# ALL-DIGITAL FREQUENCY SYNTHESIZER IN DEEP-SUBMICRON CMOS

# ALL-DIGITAL FREQUENCY SYNTHESIZER IN DEEP-SUBMICRON CMOS

**ROBERT BOGDAN STASZEWSKI**
Texas Instruments

**PORAS T. BALSARA**
University of Texas at Dallas

A JOHN WILEY & SONS, INC., PUBLICATION

Copyright © 2006 by John Wiley & Sons, Inc. All rights reserved.

Published by John Wiley & Sons, Inc., Hoboken, New Jersey.
Published simultaneously in Canada.

No part of this publication may be reproduced, stored in a retrieval system, or transmitted in any form or by any means, electronic, mechanical, photocopying, recording, scanning, or otherwise, except as permitted under Section 107 or 108 of the 1976 United States Copyright Act, without either the prior written permission of the Publisher, or authorization through payment of the appropriate per-copy fee to the Copyright Clearance Center, Inc., 222 Rosewood Drive, Danvers, MA 01923, (978) 750-8400, fax (978) 750-4470, or on the web at www.copyright.com. Requests to the Publisher for permission should be addressed to the Permissions Department, John Wiley & Sons, Inc., 111 River Street, Hoboken, NJ 07030, (201) 748-6011, fax (201) 748-6008, or online at http://www.wiley.com/go/permission.

Limit of Liability/Disclaimer of Warranty: While the publisher and author have used their best efforts in preparing this book, they make no representations or warranties with respect to the accuracy or completeness of the contents of this book and specifically disclaim any implied warranties of merchantability or fitness for a particular purpose. No warranty may be created or extended by sales representatives or written sales materials. The advice and strategies contained herein may not be suitable for your situation. You should consult with a professional where appropriate. Neither the publisher nor author shall be liable for any loss of profit or any other commercial damages, including but not limited to special, incidental, consequential, or other damages.

For general information on our other products and services or for technical support, please contact our Customer Care Department within the United States at (800) 762-2974, outside the United States at (317) 572-3993 or fax (317) 572-4002.

Wiley also publishes its books in a variety of electronic formats. Some content that appears in print may not be available in electronic formats. For more information about Wiley products, visit our web site at www.wiley.com.

*Library of Congress Cataloging-in-Publication Data:*

Staszewski, Robert Bogdan, 1965–
    All-digital frequency synthesizer in deep-submicron CMOS/Robert Bogdan Staszewski, Poras T. Balsara.
      p. cm.
    "Wiley-Interscience."
    Includes bibliographical references and index.
    ISBN-13: 978-0-471-77255-2
    ISBN-10: 0-471-77255-0
    1. Frequency synthesizers–Design and construction. 2. Wireless communication systems–Equipment and supplies–Design and construction. 3. Metal oxide semiconductors, Complementary–Design and construction. I. Balsara, Poras T., 1961– II. Title

TK7872.F73S73 2006
621.3815'486–dc22
    2006040508

Printed in the United States of America
10  9  8  7  6  5  4  3  2  1

To my parents, Kazimierz and Irena,
and my wife, Sunisa
*Robert Bogdan Staszewski*

To my parents, Roshan and Tehmurasp
To my friend, philosopher, and guide, KM
To my students and teachers
To Pearl, Farah, and Burzin
*Poras T. Balsara*

# CONTENTS

**PREFACE** xiii

**1 INTRODUCTION** 1

    1.1 Frequency Synthesis / 1
        1.1.1 Noise in Oscillators / 2
        1.1.2 Frequency Synthesis Techniques / 5
    1.2 Frequency Synthesizer as an Integral Part of an RF Transceiver / 9
        1.2.1 Transmitter / 10
        1.2.2 Receiver / 11
        1.2.3 Toward Direct Transmitter Modulation / 12
    1.3 Frequency Synthesizers for Mobile Communications / 16
        1.3.1 Integer-$N$ PLL Architecture / 17
        1.3.2 Fractional-$N$ PLL Architecture / 18
        1.3.3 Toward an All-Digital PLL Approach / 23
    1.4 Implementation of an RF Synthesizer / 25
        1.4.1 CMOS vs. Traditional RF Process Technologies / 25
        1.4.2 Deep-Submicron CMOS / 25
        1.4.3 Digitally Intensive Approach / 26
        1.4.4 System Integration / 27

1.4.5 System Integration Challenges for Deep-Submicron CMOS / 29

## 2 DIGITALLY CONTROLLED OSCILLATOR — 30

2.1 Varactor in a Deep-Submicron CMOS Process / 31
2.2 Fully Digital Control of Oscillating Frequency / 33
2.3 LC Tank / 35
2.4 Oscillator Core / 37
2.5 Open-Loop Narrowband Digital-to-Frequency Conversion / 39
2.6 Example Implementation / 45
2.7 Time-Domain Mathematical Model of a DCO / 47
2.8 Summary / 51

## 3 NORMALIZED DCO — 52

3.1 Oscillator Transfer Function and Gain / 52
3.2 DCO Gain Estimation / 53
3.3 DCO Gain Normalization / 54
3.4 Principle of Synchronously Optimal DCO Tuning Word Retiming / 55
3.5 Time Dithering of DCO Tuning Input / 56
    3.5.1 Oscillator Tune Time Dithering Principle / 56
    3.5.2 Direct Time Dithering of Tuning Input / 57
    3.5.3 Update Clock Dithering Scheme / 59
3.6 Implementation of PVT and Acquisition DCO Bits / 60
3.7 Implementation of Tracking DCO Bits / 64
    3.7.1 High-Speed Dithering of Fractional Varactors / 64
    3.7.2 Dynamic Element Matching of Varactors / 70
    3.7.3 DCO Varactor Rearrangement / 71
3.8 Time-Domain Model / 73
3.9 Summary / 74

## 4 ALL-DIGITAL PHASE-LOCKED LOOP — 76

4.1 Phase-Domain Operation / 77
4.2 Reference Clock Retiming / 79
4.3 Phase Detection / 81
    4.3.1 Difference Mode of ADPLL Operation / 85
    4.3.2 Integer-Domain Operation / 86
4.4 Modulo Arithmetic of the Reference and Variable Phases / 86
    4.4.1 Variable-Phase Accumulator (PV Block) / 89

4.5 Time-to-Digital Converter / 91
    4.5.1 Frequency Reference Edge Estimation / 93
4.6 Fractional Error Estimator / 94
    4.6.1 Fractional-Division Ratio Compensation / 96
    4.6.2 TDC Resolution Effect on Estimated Frequency Resolution / 97
    4.6.3 Active Removal of Fractional Spurs Through TDC (Optional) / 98
4.7 Frequency Reference Retiming by a DCO Clock / 100
    4.7.1 Sense Amplifier–Based Flip-Flop / 102
    4.7.2 General Idea of Clock Retiming / 103
    4.7.3 Implementation / 104
    4.7.4 Time-Deferred Calculation of the Variable Phase (Optional) / 107
4.8 Loop Gain Factor / 109
    4.8.1 Phase-Error Dynamic Range / 111
4.9 Phase-Domain ADPLL Architecture / 112
    4.9.1 Close-in Spurs Due to Injection Pulling / 114
4.10 PLL Frequency Response / 115
    4.10.1 Conversion Between the $s$- and $z$-Domains / 119
4.11 Noise and Error Sources / 119
    4.11.1 TDC Resolution Effect on Phase Noise / 120
    4.11.2 Phase Noise Due to DCO $\Sigma\Delta$ Dithering / 122
4.12 Type II ADPLL / 127
    4.12.1 PLL Frequency Response of a Type II Loop / 130
4.13 Higher-Order ADPLL / 133
    4.13.1 PLL Stability Analysis / 136
4.14 Nonlinear Differential Term of an ADPLL / 139
    4.14.1 Quality Monitoring of an RF Clock / 140
4.15 DCO Gain Estimation Using a PLL / 141
4.16 Gear Shifting of PLL Gain / 142
    4.16.1 Autonomous Gear-Shifting Mechanism / 143
    4.16.2 Extended Gear-Shifting Scheme with Zero-Phase Restart / 148
4.17 Edge Skipping Dithering Scheme (Optional) / 154
4.18 Summary / 155

## 5 APPLICATION: ADPLL-BASED TRANSMITTER    156

5.1 Direct Frequency Modulation of a DCO / 157
    5.1.1 Discrete-Time Frequency Modulation / 158
    5.1.2 Hybrid of Predictive/Closed PLL Operation / 158
    5.1.3 Effect of FREF/CKR Clock Misalignment / 163
5.2 Just-in-Time DCO Gain Calculation / 164
5.3 GFSK Pulse Shaping of Transmitter Data / 167
    5.3.1 Interpolative Filter Operation / 172
5.4 Power Amplifier / 175
5.5 Digital Amplitude Modulation / 177
    5.5.1 Discrete Pulse-Slimming Control / 180
    5.5.2 Regulation of Transmitting Power / 181
    5.5.3 Tuning Word Adjustment / 182
    5.5.4 Fully Digital Amplitude Control / 183
5.6 Going Forward: Polar Transmitter / 183
    5.6.1 Generic Modulator / 186
    5.6.2 Polar TX Realization / 187
5.7 Summary / 188

## 6 BEHAVIORAL MODELING AND SIMULATION    189

6.1 Simulation Methodology / 190
6.2 Digital Blocks / 191
6.3 Support of Digital Stream Processing / 192
6.4 Random Number Generator / 192
6.5 Time-Domain Modeling of DCO Phase Noise / 192
    6.5.1 Modeling Oscillator Jitter / 192
    6.5.2 Modeling Oscillator Wander / 194
    6.5.3 Modeling Oscillator Flicker ($1/f$) Noise / 195
    6.5.4 Clock Edge Divider Effects / 200
    6.5.5 VHDL Model Realization of a DCO / 201
    6.5.6 Support of Physical $K_{DCO}$ / 202
6.6 Modeling Metastability in Flip-Flops / 203
6.7 Simulation Results / 206
    6.7.1 Time-Domain Simulations / 206
    6.7.2 Frequency-Deviation Simulations / 207
    6.7.3 Phase-Domain Simulations of Transmitters / 209
    6.7.4 Synthesizer Phase-Noise Simulations / 209
6.8 Summary / 212

## 7 IMPLEMENTATION AND EXPERIMENTAL RESULTS — 213

- 7.1 DSP and Its RF Interface to DRP / 213
- 7.2 Transmitter Core Implementation / 214
- 7.3 IC Chip / 216
- 7.4 Evaluation Board / 218
- 7.5 Measurement Equipment / 218
- 7.6 GFSK Transmitter Performance / 219
- 7.7 Synthesizer Performance / 221
- 7.8 Synthesizer Switching Transients / 224
- 7.9 DSP-Driven Modulation / 225
- 7.10 Performance Summary / 226
- 7.11 Summary / 227

## APPENDIX A: SPURS DUE TO DCO SWITCHING — 228

- A.1 Spurs Due to DCO Modulation / 229

## APPENDIX B: GAUSSIAN PULSE-SHAPING FILTER — 232

## APPENDIX C: VHDL SOURCE CODE — 237

- C.1 DCO Level 2 / 237
- C.2 Period-Controlled Oscillator / 239
- C.3 Tactical Flip-Flop / 241
- C.4 TDC Pseudo-Thermometer Output Decoder / 243

## REFERENCES — 247

## INDEX — 253

# PREFACE

Design flow and circuit techniques of contemporary transceivers for multigigahertz mobile radio-frequency (RF) wireless applications are typically quite analog intensive and utilize process technologies that are incompatible with a digital baseband (DBB) and application processor (AP). Nowadays, the DBB and AP designs constantly migrate to the most advanced deep-submicron digital CMOS process available, which usually does not offer any analog extensions and has very limited voltage headroom. The aggressive cost and power reductions of high-volume mobile wireless solutions can realistically only be achieved by the highest level of integration, and this favors a digitally intensive approach to conventional radio-frequency (RF) functions in the most advanced deep-submicron process.

Given the task of designing highly integrated RF circuits in the digital deep-submicron process environment, we have realized that we are facing a *new paradigm*: *In a deep-submicron CMOS process, time-domain resolution of a digital signal edge transition is superior to voltage resolution of analog signals.* This is in clear contrast with the older process technologies, which rely on a high supply voltage (originally, 15 V, then 5 V, and finally, 3.3 and 2.5 V) and a stand-alone configuration with few extraneous noise sources in order to achieve a good signal-to-noise ratio and resolution in the voltage domain, often at the cost of a long settling time. In a deep-submicron process, with its low supply voltage (at and below 1.5 V), relatively high threshold voltage (0.6 V and often higher due to the MOSFET body effect), the available voltage headroom is quite small for any sophisticated analog functions. Moreover, considerable switching noise of substantial surrounding digital circuitry makes it harder to resolve signals in the voltage domain. On the positive side, the switching characteristics of a MOS transistor, with rise and fall times on the order of tens of picoseconds, offer excellent timing accuracy at high frequencies, and the fine lithography offers precise control of capacitor ratios. Hence, we exploit this new paradigm by leveraging these advantages while avoiding the weaknesses.

In this book we describe techniques for the design and implementation of an all-digital RF frequency synthesizer using deep-submicron CMOS technology. Frequency synthesizers are used as part of a modern wireless transceiver (transmitter/receiver) front end of a mobile communications channel. It can be deployed as a local oscillator in both a transmitter path (such as in Fig. 1.11) and a receiver path (such as in Fig. 1.13) RF synthesizers remain one of the most challenging tasks in mobile RF systems because they must meet very stringent requirements of a low-cost, low-power, and low-voltage monolithic implementation while meeting the phase noise and switching transient specifications. The techniques presented here were developed as part of a recent investigation [1] conducted to design a synthesizer architecture that exploits strong advantages of a deep-submicron digital CMOS process technology as well as advances in digital VLSI (very large scale integration) design. The underlying theme of the techniques presented here is to maximize digitally intensive implementation by operating in a synchronous phase domain taking full advantage of the incredible digital gate density (150,000 equivalent gates per square millimeter in 130-nm CMOS) and still-untapped capabilities of the latest deep-submicron CMOS processes. The primary benefit obtained with this architecture is to allow integration of the RF front end with the digital back end onto a single silicon die using a standard ASIC design methodology.

The synthesizer described here has a natural capability to perform frequency/phase modulation that would greatly simplify the transmitter structure by eliminating the analog-intensive I/Q mixing modulator with digital-to-analog converters and antialiasing filters as presented in Fig. 1.11. The ideas developed here have been implemented in a commercial deep-submicron CMOS process and demonstrated in a working silicon BLUETOOTH transmitter for short-range communications and are also being incorporated in other single-chip mobile phone systems.

Figure P.1 shows the scope of the book within the framework of a BLUETOOTH transmitter. The main focus is RF synthesizer design. However, to demonstrate its use, the book also covers relevant aspects of the entire RF front end that starts with the physical-layer 1-Mb/s data bit stream and ends with the RF signal

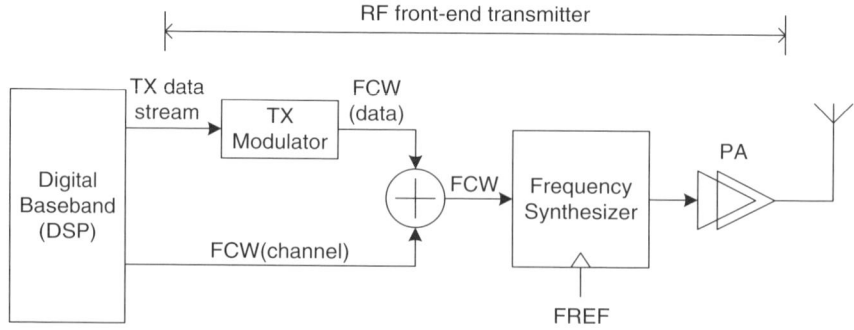

**Figure P.1** RF front end of a synthesizer-based transmitter.

feeding the antenna. This digitally intensive transmitter was integrated onto the same silicon die with Texas Instruments' C54X DSP processor (used in cellular phones) to demonstrate the feasibility of the single-chip radio as illustrated in Fig. 1.29. It should be noted that the same transmitter architecture could be used for GSM cellular phones with the differences of a slower data rate (about 280 kb/s), more stringent RF performance specifications, and a required external power amplifier, since at 1.5 V it is not possible to drive signal power on the order of 2 W.

In summary, the ideas presented in this book have culminated in a silicon demonstration of a novel RF synthesizer optimized for short-range wireless communication transceivers with the following characteristics and constraints:

- *2.4-GHz ISM band*: FCC unlicensed band providing great potential for consumer application of short-range wireless devices, such as BLUETOOTH or 802.11b WLAN.
- *Low area*: mass-production deployment in short-range wireless-enabled devices and cellular phones.
- *Low power*: battery-operated mobile communication units.
- *Fully monolithic implementation with minimal count of external components*: internal voltage-controlled oscillator, no charge-pump capacitors, no analog filters.
- *Integrateable*: utilizing the deep-submicron CMOS process technology in order to integrate with digital baseband and take advantage of high digital gate density.
- *Fully digital*: to take advantage of the incredible digital gate density of advanced CMOS processes, fast design turnaround time, and portability. Minimize analog and RF circuit content.
- *Naturally capable of performing direct* Gaussian frequency shift keying *modulation*: uses transmittal data without I/Q image reject mixers, which are analog in nature.
- *Fits well with the receiver and transmitter architecture*: acting as an optimized local oscillator capable of performing channel hopping, with fast switching times and low phase noise.
- *Top-down modeling and simulation methodology*: designing the entire system in the VHDL high-level hardware description language, as described in reference [2].

## Overview of Contents

A bottom-up rather than top-down approach was chosen to present the development of ideas and to guide the reader through the design steps. The organization of the book is as follows:

- In Chapter 1 we give necessary background by introducing basic concepts of frequency synthesis and its possible realizations in a modern wireless transceiver.

- In Chapter 2, the frequency synthesizer design starts with a raw digitally controlled oscillator (DCO), which is the foundation of a novel digital architecture. A time-domain model, which is used extensively for analysis and VHDL simulation, is introduced.
- In Chapter 3, a hierarchical layer of arithmetic abstraction is added over the DCO which makes it easier to operate algorithmically. The main task of this overlay block is to perform DCO calibration and normalization such that the normalized DCO transfer function is largely independent of the process and environmental factors. Other improvements, such as increasing the frequency resolution through $\Sigma\Delta$ dithering and dynamic element matching, are also introduced.
- In Chapter 4, a phase correction mechanism is built around the normalized DCO oscillator such that the frequency drift or wander performance of the system is as good as that of the stable external frequency reference.
- In Chapter 5, an application of the all-digital RF synthesizer is presented. A frequency modulation capability is added to the synthesizer core to enable it to perform transmitter data modulation efficiently. A full transmitter is completed with the addition of two blocks. The first of these blocks is the transmitter pulse filter operating in baseband. The second block is a class E power amplifier that is capable of emitting an RF signal with a power level of several milliwatts.
- In Chapter 6, the behavioral modeling and simulation methodology used in the design are described.
- Finally, the implementation details of a full transmitter and experimental results are presented in Chapter 7.

## Acknowledgments

The work described in this book originated as an exploratory research project at Texas Instruments and subsequently culminated as the foundation of the Digital RF Processor (DRP$^{TM}$) technology. The DRP is a key component in several commercial integrated circuits for single-chip BLUETOOTH and GSM transceivers. This success would not be possible without the early help of several people. We would like especially to thank:

- Bill Krenik, for sponsoring the project at Texas Instruments and providing very useful feedback from the day the idea was conceived.
- Dirk Leipold, a physicist and engineer at Texas Instruments, for many long and fruitful discussions leading to refinements of the digital RF architecture. His orthogonal thinking and expertise in process technology are unparalleled.
- Chih-Ming Hung, for his detailed circuit design and layout supervision of the digitally controlled oscillator and the RF power amplifier.
- Ken Maggio, manager of the RF-CMOS group at Texas Instruments, for his feedback and day-to-day activities to make the test chip a success.

- Stanley Goldman, a recognized phase-locked-loop expert at Texas Instruments, for valuable technical discussions.
- Roman Staszewski, John Wallberg, Tom Jung, and Khurram Muhammad, for their help with the digital circuit design and design flow as well as DSP integration.

*Dallas, Texas*
ROBERT BOGDAN STASZEWSKI
PORAS T. BALSARA

# CHAPTER 1

# INTRODUCTION

With the explosive growth of the wireless communication industry, research related to communication circuits and architectures has received a great deal of attention. The major issues being addressed are low-cost, low-voltage, and low-power designs, which combine necessary performance with the ability to be manufactured economically in high volumes. Recently, there has been an additional emphasis on *integration* of heterogeneous parts that constitute a communication transceiver. Modern transceivers are expected to operate over a wide range of frequencies. Although crystal oscillators offer high spectral purity, they cannot be tuned over a wide range of frequencies. Hence, some form of frequency synthesis is employed by these transceivers.

## 1.1 FREQUENCY SYNTHESIS

The term *frequency synthesizer* generally refers to an active electronic device (Fig. 1.1) that accepts some *frequency reference* (FREF) input signal of a very stable frequency $f_{ref}$ and then generates frequency output as commanded by the *frequency command word* (FCW), whereby the stability, accuracy, and spectral purity of the output correlate with the performance of the input reference. The desired value of the output frequency is an FCW multiple (generally, a real number) of the reference frequency according to the equation.

$$f_{out} = \text{FCW} \cdot f_{ref} \qquad (1.1)$$

---

*All-Digital Frequency Synthesizer in Deep-Submicron CMOS*, by Robert Bogdan Staszewski and Poras T. Balsara
Copyright © 2006 John Wiley & Sons, Inc.

## 2  INTRODUCTION

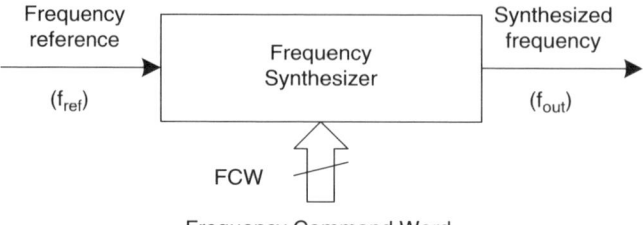

**Figure 1.1**  Frequency synthesis.

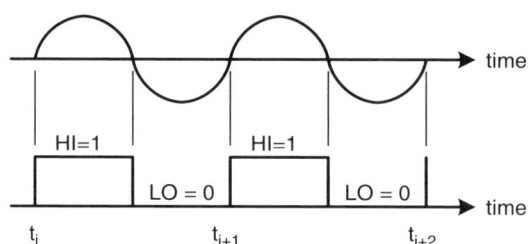

**Figure 1.2**  Possible outputs of a synthesizer: sinusoidal and digital waveforms. (From [3], © 2005 IEEE.)

Interestingly, the definition above does not specify the shape of the synthesized output. It could be a sinusoid or a rectangular signal (Fig. 1.2). The frequency and phase information is preserved in either a continuous-time waveform fit to the ideal sinusoid or in edge transition times, respectively. A clear advantage of the rectangular digital signal is that it is more useful for digital CMOS process technology.

### 1.1.1  Noise in Oscillators

For an ideal oscillator operating at frequency $\omega_c$, all its power is concentrated at a single frequency, $\omega_c$, as shown in Fig. 1.3. In a practical oscillator, the spectrum spreads into nearby frequencies around $\omega_c$ (Fig. 1.3). In oscillators this spreading is referred to as *phase noise*. Phase noise in a transmitter can cause interference in adjacent bands. In a receiver's local oscillator, it can reduce selectivity.

Phase noise is normally characterized in the frequency domain [4,5]. For an ideal oscillator operating at frequency $\omega_c$, the voltage output can be expressed as $v(t) = A\cos(\omega_c t + \phi)$, where $A$ is an amplitude and $\phi$ is an arbitrary but fixed phase reference. The power is concentrated at a single frequency $\omega_c$. Equivalently, its power spectrum is $S_v(\omega) = (A^2/2)\delta(\omega - \omega_c)$, where $\delta$ is a unit impulse or Dirac delta function (Fig. 1.3). In a practical oscillator, however, both the amplitude and the phase are time-varying fluctuations and the spectrum will exhibit "skirts" around the carrier frequency and spread into nearby frequencies. In most cases, the disturbance in the amplitude is negligible or unimportant, since it could easily be removed by a limiter circuit, and therefore only a random deviation of the

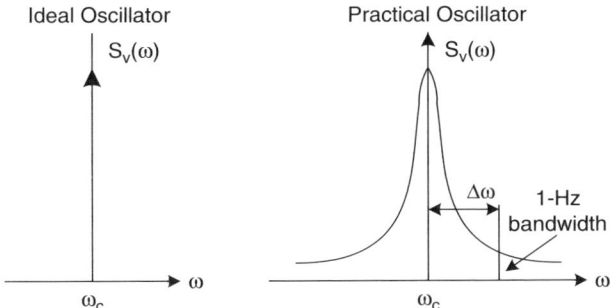

**Figure 1.3** Output spectrum of ideal and practical oscillators.

phase is considered:

$$v(t) = A \cos[\omega_c t + \phi(t)] \quad (1.2)$$

where $\phi(t)$ is a small random excess phase representing variations in the period and is commonly called *phase noise*. For a small value of the phase-noise fluctuation, $|\phi(t)| \ll 1$ rad,[1] Eq. 1.2 could be simplified to

$$v(t) \approx A \cos \omega_c t - A\phi(t) \sin \omega_c t \quad (1.3)$$

which means that the spectrum of $\phi(t)$ is frequency-translated to $\pm \omega_c$.

This phase noise can be quantified by considering a 1-Hz unit bandwidth at an offset of $\Delta\omega$ from the carrier, calculating the noise power in the band, and dividing this result by the carrier power [5]. This is the single-sided spectral noise density in units of decibel carrier per hertz (dBc/Hz). By convention, the "c" in "dBc" means "with respect to the carrier."

$$\mathcal{L}\{\Delta\omega\} = 10 \log_{10} \frac{\text{noise power in a 1-Hz bandwidth at frequency } \omega_c + \Delta\omega}{\text{carrier power}} \quad (1.4)$$

The single-sided phase noise of Eq. 1.4 is simply one-half of the phase-noise spectrum, which contains both upper- and lower-frequency components:

$$\mathcal{L}\{\Delta\omega\} = 10 \log_{10} \frac{S_\phi(\Delta\omega)}{2} \quad (1.5)$$

---

[1] $|\phi(t)| \ll 1$ rad is valid for any wireless standard. If $|\phi(t)|$ is close to 1, the oscillator has exceedingly poor phase noise or is engaged in frequency modulation.

**4** INTRODUCTION

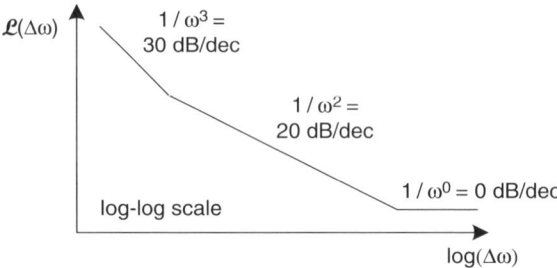

**Figure 1.4** Phase-noise spectrum of an actual oscillator.

where $S_\phi(\Delta\omega)$ is given by

$$S_\phi(\Delta\omega) = \frac{S_v(\Delta\omega)}{\text{carrier power}} \quad (1.6)$$

Figure 1.4 shows a typical oscillator phase-noise spectrum. In this log-log plot, phase noise normalized to dBc/Hz is plotted against the offset frequency $\Delta\omega$ from the carrier $\omega_c$. The phase-noise profile follows the curve shown, where it traverses through $1/\omega^3$, $1/\omega^2$, and $1/\omega^0$ slope regions. The region $1/\omega^2$ is generally referred to as the *thermal noise region* because it is caused by white or uncorrelated timing fluctuations in the period of oscillation. The $1/f$ flicker noise of electronic devices is also substantial for lower offset frequencies. It gets upconverted to the $1/\omega^3$ region. Finally, the $1/\omega^0$ region is the thermal electronic noise added outside the oscillator, such as in an output buffer, and does not affect the oscillator time base.

Let us consider the effect of a single sinusoidal tone in the phase, $\phi(t) = \phi_p \sin \omega_m t$. Equation 1.2 now becomes

$$v(t) \approx A \cos \omega_c t + A \frac{\phi_p}{2} [\cos(\omega_c + \omega_m)t - \cos(\omega_c - \omega_m)t] \quad (1.7)$$

Therefore, the *power spectral densities* (PSDs) of the oscillator output voltage and its phase noise are directly related. Using single-sided spectral densities, we have

$$S_\phi(\omega) = \frac{\phi_p^2}{2} \delta(\omega - \omega_m) \quad (1.8)$$

$$S_v(\omega) = \frac{A^2}{2} \left[ \delta(\omega - \omega_c) + \frac{1}{2} S_\phi(\omega - \omega_c) + \frac{1}{2} S_\phi(\omega_c - \omega) \right] \quad (1.9)$$

Equation 1.9 shows that the phase noise is directly shifted in frequency toward the carrier and put on either side of the synthesized frequency. It is depicted graphically in Fig. 1.5. The figure also demonstrates an *undesired* systematic fluctuation in the oscillator phase noise, giving rise to a spurious tone. Spurious tones (or spurs)

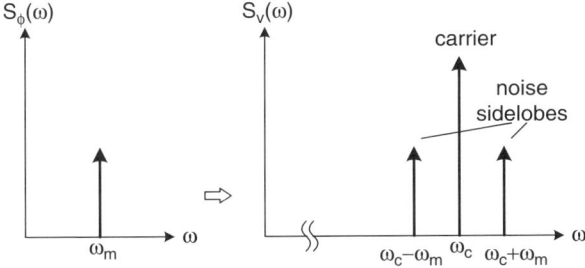

**Figure 1.5** Equivalence between PSD and single-sided phase noise.

are normally caused by the phase/frequency detector and divider circuits in classical *phase-locked-loop* (PLL)–based synthesizers. In the time domain, the presence of systematic timing fluctuations in an oscillator waveform represents a periodic timing error. In the frequency domain, it manifests as undesired tones in the frequency spectrum. Ideally, an oscillator output spectrum centers at a single frequency with no spurious tones. In reality, the presence of spurious tones causes other frequency components to appear in the oscillator output spectrum. Spurious tones are measured in dBc at a specific frequency location in the spectrum. It is simply the power difference between the carrier and the spurious tone signals in decibels.

At times, it is necessary to relate the PSD of the instantaneous frequency deviation $\Delta\omega(t)$ to the phase noise PSD $S_\phi(\omega)$ and to the single-sided phase noise $\mathcal{L}\{\omega\}$ [5] (as done in Section 6.5). Since frequency is the derivative of phase, we get

$$S_{\Delta\omega}(\omega) = \omega^2 S_\phi(\omega) = 2\omega^2 \mathcal{L}\{\omega_m\} \tag{1.10}$$

The oscillator perturbations seen in the frequency domain have the underlying cause in the time domain, where the exact time of one period of oscillation will differ from the other. The period has an average value $T_0$ and a timing error $\Delta T$. The timing error variance $\sigma_{\Delta T} = \sqrt{\overline{\Delta T^2}}$ is called *jitter*. A first-order formula is given in [7] that relates jitter in the time domain to phase noise in the frequency domain:

$$\mathcal{L}\{\Delta\omega\} = 10\log_{10}\left[\frac{2\pi\omega_c}{\Delta\omega^2}\left(\frac{\sigma_{\Delta T}}{T_0}\right)^2\right] \tag{1.11}$$

The region modeled by Eq. 1.11 is the $1/\omega^2$ up-converted thermal noise, which is the dominant noise mechanism in oscillators.

### 1.1.2 Frequency Synthesis Techniques

There are three major conventional frequency synthesis techniques:

- Direct analog mix/filter/divide
- Direct digital

- Indirect or phase-locked loop
- Hybrids: any combination of the three methods above

Each of these methods has advantages and disadvantages; hence, each application requires selection based on the most acceptable combination of compromises.

**1.1.2.1 Direct Analog Synthesis** Direct analog synthesis, also called *mix/filter/divide*, uses echelons of frequency multipliers, dividers, and other mathematical manipulations to produce the desired new frequency [8]. The process is called *direct* because the error correction process is avoided; hence, the quality of the output correlates directly with the quality of the input. Phase noise is typically excellent because the direct process and switching speed can be very fast. Unfortunately, a broadband mix/filter/divide synthesizer requires many references, which makes it extremely expensive. Because of its high cost and high power disadvantages, the direct analog synthesis method is used primarily in instrumentation and is not practical for low-power portable applications such as mobile communication terminals.

**1.1.2.2 Direct Digital Synthesis** *Direct digital frequency synthesis* (DDFS) is the most recently developed frequency synthesis technique, dating from the early 1970s [9]. A DDFS system uses logic and memory to construct the desired output signal digitally, and a data conversion device [a digital-to-analog converter (DAC)] to convert it from the digital to the analog domain, as shown in Fig. 1.6. Therefore, the DDFS method of constructing a signal is almost entirely digital, and the precise amplitude, frequency, and phase are known and controlled at all times. For these reasons, the switching speed is extremely high, but the power consumption could be excessive at high clock frequencies. The DDFS method is not entirely digital in the true sense of the word since it requires a DAC and a low-pass filter (LPF) to attenuate spurious frequencies produced by the digital switching. In addition, a very stable frequency reference clock of at least three times the output frequency is required. Considering this and the fact that the DAC and LPF might be difficult to build and would consume excessive amount of power at gigahertz operational frequency, the DDFS solution is not acceptable for radio-frequency (RF) applications such as mobile communication terminals.

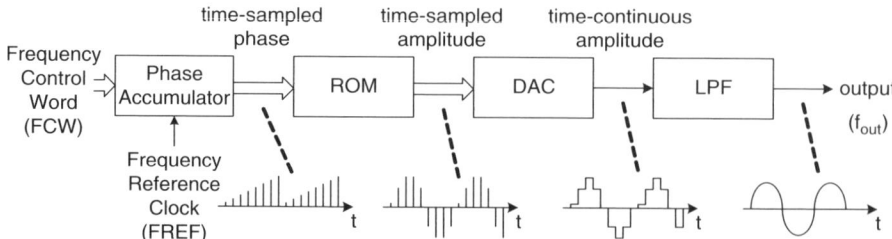

**Figure 1.6** Direct digital frequency synthesis.

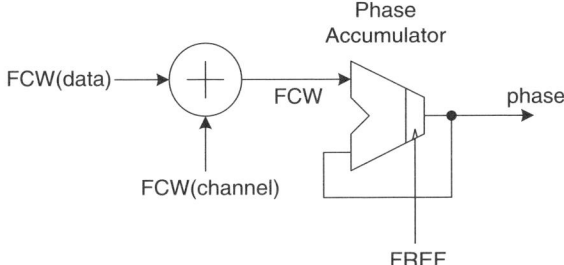

**Figure 1.7** Phase accumulator front end of a DDFS system with frequency modulation capability.

Due to its digital waveform reconstruction nature, the DDFS technique is best suited for implementing wideband transmit modulation as well as fast channel-hopping schemes [10]. As an example, Fig. 1.7 shows the phase accumulator front end of a DDFS system with an arithmetic adder that combines the FCW components of the channel selected and the frequency-modulating data.

Let the wordlength of the accumulator be $W$. For a given frequency command word FCW and clocking (FREF) frequency $f_{\text{ref}}$, the output frequency $f_{\text{out}}$ of the synthesizer is given by

$$f_{\text{out}} = \frac{f_{\text{ref}}(\text{FCW})}{2^W} \tag{1.12}$$

and the frequency resolution is

$$\Delta f = \frac{f_{\text{ref}}}{2^W} \tag{1.13}$$

Because it is very costly to implement a DDFS system at frequencies of interest for wireless communications (multi-GHz range), to date this technique has been used directly only in military applications. From yet another perspective, the DDFS method is fundamentally not the best choice for generating RF signals. As explained previously, in a deep-submicron process technology, the digital clock of Fig. 1.2 is preferred over the sinusoidal signal, which the DDFS technique attempts to produce. The complete digital *reconstruction* of the entire waveform is simply too wasteful if ultimately, only the zero crossings are what is needed.

### 1.1.2.3 Indirect Synthesis Using Phase Locking

Indirect synthesis using a PLL compares the output phase of an oscillator, such as a *voltage-controlled oscillator* (VCO), with a phase of a reference signal FREF ($f_{\text{ref}}$) [11,12], as shown in Fig. 1.8. As the output drifts, detected errors produce correction commands to the oscillator, which responds in a negative-feedback manner. Phase and frequency deviation monitoring occurs in the *phase/frequency detector* (PFD), which adds phase

**8** INTRODUCTION

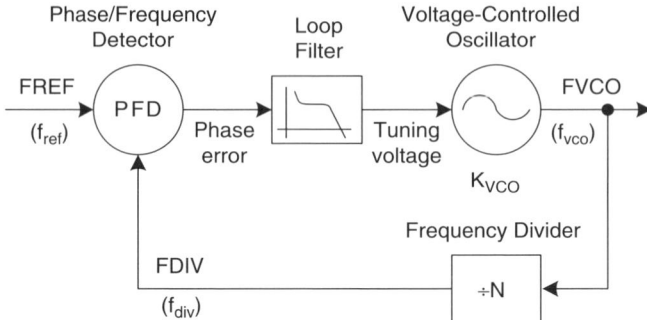

**Figure 1.8** Phase-locked loop.

noise close to the carrier. However, a PLL can outperform direct synthesis at larger frequency offsets. Fine frequency steps degrade phase noise, and fast switching is difficult to achieve with a PLL design even when using aggressive VCO pretuning techniques [13].

In general, the indirect synthesizer uses a PLL and a programmable fractional-$N$ divider, which multiplies the stable reference frequency $f_{ref}$. In the loop, a *loop filter* (LF) is present so as to suppress spurs produced in the phase detector so that they do not cause unacceptable frequency modulation in the VCO (see Section 1.3). However, the filter causes a degradation in transient response, which limits the switching time. Therefore, the requirements for frequency switching time and suppression of spurs are in conflict.

The classical PLL-based frequency synthesizers are suitable only for narrowband frequency modulation schemes, in which the modulating data rate is well within the PLL bandwidth.

#### 1.1.2.4 Frequency Synthesis Based on Hybrid Structure
In certain applications it is necessary to combine two (rarely, three) major synthesis techniques such that the best features of each basic method are emphasized. Generally, it is a hybrid of DDFS and PLL structures that is used in certain wireless devices. As shown in Fig. 1.9, the wideband modulation and fast channel-hopping capability of the DDFS method, which now operates at lower frequency, could be combined with the frequency multiplication property of a PLL that up-converts it to the RF band.

As another example of a hybrid approach, Hafez and Elmasry [14] describe a 900-MHz band hybrid synthesizer structure that uses a 1.10 to 1.85-MHz

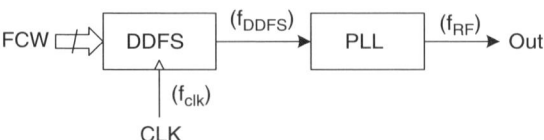

**Figure 1.9** DDFS–PLL hybrid.

low-frequency DDFS to generate a stable frequency reference to the main PLL. Instead of using a conventional digital frequency divider or prescaler, a PLL uses subsampler mixing to translate the RF frequency down to $f_{\text{ref}}$. The frequency resolution of the synthesizer is established by the DDFS system, and the PLL is used primarily as a frequency integer multiplier. Since the DDFS system operates at low frequency, its major limitation of high-power dissipation is not a concern. Unfortunately, the subsampling process introduces excessive noise.

## 1.2 FREQUENCY SYNTHESIZER AS AN INTEGRAL PART OF AN RF TRANSCEIVER

The design of RF synthesizers remains one of the most challenging tasks in mobile RF systems because these synthesizers must meet the very stringent requirements of low-cost, low-power, low-voltage monolithic implementation while also meeting the phase noise and switching transient specifications. We can generally evaluate a synthesizer design by considering the following criteria (in order of importance):

- *Phase-Noise Performance.* As for any analog circuit, oscillators are susceptible to noise, which causes adverse affects in system performance during receiving and transmitting.
- *Discrete Spurious Noise Performance.* Unwanted frequency components appear in the oscillator output spectrum.
- *Switching Speed.* The speed of switching is very important in modern communications systems, which utilize channel or frequency hopping (see Fig. 1.10) to combat various wireless channel impairments (e.g., fading, interference). Since the system switches carrier frequency often (as often as once every 1.6 ms in BLUETOOTH or GSM), a fast-switching, stable frequency synthesizer is essential for proper operation. Switching speed is also important in fixed-channel *time-division multiple access* (TDMA) systems for quick handoff between neighboring cells.

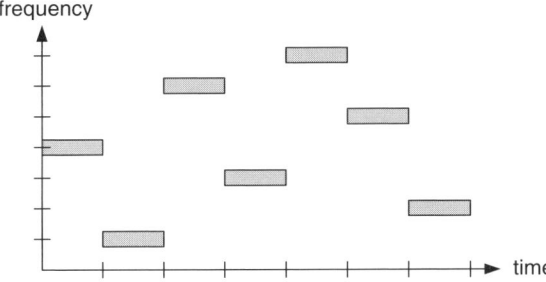

**Figure 1.10** Channel hopping of a transmitted signal.

- *Frequency and Tuning Bandwidth.* The frequency range has to be wide enough to cover the operational band, with an extra margin to account for *process–voltage–temperature* (PVT) variations.
- *Power Consumption.* How much power is consumed is important for battery-operated mobile communication units.
- *Size.* Size is important for mass production deployment.
- *Integrateability.* Deep-submicron CMOS process technology is chosen to integrate with digital baseband and application processor.
- *Cost.* No extra cost is added to the process. Only a minimal number of external components are required.
- *Portability.* The ability to transfer a design from one application to another and from one process technology node to the next is vital, especially in digital VLSI (very large scale integration) and for *intellectual property*–based applications. Designs described in a hardware description language (HDL) are very portable.

### 1.2.1 Transmitter

An RF frequency synthesizer is employed as a *local oscillator* (LO) in a transmitter to perform frequency translation from baseband to RF. Figure 1.11 shows a conventional direct up-conversion transmitter. The *in-phase* (I) and *quadrature* (Q) pulse-shaped digital baseband signals are converted into the analog domain with DACs. Due to their discrete-time nature, DAC outputs contain strong sampling-time harmonics and switching noise and have to be conditioned with *low-pass filters* before being up-converted to the RF carrier by a modulator, which is a critical RF/analog block. A *power amplifier* (PA) is the last stage in the transmitter path. It performs antenna impedance matching and brings the signal emitted to the power level required. A major weakness of this mixer-based transmitter architecture

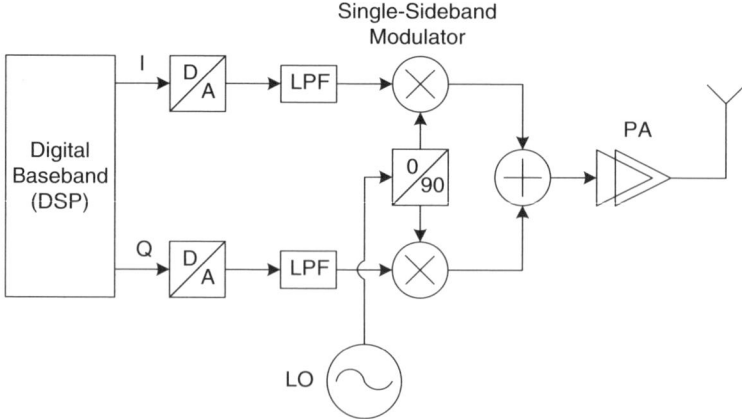

**Figure 1.11** Conventional direct up-conversion transmitter.

## 1.2 FREQUENCY SYNTHESIZER AS AN INTEGRAL PART OF AN RF TRANSCEIVER

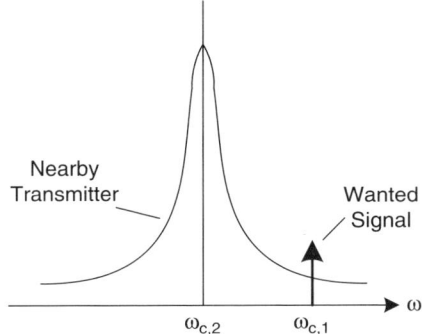

**Figure 1.12** Effect of LO phase noise in a transmitter.

is that even a small mismatch in phase shift or amplitude gain between the I and Q paths can impair system performance significantly. Furthermore, because of a certain amount of inherent frequency shift between the modulator input and output (it performs frequency translation by design), the strong power amplifier signal can cause frequency pulling of the oscillator through injection locking. This mechanism finds parasitic paths, such as substrate, power, and ground lines as well as electromagnetic radiation, to feed a strong PA signal into the most sensitive parts of the oscillator, causing its resonant frequency to be shifted in an injection-pulling manner.

The phase noise generated by the LO in the transmitter must be minimized. When a noiseless receiver must detect a weak desired signal at frequency $\omega_{c,1}$ in the presence of a powerful nearby transmitter generating a signal at frequency $\omega_{c,2}$ with substantial phase noise or spurs, the problem shown in Fig. 1.12 would occur. The signal desired will be corrupted by the phase-noise tail of the transmitter. This leads to a challenging noise requirement for the noise skirt of an RF LO. The tough standards for modern wireless communication systems result in very tight specifications on the close-in and far-out phase noise of synthesizers [15].

### 1.2.2 Receiver

An RF frequency synthesizer is used as an LO in a receiver to perform frequency translation and channel selection. Figure 1.13 shows a modern zero-IF receiver architecture in which the LO is used to down-convert the RF signal directly into baseband without *intermediate-frequency* (IF) stages. Alternatively, due to dc-offset issues plaguing true zero-IF architecture [16], the conversion could preferably be near-zero IF. In this case, the IF signal does not get demodulated exactly to dc but to a fraction of the signal bandwidth away.

The power level of an RF signal received at the antenna output port is immediately strengthened by a *low-noise amplifier* (LNA). It is then down-converted by an image-reject mixer into baseband. *Low-pass filters* are used to keep the unwanted frequency components from being fed further in the receiving chain. Variable-gain

**12** INTRODUCTION

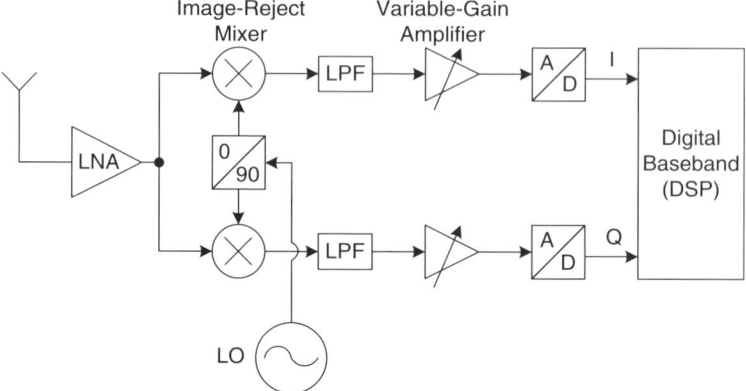

**Figure 1.13** Zero-IF receiver topology.

amplifiers bring the signal to the required level before it gets converted into digital form by an *analog-to-digital converter* (ADC). The digital baseband processes both the *in-phase* and *quadrature* parts and performs detection and other DSP processing.

This direct down-conversion architecture is considered to be best suited for on-chip integration since it does not require external high-$Q$ tuning circuits for IF filtering. Unfortunately, it suffers from dc-offset problems due to LO leakage as well as self-mixing. Various DSP-based cancellation methods have been developed, and this remains an area of active research [17].

Modern wireless receiver specifications impose stringent requirements on an RF local oscillator. For instance, the specifications for a BLUETOOTH receiver states that a $-70$-dBm signal power in the 1-MHz bandwidth must be detected with a *bit error rate* (BER) of less than $10^{-3}$. This translates into a requirement that the *signal-to-noise ratio* (SNR) at the noncoherent detector be no less than 17 dB. The system must also be able to reject $-10$ dBm out-of-band single-tone blocking signals with a signal level of $-67$ dBm. The system must also pass an intermodulation test with a $-39$ dBm single-tone signal close to a $-39$ dBm BLUETOOTH-modulated interferer when the signal level is at $-64$ dBm. These impose stringent requirements on the RF LO. Figure 1.14 illustrates the effect of LO purity in a receiver. Since the wanted signal power is small and the interferer is large, the noise skirt of the LO must be low so that the wanted signal would remain recoverable after down-conversion.

### 1.2.3 Toward Direct Transmitter Modulation

Figure 1.15 is a general block diagram of a transmitter *quadrature amplitude modulation* (QAM) using complex signals. It graphically describes an arbitrary modulation process. The incoming bit stream $b_k$ is fed to a coder, which converts the 0 or 1 digital bits into a stream of symbols $a_m$. A symbol assumes values from an alphabet. Since the coder may map multiple bits into a single data symbol, a

## 1.2 FREQUENCY SYNTHESIZER AS AN INTEGRAL PART OF AN RF TRANSCEIVER

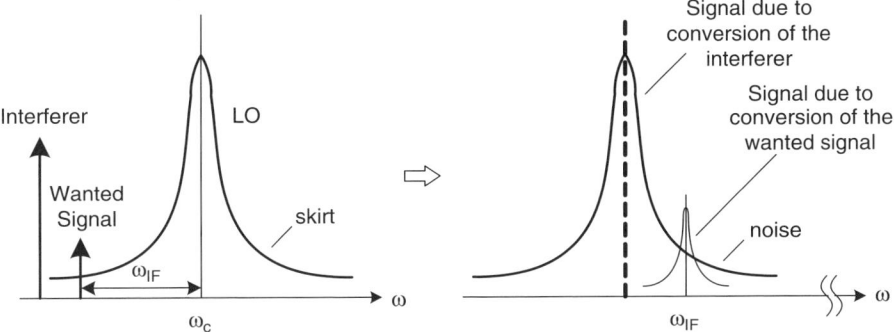

**Figure 1.14** Effect of LO phase noise in a receiver.

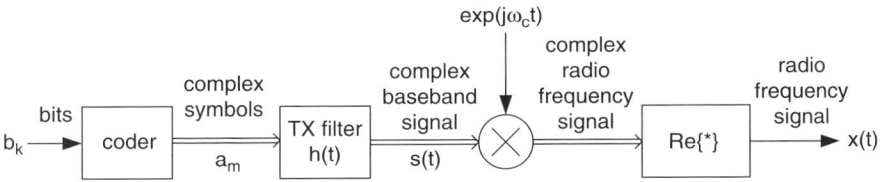

**Figure 1.15** QAM modulation with complex signals.

distinction must be made between the symbol rate and the bit rate. In BLUETOOTH and GSM there is a one-to-one correspondence between the bits and symbols: $\{0, 1\} \rightarrow \{-1, +1\}$. In contrast, more advanced encoding schemes, such as QPSK or 8PSK, pack 2 or 3 bits into a symbol, respectively.

A sequence of symbols is applied to a transmitter filter, which produces a continuous-time signal for transmission over the continuous-time channel. The main purpose of employing a baseband transmitter filter is to properly constrain the bandwidth occupied by the modulated RF spectrum. When rectangular pulses are passed through a bandlimited channel, the pulses will spread in time, and the pulse for each symbol will spread into the time intervals of succeeding symbols [18]. This causes *intersymbol interference* (ISI) and leads to increased probability that the receiver will make an error in detecting a symbol. Out-of-band radiation in the adjacent channel in a mobile system should generally be 40 to 80 dB below that in the desired passband. Since it is difficult to manipulate the transmitter spectrum directly at RF frequencies, spectral shaping is done in baseband.

The impulse response $h(t)$ of the transmitter filter, called the *pulse shape*, could be raised-cosine or Gaussian [18]. The raised-cosine rolloff filter belongs to the class of filters that satisfy the Nyquist criterion of zero ISI at sampling instances. Gaussian filters, on the other hand, have a smooth transfer function but do not satisfy the Nyquist criterion and allow for a certain amount of ISI at zero crossings. However, they can employ power-efficient nonlinear amplifiers and are commonly used with frequency-modulated signals.

**14** INTRODUCTION

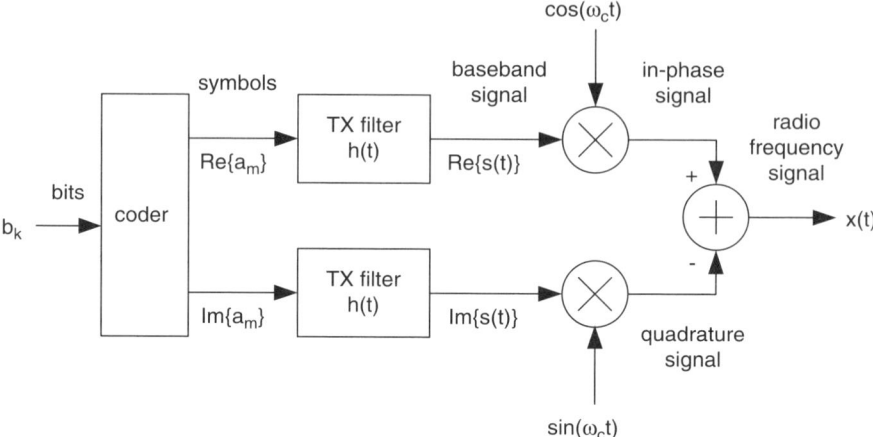

**Figure 1.16** QAM modulation with I and Q baseband signals.

In modern implementations, the pulse shape is oversampled by a *"chip" clock*, which here is an integer multiple of the symbol clock and represented digitally throughout the pulse filtering process, even though usually the filter output $s(t)$ is ultimately brought back to the continuous-time domain by performing a digital-to-analog conversion and subsequent low-pass filtering.

The digital baseband data bits $b_k$ are synchronous to the baseband clock, whereas the digital filter output samples are synchronous to the "chip" clock, which is conventionally a multiple of the data rate.

Complex signal representation requires two physical wires that carry both real-valued parts of a complex number. Figure 1.16 shows a block diagram of QAM transmitter modulation using *in-phase* (I) and *quadrature* (Q) signals that represents a natural progression toward a more physically realizable representation. This realization is the basis for the conventional transmit modulator introduced in Section 1.2 and can handle a wide range of modulation schemes. However, its I/Q imbalance and carrier feedthrough usually lead to poor sideband suppression.

Figure 1.17 shows a block diagram of a QAM transmit modulation using a polar alternative in a form of direct amplitude and phase modulation. The direct phase

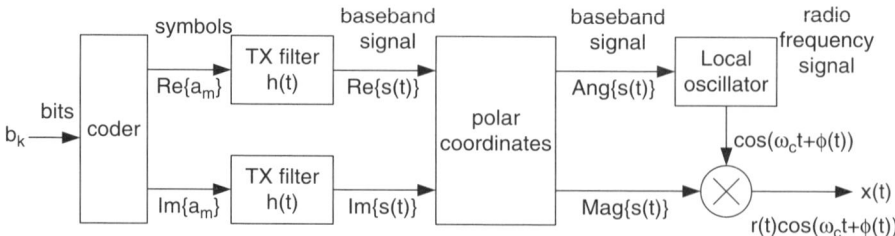

**Figure 1.17** QAM modulation with direct phase and amplitude modulation.

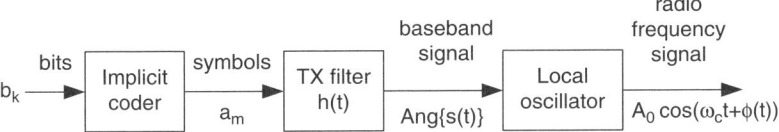

**Figure 1.18** GFSK modulation as QAM modulation with a constant envelope.

modulation is performed conventionally by modulating the oscillator tuning input in a feedforward manner with a possible PLL compensation method, as shown in [19] and [20]. The direct amplitude modulation might be performed by a conventional method of regulating the supply voltage to a saturation-mode power amplifier. The QAM polar method is clearly the best choice for digital integration of mobile RF transceivers because it does not use the traditional RF/analog-intensive up-conversion mixer of Section 1.2.

Constant-envelope transmitter modulation schemes such as *Gaussian frequency shift keying* (GFSK), which is used in BLUETOOTH and GSM (Global System for Mobile Communications)[2] standards, permits the use of nonlinear power amplifiers that have better power efficiency than that of their linear counterparts. This is important in keeping the total system power consumption to a minimum, since the RF power amplifier consumes the lion's share of an overall power budget. With constant-envelope modulation, dynamic amplitude control is not required and only a slowly varying output power regulation might need to be implemented. Figure 1.18 shows the GFSK modulation as being a special case of Fig. 1.17 with the constraint of having a constant envelope. This architecture is selected as the preferred embodiment for the BLUETOOTH transmitter (see Section 5.1).

The RF output of an angle (i.e., phase or frequency)-modulated system can be expressed as

$$s(t) = A_c \cos[\omega_c t + \phi(t)] \quad (1.14)$$

where $A_c$ is a carrier amplitude, $\omega_c$ is an angular carrier frequency (rad/s), and $\phi(t)$ is a modulating phase (rad). In GFSK, the information-bearing quantity is frequency; that is, the instantaneous frequency $f(t)$ of the carrier signal is proportional to the modulating signal $y(t)$:

$$f(t) = k_f y(t) \quad (1.15)$$

where $k_f$ is the *frequency modulation* (FM) proportionality constant. Since phase, $\phi(t)$, is an integral of frequency $f$,

$$\phi(t) = 2\pi \int_{-\infty}^{t} f(\tau) \, d\tau \quad (1.16)$$

---

[2]GSM actually uses *Gaussian minimum shift keying* (GMSK), which is a special case of GFSK.

Equation 1.14 could be rewritten as

$$s(t) = A_c \cos\left[\omega_c t + 2\pi \int_{-\infty}^{t} f(\tau)\,d\tau\right] \quad (1.17)$$

$$= A_c \cos\left[\omega_c t + 2\pi \int_{-\infty}^{t} k_f y(\tau)\,d\tau\right] \quad (1.18)$$

More complex modulation schemes, such as *quadrature-phase shift keying* (QPSK) and *eight-phase shift keying* (8PSK), require full symbol-rate RF amplitude control. The polar QAM method is still likely to be preferred in this case over the I/Q scheme. Here, a stripped-down version of the *envelope elimination and restoration* (EER) method (without the envelope detection and amplitude limitation parts) could be used and is based on power-efficient nonlinear saturation-mode *power amplifiers* (PAs). In this method, the supply voltage of a nonlinear PA is adjusted to achieve the desired amplitude of the output, while the input signal has a constant envelope with a duty cycle not far from 50%. Another saturation-mode PA method could be used: *Linear amplification with nonlinear components* (LINC) adds two constant-envelope PA outputs of the properly phase-shifted signals. However, it is a more area-intensive and less power-efficient technique than the EER and is currently used mainly in base stations. A method described in this book uses a novel digital pulse slimming circuit instead.

## 1.3 FREQUENCY SYNTHESIZERS FOR MOBILE COMMUNICATIONS

A great majority of RF wireless synthesizers for mobile applications are based on a charge-pump PLL structure [21]. Under locked conditions the average output frequency of a PLL bears an exact relationship to the reference input frequency, so the frequency accuracy is extremely high. Unfortunately, its acquisition time is rather long since the phase/frequency detector evaluates the frequency difference between the reference and generated clocks by means of the phase difference. In modern wireless applications, the fast acquisition characteristic of the frequency synthesizer is crucial (e.g., in channel hopping). The acquisition time is directly proportional to the initial frequency difference $\Delta f_0$ and inversely proportional to the loop bandwidth $f_{BW}$ [22]. Consequently, to reduce the acquisition time, a small initial frequency difference and wide loop bandwidth are desirable. However, it is not always possible to achieve a small $\Delta f_0$ value at the design stage, due to the presence of process, voltage, and temperature (PVT) variations; thus, a self-calibrating mechanism would be preferred [23]. In addition, an attempt to enhance the acquisition time through a wide loop bandwidth increases the contribution of the reference phase noise and is therefore rarely possible.

The frequency difference could be measured directly, as in [24], to reduce the acquisition time significantly, to just a few clock cycles independent of the initial

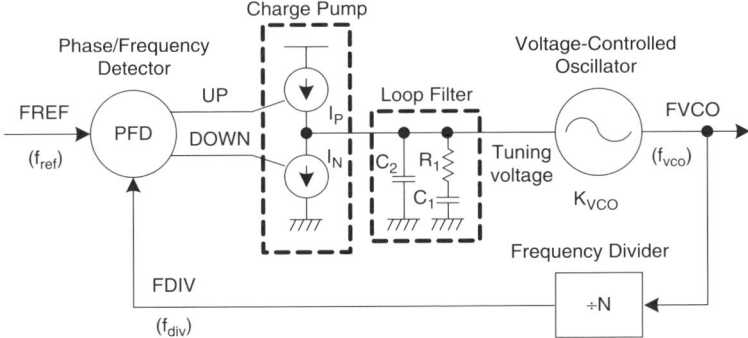

**Figure 1.19** Typical charge-pump-based PLL for RF wireless applications.

frequency difference. The technique presented in this book is capable of performing simultaneous phase and frequency estimation and nearly instantaneous correction; therefore, the acquisition performance should be even better.

As shown in Fig. 1.19, the PFD estimates the phase difference between the frequency reference FREF input and the divided-by-$N$ VCO clock FDIV by measuring time between their respective closest edges and generates either an UP or a DOWN pulse whose width is proportional to the time difference measured. This signal, in turn, produces a current pulse $I_P - I_N$ with the proportional duty cycle in a charge-pump block. At the loop filter, this current is converted into a VCO tuning voltage. $C_2$ is an integrating capacitor and introduces a pole at dc, thus giving rise to a type II loop. Its main task is to suppress the glitch generated by the charge pump on every phase comparison instant. The glitch arises from mismatches between the width of UP and DOWN pulses produced by the PFD as well as charge injection and clock feedthrough mismatches between PMOS and NMOS devices in the charge pump. This periodic glitch will modulate the VCO output frequency, thus giving rise to frequency spurs. The primary sidebands will be located on both sides of the carrier at the comparison frequency offset. The double integrator configuration is only marginally stable, and it is necessary to stabilize the loop with a pole–zero combination introduced by $R_1$ and $C_1$.

### 1.3.1 Integer-$N$ PLL Architecture

Conventional PLL has an integer divider ratio of $N$ such that $f_{vco} = Nf_{ref}$. The letter $N$ is used historically in the PLL field to designate its frequency multiplication property and is equivalent to FCW as defined by Eq. 1.1. Resolution is equal to the reference frequency, which is usually selected to be the same as the channel spacing. Narrow loop bandwidths are undesirable because of long switching times, inadequate suppression of he VCO phase noise, and susceptibility to supply and substrate noise.

The PLL building blocks for the RF applications require extreme care due to the high frequency of operation, matching, and linearity. For example, it is impractical

## 18 INTRODUCTION

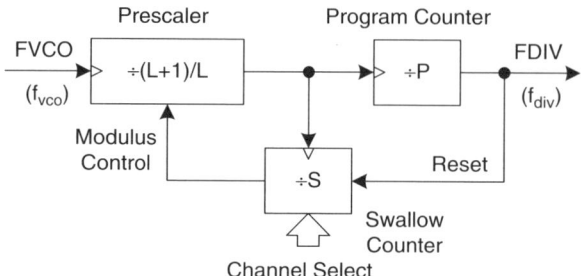

**Figure 1.20** Pulse swallow frequency divider.

to build the programmable frequency divider of Fig. 1.8 directly at the RF frequency. The pulse swallower structure shown in Fig. 1.20 is commonly used instead. It consists of a high-frequency prescaler that divides by either $L$ or $L+1$, where $L$ is usually a low-power-of-2 number (8, 16, or 32 are used in practice), and a program counter and swallow counter. The program counter always divides the prescaler output by $P$, whereas the swallow counter divides the prescaler output by a selectable value of $S$ that determines the channel desired. Both counters operate at lower frequency, $f_{vco}/L$. It could easily be shown that if $S < P$, the resulting division ratio is $N = PL + S$. The $S$ counter value controls the channel selection by an integer multiple of $f_{ref}$.

Unfortunately, the charge-pump PLL structure does not lend itself easily to silicon integration. Because of the spur reduction requirements, the loop filter, usually realized as a charge pump as in Fig. 1.19, requires large resistors and capacitors, most likely external to the IC chip, to achieve a low PLL bandwidth of several kilohertz. Realizing a monolithic capacitance on the order of a few hundred picofarads would require a prohibitively large area if implemented as a high-quality *metal–insulator–metal* (MIM) *capacitor*. Implementing it as a MOS capacitor would take less area, but it would probably be unacceptable because of its high leakage current and nonlinearity. Another major disadvantage of this analog-intensive synthesizer is lack of portability from one process technology to another.

### 1.3.2 Fractional-N PLL Architecture

In fractional-$N$ synthesizers, the output frequency can increment by fractions of the reference frequency, allowing the latter to be much greater than the channel spacing required. This allows wide loop filter design at the expense of fractional spurs, resulting in improved loop dynamics and attenuation of the oscillator-induced noise [4]. The PLL bandwidth is usually set at roughly 10% of the reference frequency to avoid any significant feedthrough of the reference tone and may now span several channels. In response to a change in a frequency control word, the PLL output frequency settles to the programmed value with a time-constant inversely related to the loop bandwidth.

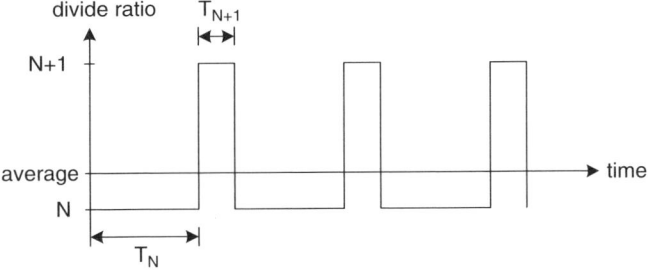

**Figure 1.21** Alternating divide ratio of fractional-$N$ PLL.

Fractional-$N$ PLL can achieve arbitrarily fine time-averaged frequency-division ratio of $(N.f)$ by modulation of the instantaneous integer division ratio of $N$ and $N+1$. (In practice, a multibit modulus could be used, as demonstrated in [25].) Figure 1.21 reveals the principle in which the integer division is periodically altered from $N$ to $N+1$. The resulting average divide ratio will be increased from $N$ by the duty cycle of the $N+1$ division:

$$N_{avg} = \frac{NT_N + (N+1)T_{N+1}}{T_N + T_{N+1}} = N + \frac{T_{N+1}}{T_N + T_{N+1}} = N + (.f) = (N.f) \quad (1.19)$$

where $f$ corresponds to the fractional part of the frequency-division ratio. Figure 1.22 shows details of various signals (active edges only) for $N = 2.25$. The timestamps shown on FREF edges are given in terms of VCO edges (each FREF cycle comprises 2.25 VCO cycles). The phase error is given with respect to the nearest FREF edge.

The phase detector will operate at a frequency of $f_{ref} + (.f/N)f_{ref}$, and the phase error of the phase detector causes VCO fractional spurs at a multiple of the offset

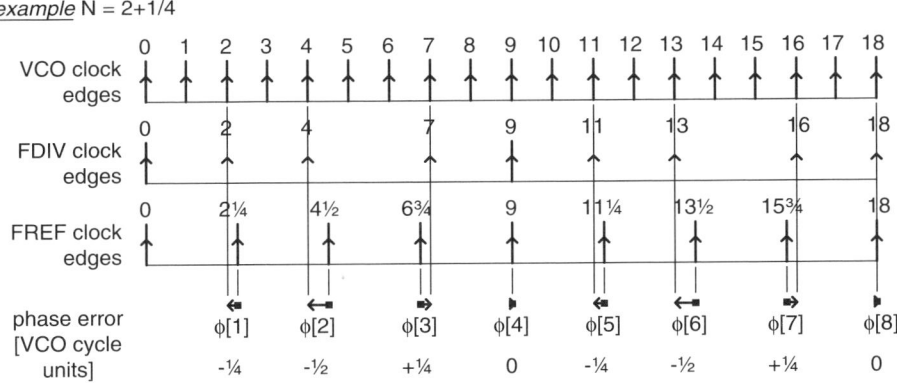

**Figure 1.22** Periodic and deterministic phase error in a fractional-$N$ PLL.

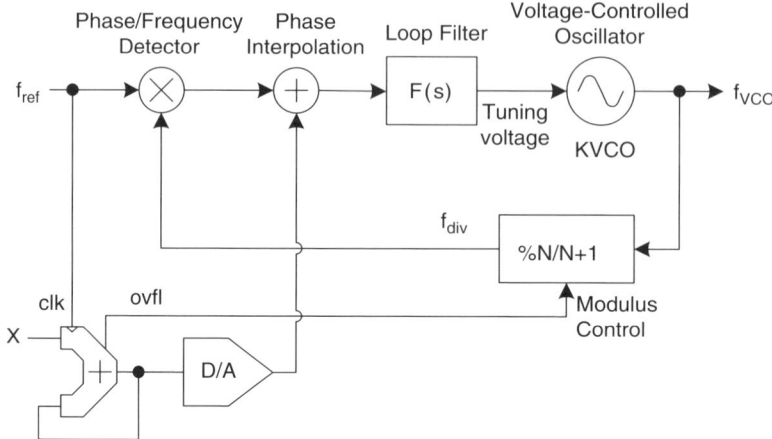

**Figure 1.23** Fractional-$N$ PLL incorporating phase interpolation.

frequency $(.f) f_{ref}$. There are several methods available to suppress the fractional spurs. A more conventional method is the analog fractional-$N$ compensation scheme that uses an accumulator and a DAC and is based on the observation that the phase-error perturbation is periodic and deterministic (Fig. 1.22) and could be canceled out by a tracking circuit. This form of correction, called *phase interpolation*, is presented in Fig. 1.23. The fractional spurs are reduced to the extent that the phase interpolation signal exactly matches the phase error. In practice, it is difficult to achieve fractional spurs lower than $-70$ dBc. This requires a precision DAC and carefully designed phase detector and sampler circuitry. Due to its analog complexity, the interpolation scheme is not suitable in most applications.

The second method uses a $\Sigma\Delta$-modulated clock divider as described by Miller and Conley [26] and Riley et al. [27] and is shown in Fig. 1.24. This solution is more digital in nature since it does not rely on precise analog component matching of the previous technique. It trades the reduction in fractional spurs for the increase in the noise floor [as shown in Fig. 1.26 (left)].

Figure 1.25 shows the third-order $\Sigma\Delta$ digital modulator divider described in [26]. It uses three accumulator stages, in which the storage is performed in the accumulator feedback path. The modulator input is a fractional fixed-point number and its output is a small integer stream. It is shown in reference [26] that its transfer function is

$$N_{\text{div}}(z) = .f(z) + (1 - z^{-1})^3 E_{q3}(z) \qquad (1.20)$$

where $E_{q3}$, the quantization noise of the third stage, equals the output of the third-stage accumulator. The first term is the desired fractional frequency, and the second term represents noise due to fractional division.

1.3 FREQUENCY SYNTHESIZERS FOR MOBILE COMMUNICATIONS    21

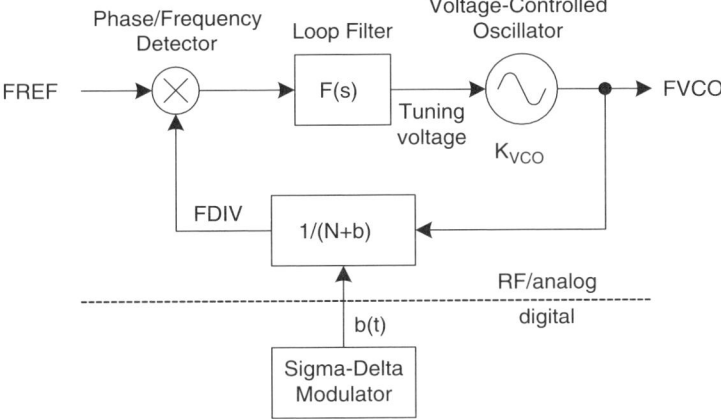

**Figure 1.24** Fractional-$N$ synthesizer using a $\Sigma\Delta$-modulated clock divider.

Figure 1.26 plots the spectrum of the output signal FVCO. It shows quantization noise-shaping properties of the first-, second-, and third-order $\Sigma\Delta$ modulation of the clock division ratio. The first order of $\Sigma\Delta$ operation turns out to be equivalent to the conventional but uncompensated fractional-$N$ architecture and exhibits systematic division ratio patterns that produce unacceptably large frequency tones. In this case, the quantization noise is concentrated at discrete frequencies rather than being spread continuously and merged into the noise floor as in second and higher $\Sigma\Delta$ orders. Figure 1.26 (left) shows the power spectral density ($S_\nu$ as defined in Section 1.1.1) of the divided clock and is centered around $f_{\text{div}}$. As shown, the third order of $\Sigma\Delta$ dithering introduces enough randomness to eliminate completely any frequency spurs that are clearly shown with the second- and first-order $\Sigma\Delta$ dithering. The right plot shows the PSD of the divided clock phase ($S_\phi$ as defined in Section 1.1.1), and it demonstrates that the second-order $\Sigma\Delta$ dithering

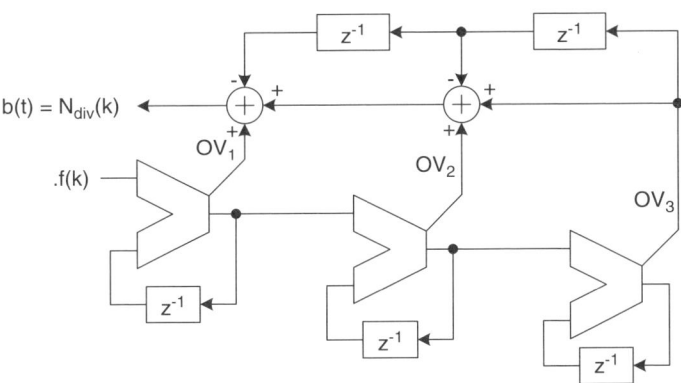

**Figure 1.25** MASH-3 $\Sigma\Delta$ digital modulator divider.

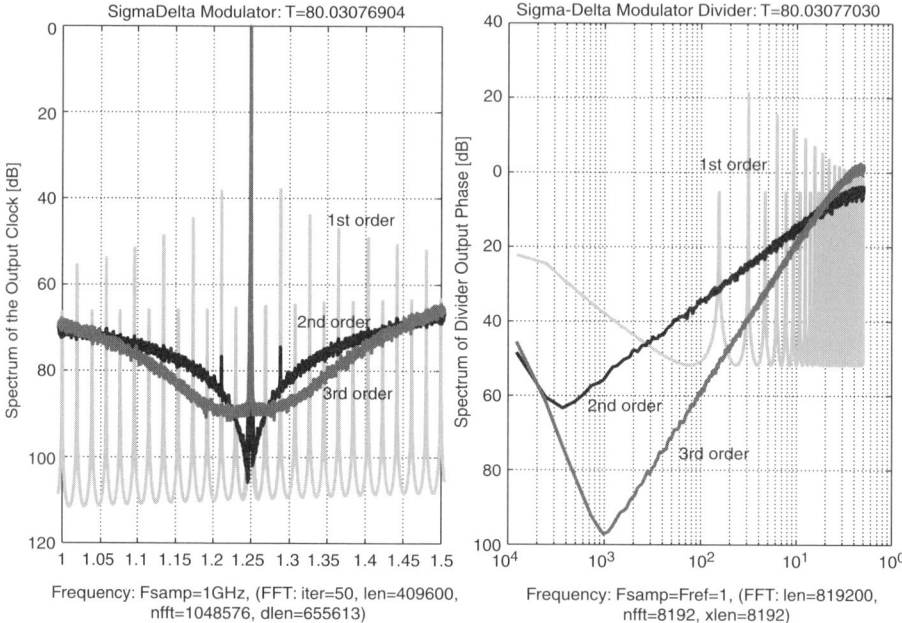

**Figure 1.26** ΣΔ-divided clock: clock output spectrum (left), phase spectrum (right).

performs high-frequency shaping of the division ratio quantization noise of 20 dB/decade, whereas the third order produces 40 dB/decade. The noise at higher frequencies then gets filtered out in the loop filter. While the phase spectrum $S_\phi$ is important in determining the noise floor, it is less convenient in dealing with frequency spurs.

The ΣΔ modulators are usually built as a multistage structure of single-bit modulators [28]. As derived in reference [26], the quantization noise shaping is related to the number of modulator stages $m$ by the following formula and is given in rad$^2$/Hz:

$$\mathcal{L}\{f\} = \frac{(2\pi)^2}{12 f_{\text{ref}}} \left(\frac{f}{f_{\text{ref}}/2\pi}\right)^{2(m-1)} \quad (1.21)$$

The fractional-$N$ frequency synthesizer architecture lends itself well to an indirect narrowband frequency modulation which could be implemented entirely in a digital manner. As long as the modulation data rate is lower than the PLL bandwidth, the average division ratio $N$ digital command word, which corresponds to a desired channel, could be augmented by the instantaneous value of the modulation frequency deviation. There has been some research to increase the data rate by compensating for the PLL high-frequency attenuation by boosting the high-frequency components of the modulation signal as shown in reference [29]. After the equalized

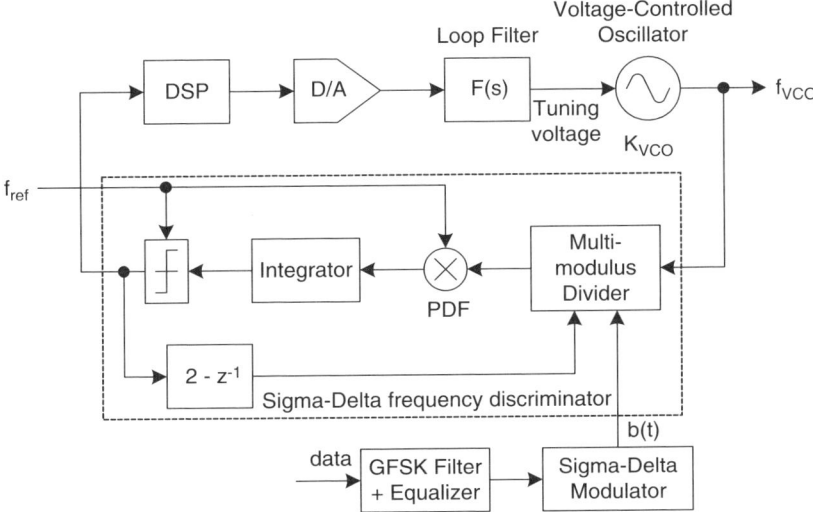

**Figure 1.27** Modulating a wideband fractional-$N$ synthesizer.

modulation signal passes through the PLL, the modulation spectrum could be restored to its original form. The digital equalizer could be embedded in the GFSK filter with little extra overhead. However, the precise loop compensation requirement makes this architecture not very practical for manufacturing.

The problem of mismatch between the digital compensation filter and the analog PLL got addressed by Bax and Copeland [30]. The fractional-$N$ PLL is re-architectured there to place a $\Sigma\Delta$ frequency discriminator, whose transfer function is set digitally and well controlled, in the feedback path (Fig. 1.27). As a result, the only analog component left that requires a substantial amount of matching is the VCO.

### 1.3.3 Toward an All-Digital PLL Approach

At the time the all-digital techniques described in this book were introduced, there was no reported research on successful low-cost and low-power synthesizer architectures for mass-market RF mobile communications that would employ an *all-digital* approach. There were, however, several reports on all-digital PLL (ADPLL) synthesizers [31–35] used for clock recovery applications in wireline communication circuits (i.e., Ethernet, asynchronous transfer mode, fiber optics, etc.), as well as clock generation in microprocessors and DSPs. In most ADPLLs, the oscillator is controlled by digital commands as opposed to analog tuning voltage. The rest of the controller parts can be designed at the HDL level. However, intensive SPICE simulation still has to be carried out to ensure that the target frequency band remains achievable under PVT variations. Specific transistor sizing and layout design of the oscillator have to be "handcrafted" as the design specification or process changes [32].

**24** INTRODUCTION

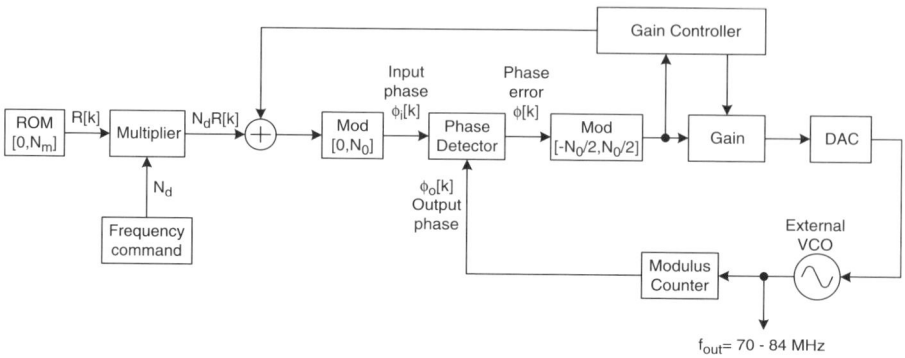

**Figure 1.28** Phase-domain PLL. (Based on [36], © 1992 IEEE.)

The applications described above have much more relaxed requirements on the phase noise and spurious content than those of RF wireless applications. In fact, they are *all* based on a ring oscillator structure which inherently features relatively poor phase-noise characteristics. Another limitation is their integer-only multiplication ratio $N$.

A 1992 paper by Kajiwara and Nakagawa [36] is worth mentioning here. The authors describe a digitally intensive architecture (shown in Fig. 1.28) that is suitable only for lower-frequency synthesizers. It addresses the chief limitation of traditional PLL-based frequency synthesizers, the slow frequency switching times, which make them less desirable for advanced portable wireless applications that use spread-spectrum and frequency-hopping techniques. On the other hand, the direct digital synthesizers, whose switching time is extremely fast, cannot be used in a straightforward manner at wireless frequencies. The paper proposes a *digital signal processor* (DSP)–based phase-domain PLL structure that features fast switching times. This architecture, however, has a major limitation of handling only integer-$N$ division ratios.

The authors make a very important observation that since the reference phase and oscillator phase are in a linear form, their difference produced by the phase detector is also linear with no spurs, and the loop filter is not needed. This is in contrast with conventional PLLs with a correlation detector, especially those based on a charge pump, which generate a significant amount of spurs and which require a filter that degrades the transients and limits the switching time (Fig. 1.19). The settling time is only several reference clock cycles for the larger loop bandwidth setting. The acquisition time of this synthesizer is barely influenced by the step frequency, due to the lack of a loop filter that accumulates the phase error.

The metastability problem that arises due to asynchronous timing between the two digital-phase detector inputs is handled by converting the terminating path into an analog domain through the DAC, which does not require resampling. For this reason, the architecture is not entirely digital. In this book we describe techniques for the design of a low-cost low-power all-digital synthesizer suitable for mass-market RF wireless communications.

## 1.4 IMPLEMENTATION OF AN RF SYNTHESIZER

### 1.4.1 CMOS vs. Traditional RF Process Technologies

RF front-end circuits have traditionally been implemented by gallium arsenide (GaAs) MESFET, Si-bipolar junction transistors (BJTs), III–V heterojunction bipolar junction transistors (HBTs) and silicon–germanium (SiGe) HBTs, whereas baseband *digital signal processing* (DSP) and analog circuits are being implemented exclusively using CMOS technologies. As the emphasis of wireless applications evolves toward *personal communication systems* (PCSs) and *wireless local area networks* (WLANs) as well as wireless entertainment electronics, light weight, small dimensions, low cost, low power, and a higher level of integration are becoming ever more critical. These have spurred interest in wide use of low-cost CMOS technologies, which are typically used for *application-specific integrated circuits* (ASICs). RF wireless systems using CMOS technologies are being investigated intensively, due mainly to their low cost, high yield, and higher level of integration, with baseband circuits.

On the device physics side, there is a fundamental limitation in bipolar technology that prevents it from being used in low-voltage applications. Operation at a 1-V power supply is difficult to achieve since the *base–emitter voltage* $V_{BE}$ of bipolar transistors lies around 0.7 V and the base-to-collector junction must be reverse biased. By contrast, an MOS transistor can operate with a *drain–source voltage* $V_{DS}$ lower than the *gate–source voltage* $V_{GS}$, which in turn can be reduced by lowering the threshold voltage of the MOS device. Therefore, if the saturation voltage is kept at a few hundred millivolts, there is enough room for some output dynamic range.

### 1.4.2 Deep-Submicron CMOS

As the feature size of the digital CMOS technology becomes smaller, the frequency at which CMOS transistors can operate while delivering acceptable performance becomes higher [37–40]. In terms of CMOS transistor application for RF circuits, there are several *figure-of-merit* (FOM) parameters: cutoff frequency $f_T$, maximum oscillation frequency $f_{max}$, minimum noise figure $F_{min}$, flicker noise $1/f$, power-added efficiency PAE, and power gain $G_A$. A conventional deep-submicron CMOS technology designed for logic applications has been shown to exhibit useful RF device characteristics [39].

Scaling of CMOS devices increases the $1/f$ flicker noise, which is caused by carrier trapping near the thin oxide–silicon interface. This phenomenon is absent in bipolar transistors. The flicker noise could be converted into close-in phase noise of *voltage-controlled oscillators* (VCOs) and appear at the output of down-conversion mixers. Fortunately, as shown in [37], this noise frequency translation mechanism can be avoided through half-circuit topological symmetry.

Both $f_T$ and $f_{min}$ peak values already exceed 100 GHz for sub−0.1-μm deep-submicron CMOS logic devices and are predicted to double every three years [37] for the foreseeable future. Despite the continual RF performance lag of CMOS

**26** INTRODUCTION

relative to the latest SiGe bipolar processes, at this point CMOS performance is adequate for wireless communication in frequency bands up to 5 GHz.

This has opened the possibility for CMOS RF circuits that meet the stringent requirements of communication systems. Efforts have been made to show the feasibility of CMOS front-end circuits [41–43], and the performance has become comparable to those of BJT circuits. The ultimate goal is to take advantage of the CMOS process to implement a *single-chip radio* by integrating RF front end *intermediate-frequency* (IF) modulation/demodulation circuits and analog baseband signal-processing circuits with the digital baseband section.

In this book, the following fundamental property of the digital deep-submicron process is utilized as basis for a *new paradigm*:

> In a deep-submicron CMOS process, time-domain resolution of a digital signal edge transition is superior to voltage resolution of an analog signal.

This is in clear contrast with the older process technologies, which rely on a large supply voltage (originally, 15 V, then 5 V, and finally, 3.3 V and 2.5 V) and a stand-alone configuration with few extraneous noise sources in order to achieve a high signal-to-noise ratio and resolution in the voltage domain, often at the cost of a long settling time. In a deep-submicron process, with its low supply voltage (at or below 1.5 V) and relatively high threshold voltage (0.6 V and often higher, due to the body effect), the voltage headroom available is quite small. Moreover, the massive switching noise of a large digital circuitry around makes it harder to resolve signals in the voltage domain. At the same time, the switching characteristics of a MOS transistor are excellent with rise and fall times on the order of tens of picoseconds or less.

### 1.4.3 Digitally Intensive Approach

Rapid reduction in the transistor feature size in recently developed deep-submicron CMOS processes shifts the design paradigm toward more digitally intensive or digitally assisted techniques [44]. In a monolithic implementation, the manufacturing cost of a design is measured not in terms of a number of devices used but rather, in terms of the silicon area occupied and is only marginally dependent on the actual circuit complexity. The testing part of the overall cost does indeed depend on the circuit complexity, but a large number of digital gates typically have higher test coverage at lower testing cost than even a small analog circuit.

Each new digital CMOS process node occurs roughly every 18 months, resulting in an increase in the digital gate density by a factor of 2 (known as Moore's law). A typical digital cellular phone on the market today contains over 4 million transistors. In contrast, analog and RF circuits do not scale down very well. For example, one of the latest commercial CMOS processes [45] (summarized in Table 1.1) which has a 0.13-$\mu$m feature size, achieves an astonishing digital gate density of 150K equivalent (two-input NAND) gates per mm$^2$, which is an order of magnitude greater than is achievable with more traditional RF BiCMOS process

**Table 1.1 Technology Parameters of Texas Instruments' 0.13-μm CMOS Process**

| | |
|---|---|
| Interconnection material | Copper |
| Top metal resistivity | $\leq 10$ mΩ/square |
| Minimum metal pitch | 0.35 μm |
| M1 routing pitch | 0.425 μm |
| Transistor nominal voltage | 1.5 V |
| $L_{\text{drawn}}$ | 0.11 μm |
| $L_{\text{effective}}$ | 0.08 μm |
| Gate oxide | 29 Å |
| Substrate resistivity | $\leq 50$ Ω · cm |

technologies. A typical inductor for an integrated LC oscillator occupies about 0.5 mm² of silicon area! A low-noise charge pump or a low-distortion image-reject mixer (both good examples of classical RF transceiver components) occupy roughly about the same area, which could be traded for tens of thousands of digital gates, which is a lot of DSP power! Consequently, there are numerous incentives to look for digital solutions. Furthermore, high-performance digital circuits and digital signal-processing techniques can be exploited for improving the performance of analog circuits in deep-submicron CMOS by providing "digital assistance" [44].

Migrating to the digitally intensive RF front-end architecture brings with it the advantages of conventional digital design flow, including:

- Fast design turnaround cycles using automated CAD tools (VHDL or Verilog hardware-level description language, synthesis, auto-place and auto-route with timing-driven algorithms, parasitic back-annotation, and post-layout optimization).
- Much lower parameter variability than with analog circuits.
- Ease of testability—automatic functional testing with good fault coverage.
- Smaller silicon area and less power dissipated, which can only get better with each CMOS technology advancement (also called a *process node*).
- An excellent chance of first-time silicon success. Commercial analog circuits usually require several design, layout, and fabrication iterations to meet marketing requirements.

### 1.4.4 System Integration

Integration presents a wide array of opportunities. The most straightforward way would be to merge various digital sections into a single silicon die, such as DRAM or flash memory embedded into a DSP or controller. More difficult would be integrating the analog baseband with the digital baseband. Care must be taken here to avoid coupling of digital noise into the high-precision analog section, usually through substrate or power/ground supply lines. In addition, the low-voltage headroom challenges one to find new circuit and architectural solutions.

**28**  INTRODUCTION

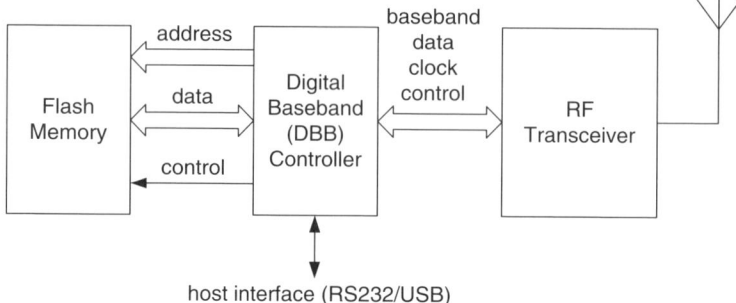

**Figure 1.29** Ultimate mobile wireless integration: single-chip BLUETOOTH radio.

Integrating the analog baseband into an RF transceiver section presents a different set of challenges.

Sensible integration of diverse sections results in the following advantages:

- A lower total silicon area. In a deep-submicron CMOS design, the silicon area is often bond-pad limited. Consequently, it is beneficial to merge various functions onto a single silicon die to maximize the core/bond pad ratio.
- A lower component count and thus lower packaging cost.
- Power reduction; there is no need to drive large external interchip connection capacitances.
- A smaller printed-circuit-board (PCB) area, thus saving precious "real estate."

The ultimate goal in mobile wireless integration is a single-chip digital radio (Fig. 1.29). The *digital baseband controller* is usually based on a *digital signal processor* (DSP) or an ARM7-like microprocessor and is responsible for taking the digital data stream to and from the RF transceiver and performing any necessary digital signal processing on it to convert the digital data stream into a stream of user data. Examples of the processing performed by the DBB controller may include digital filtering, data encoding and decoding, and error detection and correction. It also implements the GSM[3] cellular or BLUETOOTH [46] protocol layer stack, which is controlled by a software program stored in a nonvolatile flash memory. The RF transceiver module implements the physical layer by converting the information bits to and from the RF waveform. The advanced deep-submicron CMOS process total integration leads to an extremely compact and economic implementation of this sophisticated and highly functional communication system.

---

[3]GSM originally stood for *groupe spécial mobile* but for marketing reasons was later renamed *global system for mobile communications*.

### 1.4.5 System Integration Challenges for Deep-Submicron CMOS

Deep-submicron CMOS processes present new integration opportunities on the one hand, but make it extremely difficult to implement traditional analog circuits, on the other. For example, frequency tuning of a low-voltage deep-submicron CMOS oscillator is an extremely challenging task, due to its highly nonlinear frequency vs. voltage characteristics and low voltage headroom, making it susceptible to the power/ground supply and substrate noise. In a circuit with a low supply voltage, not only does the dynamic range of the signal suffer but also, the noise floor rises, thus causing even more severe degradation of the signal-to-noise ratio. At times, it is possible to find a specific solution, such as utilizing a voltage doubler [47]. Unfortunately, with each CMOS gate oxide thickness reduction, the supply voltage also needs to be scaled down, which is inevitable to avoid gate breakdown and reliability problems [48].

Moreover, the high degree of integration leads to the generation of substantial *digital switching noise*, which is coupled through a power supply network and substrate into noise-sensitive analog circuits [49]. Furthermore, advanced CMOS processes typically use low-resistance P-substrate, which provides an effective means of combating latch-up problems but exacerbates substrate noise coupling into the analog circuits. This problem gets worse with scaling down of the supply voltage. Fortunately, there is a serious effort today among major IC fabrication houses to develop CMOS processes with higher-resistivity silicon substrates.

Circuits designed to ensure proper operation of RF amplifiers, filters, mixers, and oscillators depend on circuit techniques that operate best with long-channel thick-oxide devices with a supply voltage of 2.5 V or higher. The process utilized for this book is optimized for short-channel thin-oxide devices operating as digital switches at only 1.5 V.

To address the various deep-submicron RF integration concerns, some new and radical system and architectural changes have to be discovered. In this book, alternative approaches and architectures for RF front ends are explored. This will allow easier integration of an RF section into the digital baseband.

# CHAPTER 2

# DIGITALLY CONTROLLED OSCILLATOR

As discussed in Section 1.1, the phase and frequency information of a *discrete-time* oscillator is contained not in the local waveform fit to the ideal sinusoid but in the significant (rising or falling) edge transition instances. If a transition timestamp represents a positive zero crossing of an arbitrary-amplitude sinusoid, this should provide a sufficient amount of phase information. However, for practical reasons it is necessary for the digital signal to have the same frequency as the desired output signal; therefore, falling edge transitions are naturally generated halfway between positive edge transitions.

From an information theory standpoint, this is a very efficient mechanism to represent a signal containing phase and frequency information. It is in close alignment with the fundamental strength of the digital deep-submicron CMOS processes stated in Section 1.4.2: Time-domain resolution is superior to voltage-domain resolution.

The oscillator presented is a cell with only digital *inputs/outputs* (I/Os) operating in the discrete-time domain, even though the underlying functionality is mainly continuous time and continuous amplitude in nature. This is a very important consideration since it stops the analog nature from propagating up in the hierarchy, right at the interface level. The analog design, modeling, and simulation constraints of the system are thus vastly simplified. An analogy can be drawn here to a flip-flop and its fundamental role in sequential digital circuits, even though its underlying nature and internals are analog.

A *digitally controlled oscillator* (DCO) is used here as a foundation to perform *digital-to-frequency conversion* (DFC). Its output is a periodic waveform whose

---

*All-Digital Frequency Synthesizer in Deep-Submicron CMOS*, by Robert Bogdan Staszewski and Poras T. Balsara
Copyright © 2006 John Wiley & Sons, Inc.

frequency *f* is a certain function of the input *oscillator tuning word* (OTW):

$$f = \text{f(OTW)} \quad (2.1)$$

In general, f(OTW) mapping of digital input to the frequency of oscillation is a nonlinear function. The frequency setting function is not known precisely and varies with the process spread and environmental factors (voltage and temperature). The instantaneous value of the frequency depends on power/ground and substrate noise as well as truly random phenomena such as thermal and flicker noise. The DCO building block accomplishes only the barest minimum of functionality, and it has to be conditioned by normalization circuits, described in Chapter 3. The DCO also provides necessary means for higher-level blocks to perform self-calibration. The basic DCO described later is built using an *LC* oscillator with fixed inductance and variable capacitance. This variable capacitance in a digital CMOS process can be implemented efficiently using MOS varactors.

## 2.1 VARACTOR IN A DEEP-SUBMICRON CMOS PROCESS

Frequency tuning of a low-voltage deep-submicron CMOS oscillator is quite a challenging task, due to its highly nonlinear frequency vs. voltage characteristics and low voltage headroom. Figure 2.1 shows normalized representative curves of a MOS varactor capacitance vs. control voltage ($C-V$ curve) for both a traditional CMOS process and a deep-submicron process. Previously, the large linear range of a $C-V$ curve could be exploited for precise and wide operational control of frequency. With a deep-submicron process, the linear range is now very compressed

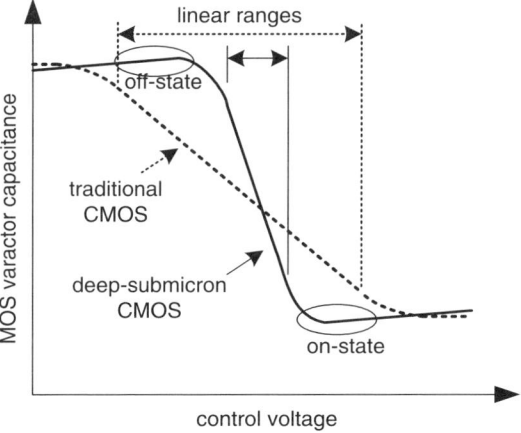

**Figure 2.1** Idealized capacitance vs. voltage curves of a MOS varactor for both a traditional and a deep-submicron CMOS process. (From [50], © 2003 IEEE.)

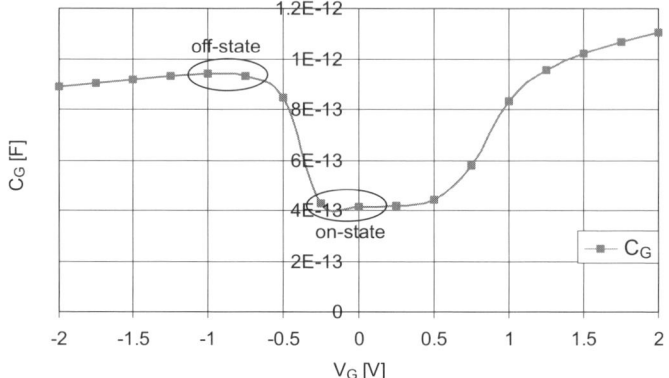

**Figure 2.2** Gate capacitance vs. gate voltage of a measured PMOS varactor: 0.13-μm CMOS process, PPOLY/NWELL, inversion type, single-contacted gate, $L = 0.5$ μm, $W = 0.6$ μm, $N = 8$ fingers × 12 × 2. (From [50], © 2003 IEEE.)

(high $\Delta C/\Delta V$) and has undesirable high gain ($K_{VCO} = \Delta f/\Delta V$), which makes the oscillator extremely susceptible to noise and operating point shifts.

An example $C-V$ curve of an actual PMOS varactor used in the DCO design for the acquisition mode is shown in Fig. 2.2. The data were measured and deembedded from a commercial IC test structure at the intended frequency of operation of 2.4 GHz. Because of the well isolation properties in this N-well process, the PMOS device (Fig. 2.3) is a better candidate for a varactor. The device has the following channel length and width dimensions and finger multiplicity: $L = 0.5$ μm, $W = 0.6$ μm, $N = 8$ fingers × 12 × 2. In this configuration, the source, the drain, and the well are all tied to ground. It was confirmed experimentally that in this process, a PPOLY/NWELL inversion-type varactor features more distinctly defined operational regions than does an accumulation-type varactor. In fact, the flat on-state region of the depletion mode and the flat off-state region of the inversion mode

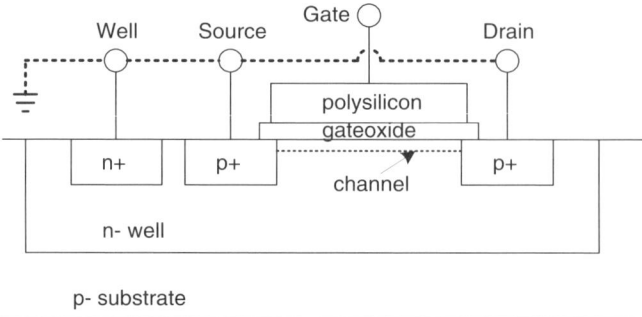

**Figure 2.3** Physical structure of a PMOS transistor used as a varactor when the source, drain, and well tie-offs are tied to ground.

(Fig. 2.2) are used as two stable binary-controlled operating points. Today's advanced CMOS process lithography makes it possible to create extremely small but well-controlled varactors. The switchable capacitance of the finest differential LSB varactor (see Fig. 2.7) is on the order of tens of attofarads.

Still referring to Figs. 2.2 and 2.3, let the gate potential $V_G$ start at $+2$ V, at the right end of the $C-V$ x-axis. The positively charged gate attracts a large number of electrons, which are the majority carriers of an n-type well. The varactor capacitance is relatively high because this structure behaves like a parallel-plate capacitor with only a silicon oxide dielectric in between. A gate conductor forms one plate of the capacitor and the high concentration of electrons in the n-well forms the second plate. This region of operations is termed the *accumulation mode*. As $V_G$ is lowered, fewer and fewer electrons are attracted to the region below the gate, and its concentration drops. This causes the effective "bottom" plate to be further separated, thus lowering the gate-to-well capacitance. As soon as the gate potential is close to zero and enters negative values, the electrons start being repelled, causing a depletion region under the gate. Now the structure is in the *depletion mode*. The capacitance gets lower and lower while the depletion region increases. Lowering $V_G$ further below the (negative) threshold level $V_t$ results in holes being attracted to the region under the gate. This gives rise to a conductive layer of holes, and this region of operation is called the *inversion mode*. Because the bottom plate of the capacitor is just below the gate oxide, the gate capacitance is high again. A strong inversion layer exists at $V_G = -2$ V.

The slight drop of capacitance in the "flat" strong inversion region in Fig. 2.2 had not been of any practical significance until the advent of deep-submicron CMOS processes. It is due to the depletion layer being created in the gate polysilicon [51], which is less doped and much thinner than in the past.

In this varactor structure, the source, drain, and backgate are tied to the same zero potential. This is very similar to the classical MOS capacitor structure, except that the latter does not have the source and drain. To create a channel, the inversion region in the MOS capacitor relies on a process of thermal regeneration of electron and hole pairs, which takes an extremely long time (on the order of microseconds). Consequently, the channel never manages to get created and destroyed for the RF range of frequencies. In the MOS varactor, on the other hand, the source and drain regions serve as vast and ready reservoirs of electrons, so this problem does not exist.

## 2.2 FULLY DIGITAL CONTROL OF OSCILLATING FREQUENCY

The digital solution to control the oscillating frequency could generally be summarized as follows: A method of weighted binary switchable capacitance devices, such as varactors, is used. An array of varactors (Fig. 2.4) could be switched into a high- or low-capacitance mode individually by a two-level digital control voltage bus, thus giving very coarse step control for the more significant bits, and less coarse step control for the less significant bits (LSBs). To achieve very fine frequency

## 34 DIGITALLY CONTROLLED OSCILLATOR

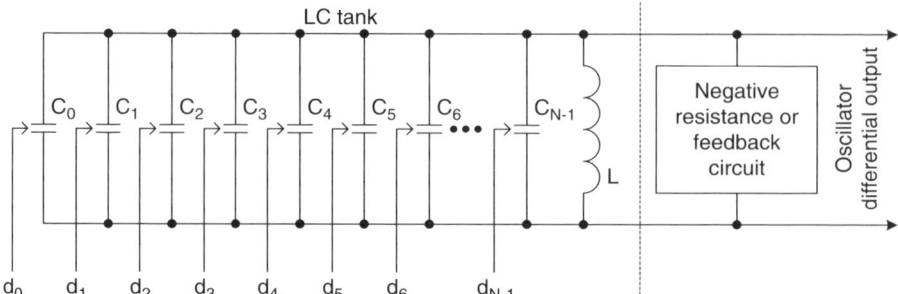

**Figure 2.4** *LC* tank–based oscillator with switchable capacitors. (From [50], © 2003 IEEE.)

resolution, the LSB could possibly be operated in analog fashion. (A similar idea is used in reference [23], which employs a hybrid of digital oscillator control for PVT and *analog* control for acquisition and tracking.) However, this requires a DAC and does not fundamentally solve the problem of the nonlinear VCO gain ($K_{VCO}$) characteristics. A better solution [50] is to dither the LSB digital control bit (or multiple bits), thus controlling its time-averaged value with a finer resolution. Consequently, each varactor could be allowed to stay in only one of the two regions where the capacitance sensitivity is the lowest and the capacitance difference between them is the highest. These two operating regions are shown by the ovals in Figs. 2.1 and 2.2.

This fully digital control of oscillator enables total integration in a deep-submicron CMOS process for the reasons expressed in Chapter 1. Due to the fact that there are several disclosures on ring oscillator–based DCOs for clock recovery and clock generation applications [31–35], where the frequency resolution and spurious tone level are quite relaxed, it seems that the latter two concerns have been an effective deterrent against digital RF synthesizers for wireless communications. The combination of circuit and architectural techniques has brought to fruition a fully digital solution that combines an extremely fine frequency resolution with low spurious content and low phase noise.

The idea of high-rate dithering of LSB capacitors is illustrated in Fig. 2.5. It is similar in principle to the fractional-*N* division ratio dithering presented earlier in conjunction with Fig. 1.21. Instead of applying a constant input that would select capacitance $C_1$ or $C_2$, where $C_2 = C_1 + \Delta C$, with $\Delta C$ being an LSB capacitor, during the entire update cycle, the selection alternates between $C_1$ and $C_2$ several times during the cycle. In the example of Fig. 2.5, $C_2$ is chosen one-eighth of the time and $C_1$ is chosen the remaining seven-eighths of the time. The average capacitance value will therefore be one-eighth of the $C_2 - C_1$ distance above $C_1$. If the dithering speed is performed at a fast enough rate, the resulting spurious tone at the oscillator output could be made vanishingly small (see Appendix A). It should also be noted that resolution of the time-averaged value relies on the dithering speed.

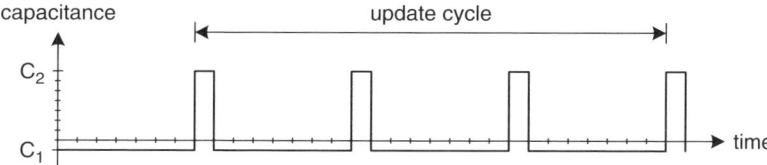

**Figure 2.5** DCO dithering by changing the discrete capacitance at a high rate.

Without any feedback[1] that would result in a supercycle, the dithering rate has to be higher than the update cycle rate times the integer value of the resolution inverse (eight in this case). Therefore, there is a proportional relationship between the frequency resolution improvement and the dithering rate.

The dithering pattern shown in Fig. 2.5 is not random at all and can create detectable spurious tones. It is equivalent to first-order $\Sigma\Delta$ modulation [26]. In Chapter 3, second- and third-order $\Sigma\Delta$ randomization is described to effectively eliminate the already-low spurious content.

## 2.3  LC TANK

The idea behind a digitally controlled *LC* tank oscillator was shown from a higher system level in Fig. 2.4. The resonant frequency of the parallel *LC* tank is established by the well-known formula

$$f = \frac{1}{2\pi\sqrt{LC}} \quad (2.2)$$

The oscillation is perpetuated by a negative-resistance device, which is normally built as a positive-feedback active amplifier network.

The frequency $f$ could be controlled by changing either the inductance $L$, the capacitance $C$, or some combination thereof. However, in a monolithic implementation it is more practical to keep the inductor fixed while changing the capacitance of a voltage-controlled device such as a varactor. Since the digital control of the capacitance $C$ is required, the total capacitance is divided into $N$ smaller digitally controlled varactors, which may or may not follow the binary-weighted pattern of their capacitance values. Equation 2.2 now becomes

$$f = \frac{1}{2\pi\sqrt{L \sum_{k=0}^{N-1} C_k}} \quad (2.3)$$

---

[1]A phase-locked loop would naturally dither the oscillator input to maintain a constant average integrated frequency error.

Digital control signifies that each of the individual capacitors (of index $k$) could be placed in either a high-capacitative state $C_{1,k}$ or a low-capacitative state $C_{0,k}$ (see Fig. 2.1). The capacitance difference between the high- and low-capacitative states of a single-bit $k$ is $\Delta C_k = C_{1,k} - C_{0,k}$ and is considered the effective switchable capacitance. Since the frequency of oscillation grows with lower capacitance, increasing the digital control value must result in the increased frequency of oscillation. Therefore, the digital control state is the opposite of the capacitative state, so the digital bits need to be *inverted* such that the $k$th capacitor could be expressed as

$$C_k = C_{0,k} + \overline{d_k} \Delta C_k$$

The bit inversion turns out to be quite convenient from an implementation point of view. Figure 2.7 reveals that it is necessary to provide a buffering scheme that would (1) isolate the "raw" varactor input from the noisy digital circuits, (2) have sufficiently low driving resistance to minimize the thermal and flicker noise, and (3) establish two stable low and high voltage levels for the best varactor operation. Equation 2.3 could be rewritten to include the digital control details:

$$f = \frac{1}{2\pi \sqrt{L \sum_{k=0}^{N-1} (C_{0,k} + \overline{d_k} \Delta C_k)}} \qquad (2.4)$$

Figure 2.6 shows a model of a single-cell binary-weighted switchable capacitor of index $k$ that is equivalent to a weight of $2^k$. The basic unit cell is created for a weight of $2^0$. The next varactor, of weight $2^1$, is created not as a single device of double the unit area but is built of two unit cells. This is done for matching purposes. It mainly ensures that the parasitic capacitance due to fringing electric fields, which is quite significant for a deep-submicron CMOS process and is extremely difficult to control and model, is well ratioed and matched. Each cell consists of double the number of unit cells in the preceding cell. Even though the total occupied silicon

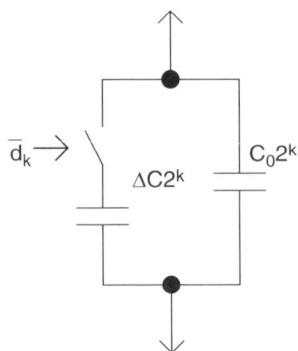

**Figure 2.6** Modeling a binary-weighted switchable capacitor of index $k$. (From [50], © 2003 IEEE.)

area of the device multiplicity method is somewhat larger than the straightforward method of progressively larger uniform devices, it makes it possible easily to achieve an economical component-matching resolution of 8 bits.

When the $d_k$ digital control bit is 1, the only capacitance seen by the oscillating circuit is $C_0$ times the weight. This capacitance is always present because the varactor can never truly be turned off. For this reason it could be treated as a "parasitic" shunt capacitance. The total sum of these contributions $C_{0,\text{tot}}$ sets the upper limit on the oscillating frequency for a given inductance $L$. When the digital control bit is 0, the $\Delta C$ capacitance times the weight is added. The index $k$ of the binary-weighted capacitance can thus be described as

$$C_k = C_{0,k} \cdot 2^k + \overline{d_k}\Delta C_k \cdot 2^k \tag{2.5}$$

making a total binary-weighted capacitance of size $N$:

$$C = \sum_{k=0}^{N-1} C_k = \sum_{k=0}^{N-1}(C_{0,k} \cdot 2^k + \overline{d_k}\Delta C_k \cdot 2^k) \tag{2.6}$$

$$= \sum_{k=0}^{N-1} C_{0,k} \cdot 2^k + \sum_{k=0}^{N-1} \overline{d_k}\Delta C_k \cdot 2^k \tag{2.7}$$

$$= C_{0,\text{tot}} + \sum_{k=0}^{N-1} \overline{d_k}\Delta C_k \cdot 2^k \tag{2.8}$$

Contributions from all the static shunt capacitances are lumped into $C_{0,\text{tot}}$, so the only adjustable components are the effective capacitances in the second term of Eq. 2.8.

## 2.4 OSCILLATOR CORE

Figure 2.7 shows an implementation of the differential varactor and the preceding driver stage [50]. The $V_{\text{tune\_high}}$ and $V_{\text{tune\_low}}$ rail supply levels of the inverter are set to correspond with the two stable operating points, the off-state and on-state, respectively, as shown in Fig. 2.1 ($V_{\text{tune\_high}} = 0.9$ V and $V_{\text{tune\_low}} = 0$ V). The varactor used in this work is a differential configuration built on the basic structure described in conjunction with Figs. 2.2 and 2.3. The balanced capacitance is between the gates of PMOS transistors M1 and M2 (Fig. 2.7), whose source, drain, and backgate connections are shorted together and tied to the M3/M4 inverter output. Since the voltage control is now applied to the backgate and source/drain, the negative and decreasing values of $V_G$ in Fig. 2.2 covering the inversion mode are of interest. Because of the differential configuration, one-half of the single PMOS capacitance is achieved, which actually enhances frequency resolution.

The circuit of Fig. 2.7 also reveals a phase noise contribution mechanism from the static tuning input. When either of the driving transistors (M3 or M4) is turned on, its

## 38  DIGITALLY CONTROLLED OSCILLATOR

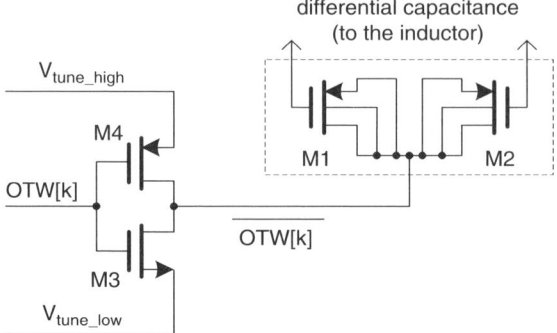

**Figure 2.7** Differential varactor and an inverting driver. (From [50], © 2003 IEEE.)

channel resistance generates thermal noise:

$$\overline{e_n^2} = 4kTR\,\Delta f \tag{2.9}$$

where $\overline{e_n}$ is the root-mean-square (rms) open-circuit noise voltage generated by the driving resistance $R$ over the bandwidth $\Delta f$ at a given temperature $T$, and $k$ is Boltzmann's constant. As an example, a 50-$\Omega$ resistance generates about 0.9 nV of rms noise over a bandwidth of 1 Hz. This noise is added to the stable control voltage, which then perturbs the varactor capacitance. This, in turn, perturbs the oscillating frequency and gives rise to the phase noise. These observations favor selection of large $W/L$ ratios of the driver-stage transistors to reduce the driving resistance and hence thermal voltage noise, and careful selection of the operational states on the $C$–$V$ curve (Fig. 2.2), which would result in the smallest possible capacitance sensitivity to the voltage noise.

Figure 2.8 shows an ideal schematic of a DCO oscillator. The inductor is connected in parallel with an array of differential varactors. NMOS transistors MN1 and MN2 comprise the first cross-coupled pair that provide negative resistance to the $LC$ tank. PMOS transistors MP1 and MP2 provide a second such pair. The current source $I_b$ limits the amount of current the oscillator is allowed to draw. The oscillator output is differential, with "outp" and "outm" pins being fed to the differential-to-complementary circuit, whose purpose is to square the nearly sinusoidal outputs and make them insensitive to the common mode level. This structure of forming the negative resistance by double cross-connection of transistor pairs may be found in the literature [5]. It was chosen for its inherent low power since the current used for amplification is utilized twice. What is novel in this DCO is the replacement of "analog" varactors with a digitally controlled varactor array. This book deals only with system-level design of the DCO and does not address the detailed component- and circuit-level design for designing an RF $LC$ tank oscillator. Interested readers should refer to references [52] and [53] for $LC$ oscillator details.

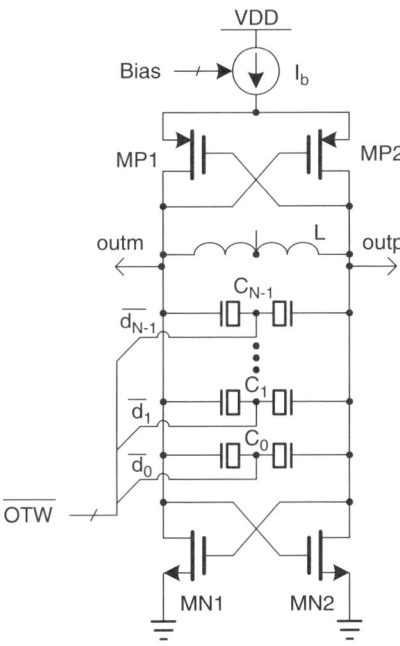

**Figure 2.8** DCO oscillator core with discrete tuning controls.

## 2.5 OPEN-LOOP NARROWBAND DIGITAL-TO-FREQUENCY CONVERSION

From the functional perspective, the operation described above can be thought of as a *digital-to-frequency conversion* (DFC), with the digital word comprising $d_k$ bits, where $k = 0, 1, \ldots, N-1$, directly controlling the output frequency $f$. To illustrate that a straightforward linear DFC conversion to the RF range is not likely to work, let's consider the following example. For a BLUETOOTH application with an oscillating frequency in the RF band of 2.4 GHz and a frequency resolution of 1 kHz, at least 22 bits of DFC resolution is required. It is clearly only with the utmost difficulty that this type of precision can be achieved, even with the most advanced component-matching techniques. The best that one could hope to achieve economically is 8 to 9 bits of capacitor-matching precision [54] without resorting to elaborate matching schemes that often require multiple iterations of time-consuming design, layout, and fabrication cycles. In fact, better than 10-bit resolution would normally require digital error correction techniques [55].

There is one aspect of digital-to-frequency conversion for wireless communications that differs significantly from general digital-to-analog conversion, and we are able to take advantage of it: namely, the narrowband nature of the wireless communication transmission. Consequently, even though frequency command steps must be very fine, the overall dynamic range required at a given instant is quite small. For example, the nominal frequency deviation of the BLUETOOTH GFSK data modulation scheme is 320 kHz. For a 1-kHz frequency resolution, 9 bits are sufficient (320 kHz/1 kHz = 320 < $2^9$). Unless the narrowband nature of BLUETOOTH is exploited, a much higher dynamic range would be necessary to cover the frequency channels of the RF band. For the BLUETOOTH band of 80 MHz, 17 bits of full 1-kHz resolution is thus required. Many additional bits would be necessary to account for the process and environmental (voltage and temperature) changes, which could reach over $\pm 20\%$ of the operational RF frequency.

A better solution to the foregoing dynamic range problem is to make the frequency resolution coarser whenever a higher dynamic range is expected. This is accomplished by traversing through the three major operational modes with progressively lower frequency range and higher resolution such that the intrinsically economical component-matching precision of 8 bits is maintained (Fig. 2.9). In the first step, a large oscillating frequency uncertainty due to *process–voltage–temperature* (PVT) variations is calibrated. After the PVT calibration, the nominal center frequency of the oscillator will be close, but not identical, to the center of the BLUETOOTH band. Since this uncertainty can reach as high as hundreds of megahertz, 1- or 2-MHz tuning steps are satisfactory. In this case, 8-bit resolution is sufficient. The second step is to acquire within the available band the operational channel requested. With 8-bit resolution, $\frac{1}{2}$-MHz steps will span over 100 MHz, which is enough to cover the 80-MHz BLUETOOTH band.

The third step gives the finest resolution but with the tightest narrowband range, and serves to track the frequency reference and to perform data modulation within

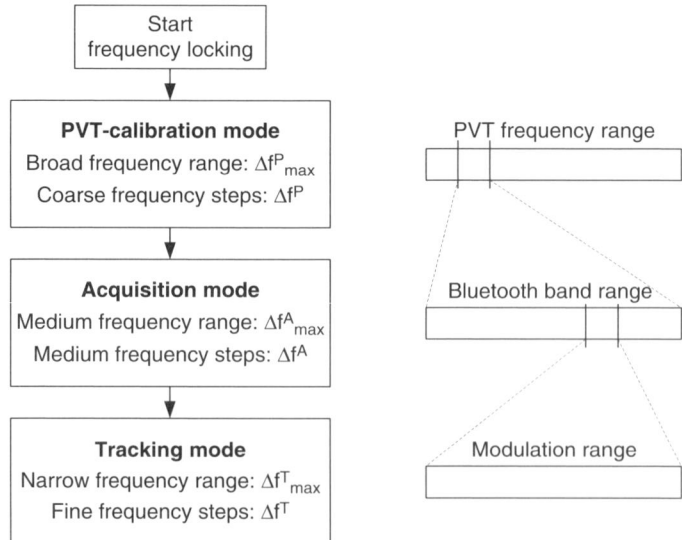

**Figure 2.9** Progression flowchart of DCO operational modes.

the channel. In this approach the 1-MHz channel spacing resolution of the BLUETOOTH band starts at the first step (PVT), but because of the very coarse frequency selection grid, possibly covering multiple channels, the best that could be achieved is to get near the neighborhood of the channel desired. It is in the second step (the acquisition mode) that the channel is approximately acquired. However, fine selection of the channel requested could only be accomplished in the third step (the tracking mode), which is the most refined of them all. Therefore, the tracking mode dynamic range has additionally to cover the resolution grid of the preceding acquisition mode. For the BLUETOOTH example, if frequency in the acquisition mode cannot be resolved to better than 500 kHz and the frequency modulation range is 320 kHz, the dynamic range of the tracking mode should be better than 10 bits [(500 kHz + 320/2 kHz)/1 kHz = 660 < $2^{10}$].

From an operational perspective, the varactor array is divided into three major groups (varactor banks) that reflect three general operational modes: process–voltage–temperature (PVT), acquisition, and tracking. The first and second groups coarsely set the desired center frequency of oscillation before the actual transmit or reception begins, and the third group controls the oscillating frequency precisely during actual operation. During PVT and acquisition, the frequency range is quite high but the precision required is relatively low; therefore, the best capacitor array arrangement here is a binary-weighted structure with a total capacitance of (based on Eq. 2.8)

$$C^P = C_0^P + \sum_{k=0}^{N^P-1} \overline{d}_k^P (\Delta C^P \cdot 2^k) \qquad (2.10)$$

$$C^A = C_0^A + \sum_{k=0}^{N^A-1} \overline{d}_k^A (\Delta C^A \cdot 2^k) \qquad (2.11)$$

where $N^P$ is the number of PVT-mode varactors, $N^A$ is the number of acquisition-mode varactors, $\Delta C^P$ and $\Delta C^A$ are the unit capacitances of the LSB varactors, and $\overline{d}_k^P$ and $\overline{d}_k^A$ are the inverted PVT and acquisition bits, respectively, of the DCO tuning word that controls the capacitance of the varactor devices.

It is important to note that at any given time, only varactors that belong to the same bank are allowed to switch. Consequently, only the varactors in each bank need to be matched. This is the key principle behind achieving an extremely fine digital frequency resolution with only 8-bit basic resolution of component matching.

The PVT correction centers the oscillating frequency of the operational band and could be performed at the time of manufacturing, on power-up, or on an as-needed basis during idle periods of BLUETOOTH operation. The channel select varactor group controls the frequency acquisition process for the transmission channel desired. Both groups are best implemented using individual binary-weighted capacitance structures, but their ranges should be slightly overlapping so that no frequencies between two groups are left out of the tuning range. There is no need to preserve binary-weighted continuity between the process/environmental and channel select

**42** DIGITALLY CONTROLLED OSCILLATOR

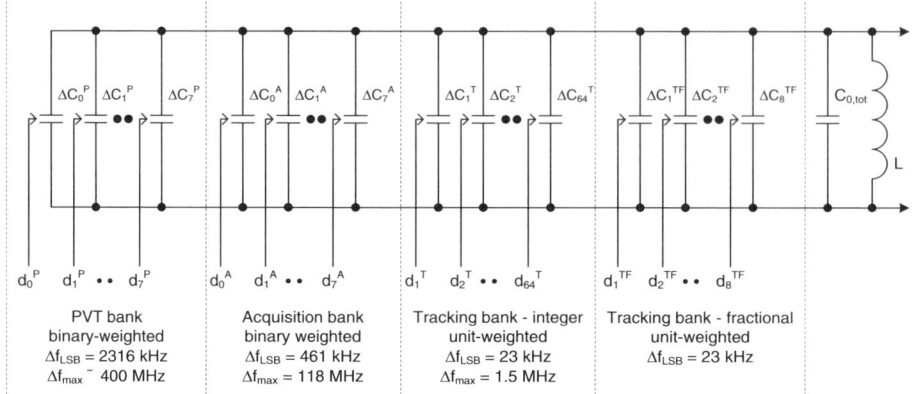

**Figure 2.10** *LC* tank with dedicated discrete capacitor banks for each of the three operational modes (for the BLUETOOTH example). (From [50], © 2003 IEEE.)

structures, due to the different origins of their respective control inputs. The PVT correction is infrequent and could usually be done through register interface (e.g., look-up table created during factory calibration), whereas channel select DCO tuning is performed dynamically and is an integral part of the synthesizer PLL. Figure 2.10 shows the dedicated capacitor banks, which are connected in parallel to create a larger quantized capacitance. Only the effective switchable capacitors are shown forming the banks. The individual shunt capacitances are indistinguishable from each other; therefore, they are lumped together as $C_{0,\text{tot}}$ (as per Eq. 2.8). Also shown in the figure is the fractional-resolution tracking varactor bank for high-speed dithering, which is discussed later.

Tracking-mode operation presents a different set of requirements. The frequency range is relatively low but the resolution required is quite high. The binary-weighted capacitance arrangement of the acquisition mode is a poor choice here, for the following reasons: binary switching noise (changing a value by 1 LSB might require many bits to toggle; for example, incrementing from decimal 31 to 32 causes 6 bits to flip), poor device matching of different-size devices (a 2× precision-matched capacitor is rarely implemented as twice the area—usually, two identical devices are in parallel next to each other), and other problems. A better structure would be an array of unit devices of fine but identical dimensions. The total capacitance is

$$C^T = C_0^T + \sum_{k=1}^{N^T} \overline{d}_k^T \Delta C^T \qquad (2.12)$$

where $N^T$ is the number of tracking-mode varactors, $\Delta C^T$ is the unit-switchable capacitance of each varactor, and the $\overline{d}_k^T$ are the inverted tracking bits of the DCO tuning word.

## 2.5 OPEN-LOOP NARROWBAND DIGITAL-TO-FREQUENCY CONVERSION

Since the relative capacitance contribution of the tracking bank is quite small compared to the acquisition bank, the frequency deviation due to the tracking capacitors could be linearized by the $df/dC$ derivative of Eq. 2.3. Consequently, the frequency resolution or granularity of the $LC$ tank oscillator is a function of the operating frequency $f$:

$$\Delta f^T(f) = -f \frac{\Delta C^T}{2C} \qquad (2.13)$$

where $\Delta C^T$ is the tracking-bank unit-switchable capacitance and $C$ is the total capacitance. The frequency resolution is a function of the operational frequency $f$, which in this implementation is established by the total capacitance $C$ of the $LC$ tank. To make it a function of only one independent variable, Eq. 2.13 is now merged with Eq. 2.2:

$$\Delta f^T(f) = -2\pi^2 L \Delta C^T f^3 \qquad (2.14)$$

In the BLUETOOTH example, $\Delta f^T = 23\,\text{kHz}$. The total tracking-bank frequency deviation from the lowest frequency is

$$f^T(f) = \Delta f^T \sum_{k=1}^{N^T} d_k^T = -f \frac{\Delta C^T}{2C} \sum_{k=1}^{N^T} d_k^T = -2\pi^2 L \Delta C^T f^3 \sum_{k=1}^{N^T} d_k^T \qquad (2.15)$$

The tracking-bank encoding is classified as a redundant arithmetic system since there are many ways to represent a number (since all bits have equal weight). The simplest encoding would be a thermometer scheme with a predetermined bit order. A less restrictive numbering scheme was chosen to facilitate a dynamic element matching—a technique to linearize the frequency vs. code transfer function. This idea is described in Chapter 3.

Further refinement of the frequency resolution is obtained by performing a time-averaged high-speed dither of one or a few of the tracking bits (Fig. 2.11). These bits belong to the separate fractional DCO part of Fig. 2.10 (i.e., bits $d_1^{TF} \cdots d_8^{TF}$). The dithering is performed by a first-, second-, or third-order digital $\Sigma\Delta$ modulator that produces a high-rate integer stream whose average value equals the lower-rate fractional input. The dithering clock is derived from DCO output by dividing its RF clock by a small integer. This method is discussed further in Chapter 3.

The DCO operational mode progression could be described mathematically in the following way. Upon power-up or reset, the DCO is set at a center or "natural" resonant frequency $f_c$ by appropriate presetting of the $d_k$ inputs. This corresponds to a state in which varactors representing half or approximately half of the total capacitance are turned on to extend maximally the operational range in both directions. The total capacitance value of the $LC$ tank is $C_c$ and the natural

## 44  DIGITALLY CONTROLLED OSCILLATOR

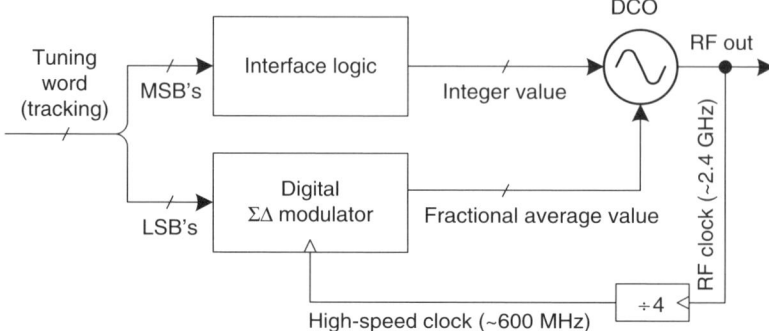

**Figure 2.11** Improving frequency resolution with $\Sigma\Delta$ dither of DCO varactors.

frequency is

$$f_c = \frac{1}{2\pi\sqrt{LC_c}} \qquad (2.16)$$

During PVT mode, the DCO will approach the desired frequency $f$ by setting the $d^P$ control bits appropriately so that the new total capacitance is $C_{tot,P} = C_c + \Delta C^P$ (where $\Delta C^P$ is a signed number). The resulting final frequency of the PVT mode is

$$f_c^P = \frac{1}{2\pi\sqrt{LC_{tot,P}}} \qquad (2.17)$$

The acquisition mode will start from a new center frequency of $f_c^P$. It will approach the desired frequency $f$ by appropriately setting the $d^A$ control bits so that the new total capacitance is $C_{tot,A} = C_c + \Delta C^P + \Delta C^A$. The resulting final frequency of the acquisition mode is

$$f_c^A = \frac{1}{2\pi\sqrt{LC_{tot,A}}} \qquad (2.18)$$

Finally, the tracking mode will begin from a new center frequency of $f_c^A$. It will reach and maintain the desired frequency $f$ by setting the $d^T$ control bits appropriately so that the new total capacitance is $C_{tot,T} = C_{0,tot} + C_c + \Delta C^P + \Delta C^A + \Delta C^T$. The resulting frequency of the tracking mode is set by Eq. 2.2.

The mode progression following the process in Fig. 2.9 contains two mode-switching events during which the center frequency is shifted rapidly closer and closer to the desired frequency. At the end of the PVT and acquisition modes, the terminating-mode capacitor state is frozen and it now constitutes a new center frequency ($f_c^P$ or $f_c^A$) from which, during the subsequent tracking mode, the frequency offsets are calculated.

## 2.6 EXAMPLE IMPLEMENTATION

The frequency tuning of a DCO is accomplished by means of the quantized capacitance of an *LC* tank oscillator. The DCO has three operational modes, with the following arithmetic word encoding, nominal frequency resolution, and range as implemented for a BLUETOOTH example:

- *Process–Voltage–Temperature (PVT)-Calibration Mode.* This mode is active during cold power-up and on an as-needed basis and places the nominal center frequency of the DCO in the middle of the BLUETOOTH band ($f_c = 2.44$ GHz). It is also possible to use this mode on a regular basis as an ultrafast acquisition before the regular acquisition mode. It uses an 8-bit binary-weighted encoding. For best matching, the binary weights are obtained by means of finger multiplicity of a unit-size varactor; for example, for 4-bit encoding, if $C_0$ represents a unit weight capacitance corresponding to a LSB with weight $2^0$, capacitances are created at bits 1, 2, and 3 using two, four, and eight copies of $C_0$, respectively. Frequency resolution $\Delta f^P = 2316$ kHz. Frequency range $\Delta f^P_{\max} \approx 400$ MHz.
- *Acquisition Mode.* This mode is active during channel select. It uses 8-bit binary-weighted encoding. Again, the binary weights are obtained by means of the finger multiplicity of a unit-size varactor. Frequency resolution $\Delta f^A = 461$ kHz. Frequency range $\Delta f^A_{\max} = 118$ MHz.
- *Tracking Mode.* This mode is active during actual transmitting and receiving. It employs 64-bit unit-weighted encoding, with 8-bit unit-weighted encoding for fractional resolution. Frequency resolution $\Delta f^T = 23$ kHz, corresponding to capacitative resolution of $\Delta C^T = 38$ aF, as governed by Eq. 2.13. Frequency range $\Delta f^T_{\max} = 1.472$ MHz.

The frequency numbers are obtained through calculation using the capacitor and inductor models obtained from RF modeling and subsequently confirmed through SPICE and Cadence SpectreRF simulations. They were finally verified through lab measurements of a working silicon chip.

The *LC* tank inductor used was an integrated planar inductor [5] realized as a center-tap octagonal structure constructed of metal in layers 3 through 5. It is the biggest single component on the entire chip and is clearly discernible in the lower left corner of the chip micrograph in Fig. 7.3. This micrograph dramatically illustrates the high cost (in terms of digital gates) of conventional RF components in high-density modern CMOS processes. This supports our contention that the number of classical RF components should be minimized with proper architectural and circuit design choices.

Figures 2.12 and 2.13 show a numerical example of frequency traversal for the DCO implemented. The numbers on the horizontal axes correspond to the decimal equivalent of $d^X$ bits for each of the $X = P, A, T$ banks. The fractional

**46** DIGITALLY CONTROLLED OSCILLATOR

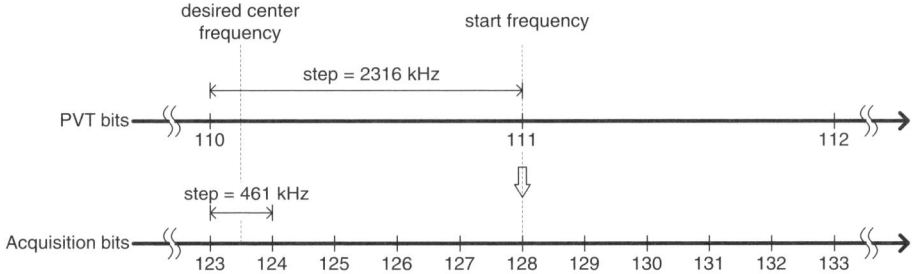

**Figure 2.12** Frequency traversal example for the DCO implemented: PVT to acquisition. PVT is calibrated to the middle of the BLUETOOTH band with code 111. (From [50], © 2003 IEEE.)

tracking bits do not contribute significantly to the gross frequency tuning, so for clarity are not included.

Assume that the PVT mode had earlier been calibrated to the middle of the BLUETOOTH band (2.44 GHz) by fixing the selection to the PVT code of $d^P = 111$. The acquisition mode therefore starts from its midpoint reset value of $d^A = 128$, where roughly half of the varactor finger units are turned on and the other half are turned off. Let's assume further that the desired center frequency lies two channels or 2 MHz lower from the center channel of the BLUETOOTH band. This translates to between four and five acquisition steps. As a result, the loop will first quickly move several steps lower from the starting code of 128. After reaching a point about 2 MHz away, it will alternate between the codes of 123 and 124, not being able to resolve any finer. In this example, transition to the tracking mode happens when the acquisition code is 123. This code stays frozen for the duration of the packet. The tracking mode always starts from the midpoint value of $d^T = 31$, as shown in Fig. 2.13. It happens that in this example the center frequency desired is located about 230 kHz higher from that point. This corresponds to 10 tracking steps. During transmission, another 160 kHz on both sides is allocated to the frequency modulation.

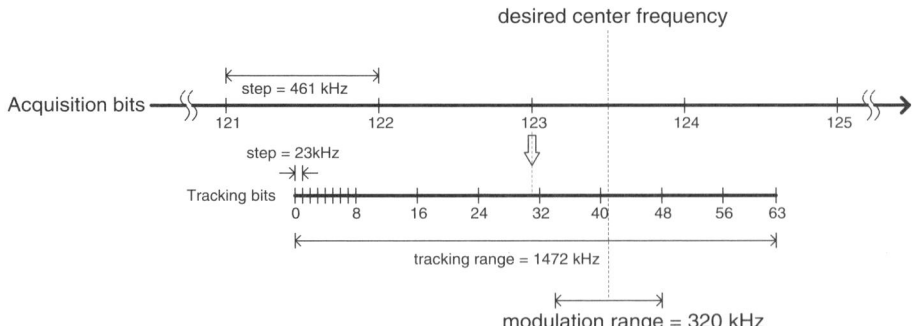

**Figure 2.13** Frequency traversal example for the DCO implemented: acquisition to tracking.

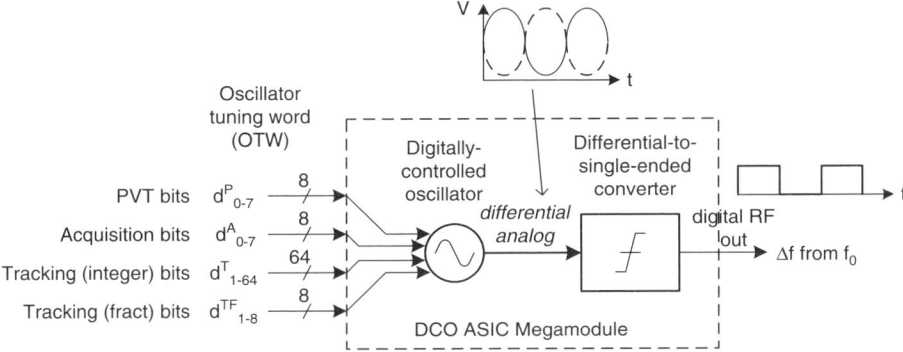

**Figure 2.14** DCO as an ASIC cell with digital I/Os. (From [56], © 2003 IEEE.)

In this implementation, the oscillator is built as an ASIC cell (Fig. 2.14) with truly digital I/Os, even at the RF output frequency of 2.4 GHz, which have rise and fall times specified to be less than 50 ps. The RF signal digitizer is a differential-to-digital converter (with complementary outputs) that transforms the analog oscillator waveform into the zero-crossing digital waveform with a high degree of common-mode rejection.

## 2.7 TIME-DOMAIN MATHEMATICAL MODEL OF A DCO

Due to the fact that the conventional RF synthesizers are based on the frequency-domain model, whereas the discrete-time architecture described here is rooted in the time domain, in this section we introduce basic time-domain equations and modeling concepts that are used in this architecture. It should be noted here that recently there have been other attempts to model RF frequency synthesizers in the time domain, such as in reference [57] for $\Sigma\Delta$ fractional-$N$ PLLs.

Let the nominal frequency of oscillation be $f_0$. Corresponding to the nominal clock period $T_0$ is $f_0 = 1/T_0$. If the clock period is shortened by $\Delta T$ (see Fig. 2.15), the new clock period will be $T = T_0 - \Delta T$. This will result in a higher frequency of oscillation of $f = f_0 + \Delta f$. Let's determine the relationship between $\Delta f$ and $\Delta T$. Expanding $f = 1/T$ results in

$$f_0 + \Delta f = \frac{1}{T_0 - \Delta T} = \frac{1/T_0}{1 - \Delta T/T_0} = \frac{f_0}{1 - \Delta T/T_0} \tag{2.19}$$

For $\Delta T/T_0 \ll 1$, using the approximation formula $1/(1-\varepsilon) \approx 1+\varepsilon$, Eq. 2.19 simplifies to

$$\Delta f \approx f_0 \frac{\Delta T}{T_0} = f_0^2 \, \Delta T = \frac{\Delta T}{T_0^2} \tag{2.20}$$

**48**  DIGITALLY CONTROLLED OSCILLATOR

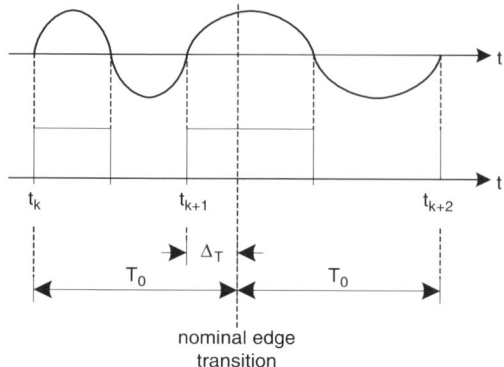

**Figure 2.15** Frequency deviation as a clock period deviation.

Therefore, for small frequency and clock period deviations, their relationship is linear by the proportionality relation

$$\frac{\Delta T}{T_0} \approx \frac{\Delta f}{f_0} \qquad (2.21)$$

The linear approximation error of Eq. 2.21 is plotted in Fig. 2.16, where $\varepsilon$ could be either $\Delta T/T_0$ or $\Delta f/f_0$. The error is negligible for reasonable frequency deviations. Even with $\Delta f = 24$ MHz, such that $\varepsilon = 24 \times 10^6/2400 \times 10^6 = 0.01$, the linear approximation error of $\Delta T$ is $-0.01\%$.

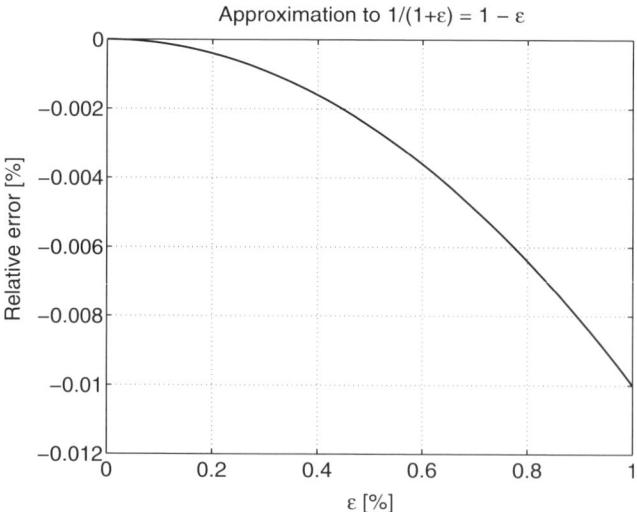

**Figure 2.16** Linear approximation error of frequency vs. period deviations, $\varepsilon = \Delta T/T_0$.

## 2.7 TIME-DOMAIN MATHEMATICAL MODEL OF A DCO

**Table 2.1  Timing Deviation vs. Frequency Deviation at Various Points in the BLUETOOTH Band Example**

| Period Deviation, $\Delta T$ (fs) | Center Frequency, $f_0$ (MHz) | Frequency Deviation, $\Delta f$ (Hz) |
|---|---|---|
| 1 | 2400 | 5760 |
| 1 | 2402 (first channel) | 5770 |
| 1 | 2440 (middle channel) | 5953 |
| 1 | 2480 (last channel) | 6159 |
| 1 | 2500 | 6250 |

Equation 2.20 is used extensively in this book as a conversion formula for system analysis and simulation. As described in Chapter 6, the simulation environment is VHDL, which being an event-driven digital simulator is foreign to the concept of frequency and operates exclusively in the native time domain.

Table 2.1 relates (based on Eq. 2.20) the DCO frequency deviation due to 1 fs (femtosecond) of a period deviation for several frequencies around the beginning and end of the BLUETOOTH band. It is obvious that a fine timing resolution is required at RF frequencies for time-domain simulation tools. In fact, it is necessary to resort to the finest timing resolution of 1 fs that the VHDL standard provides. From a physical viewpoint, it is clear that a femtosecond time deviation is quite meaningless for a single observation[2] and that only an averaged value could make sense.

Table 2.2 relates the DCO frequency resolution for various modes to the DCO period deviation at the middle of the BLUETOOTH band. For an oscillator with a time-invariant frequency excursion, a $\Delta T$ period deviation from $T_0$ will result in a $\Delta T$ deviation from ideal timing instances within one oscillator clock period, $2\Delta T$ within two clock periods, and so on, as shown in Fig. 2.17. Within $i$ oscillator clock cycles, the accumulated *timing deviation* (TDEV) will reach

$$\text{TDEV}[i] = i\Delta T = i\frac{\Delta f}{f_0^2} \qquad (2.22)$$

If the $\Delta T$ value varies for every oscillator cycle, Eq. 2.22 could be rewritten as

$$\text{TDEV}[i] = \sum_{l=1}^{i} \Delta T[l] = \sum_{l=1}^{i} \frac{\Delta f[l]}{f_0^2} \qquad (2.23)$$

Equation 2.23 states that TDEV, defined as the difference between actual and ideal timing instances, is an integral of the oscillator frequency deviation. The $\Delta T$ direction is selected toward shortening the period such that the $\Delta T$ and $\Delta f$ signs agree.

---

[2]One femtosecond is such a short interval of time that the visible light travels less than its wavelength during this interval.

**Table 2.2** Frequency Resolution and Corresponding DCO Period Deviation for the DCO Modes at the Middle of the BLUETOOTH Band, $f_0 = 2.44$ GHz

| Mode | Frequency Resolution, $\Delta f$ (kHz) | Equivalent Time Resolution, $\Delta T$ (fs) |
|---|---|---|
| PVT | 2316 | 390.7 |
| Acquisition | 461 | 77.43 |
| Tracking | 23 | 3.863 |

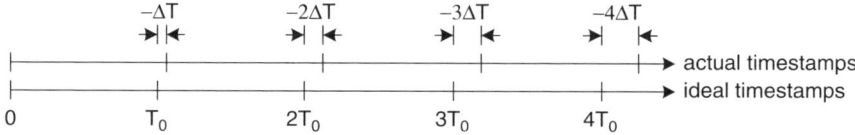

**Figure 2.17** Development of an accumulative timing deviation (TDEV).

The time-domain DCO model is presented in Fig. 2.18. The DCO input $d[i]$ is an equivalent signed representation of the *oscillator tuning word* (OTW), which corresponds to the active $d^X$ bits in Fig. 2.14, where $X$ is $P$, $A$, or $T$ and $TF$. The $d[i]$ signed input will change its operating frequency by $\Delta f[i] = d[i]K_{\text{DCO}}$ (where $\Delta f[i]$ is deviation with respect to the oscillator's natural frequency). On every rising DCO edge event, the DCO event output $\Delta f$ multiplied by a "constant" $1/f_0^2$ will be accumulated. At the end of $i$ cycles, it will accumulate the TDEV timing deviation according to Eq. 2.23. The sampled or $z$-domain representation of the accumulation, as opposed to the more commonly used $s$-domain, implies that the accumulated timing deviation is defined only at the end of the DCO clock cycles with each rising clock edge. It should be noted that because the phase is fundamentally a time integral of frequency, the DCO phase accumulation is a pure time development process and is not tied to hardware. The timing deviation is a measure of "badness" and signifies a departure from the desired timing instances that has to be corrected by a feedback loop mechanism, which is introduced in Chapter 4.

The oscillator period deviation $\Delta T$, or its inverse $\Delta f = 1/\Delta T$, can also be caused by oscillator noise. Figure 2.19 shows a time-domain DCO model that includes the additive oscillator frequency noise $\Delta f_n[i]$.

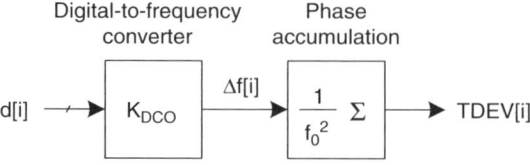

**Figure 2.18** DCO time-domain model.

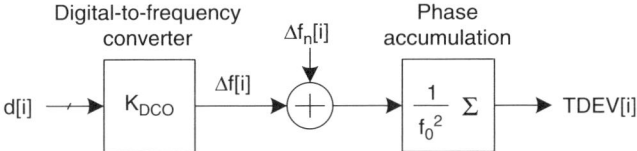

**Figure 2.19** DCO time-domain model (with noise).

The frequency resolution $\Delta f$ of each mode is referred to in subsequent chapters as the DCO gain, $K_{DCO}$, expressed as Hz/LSB (tracking-bank LSB). The oscillating frequency $f$ is referred to as the variable frequency $f_V$, for reasons that will become apparent in Chapter 4.

## 2.8 SUMMARY

In this chapter we introduced the first building block that is at the heart of the digital RF frequency synthesizer described. The digitally controlled oscillator starts a series of bottom-up progressive development levels which culminate in a full RF transmitter. The presented DCO features only a minimal but necessary set of functionalities such that it is able to be constructed as a digital I/O ASIC cell that generates a digital RF output clock in response to a digital input. The task of calibrating its transfer function is left to the higher-level blocks described in subsequent chapters. The oscillator must have a sufficiently small amount of phase noise and spurious tone content.

A novel idea is to use an array of digitally controlled varactors in place of the "analog" varactors used in conventional designs. Here we take advantage of the narrowband nature of the wireless communication system, and a progressive refinement of the effective frequency resolution is proposed. However, as shown, the resulting resolution would still be too coarse to be of any use for RF applications. To address this, LSB varactor dithering is used.

Finally, a time-domain model of the oscillator was introduced. It will be refined in subsequent chapters when more functionality is added.

# CHAPTER 3

# NORMALIZED DCO

The *digitally controlled oscillator* (DCO) of Chapter 2 provides only a raw bare minimum of functionality. In this chapter we introduce circuitry built around it for the purpose of adding the next hierarchical layer of arithmetic abstraction, which makes it easier to operate the DCO from the outside.

The oscillator frequency dependence on process spread and environmental factors such as voltage and temperature is tracked by a normalization circuit that belongs to this layer. Consequently, the oscillator described in this chapter is referred to as a *normalized DCO* (nDCO). The DCO normalization block includes circuitry to control the precise application of the tuning word, to reduce the spurious noise level. This is one of the advantages of operating the oscillator in the discrete-time domain that is not possible in conventional continuous-time designs.

As also mentioned in Chapter 2, the DCO is encapsulated at the top I/O level as a discrete-time system. Consequently, this view could be extended for a normalized DCO. This brings numerous benefits from being able to tap the rich body of knowledge from the *digital signal processing* (DSP) field.

## 3.1 OSCILLATOR TRANSFER FUNCTION AND GAIN

At the heart of the frequency synthesizer lies the digitally controlled oscillator. It generates an output with a frequency of oscillation $f_V$ that is a physically inherent

---

*All-Digital Frequency Synthesizer in Deep-Submicron CMOS*, by Robert Bogdan Staszewski and Poras T. Balsara
Copyright © 2006 John Wiley & Sons, Inc.

function of the digital *oscillator tuning word* (OTW) input. The $f_V = \text{f(OTW)}$ mapping was defined by Eq. 2.1.

In general, f(OTW) is a nonlinear function of the input. However, within a limited range of operation, it could be approximated by a linear transfer function. In this case, f(OTW) is a simple gain $K_{\text{DCO}}$. This allows us to rewrite Eq. 2.1 in the more linear form

$$f_V = f_0 + \Delta f_V = f_0 + K_{\text{DCO}} \cdot \text{OTW} \tag{3.1}$$

where $\Delta f_V$ is a deviation from a certain center frequency $f_0$. $f_0$ could be one of the mode-adjusted center frequencies described in Section 2.6. $\Delta f_V$ must be sufficiently small such that the linear approximation is satisfied.

$K_{\text{DCO}}$ is specifically defined as a frequency deviation $\Delta f_V$ (in hertz) from a certain oscillating frequency $f_V$ in response to 1 LSB of the input change. For this reason, $K_{\text{DCO}}$ is identical with the $\Delta f$ frequency resolution in Section 2.7 (an exact $K_{\text{DCO}}$ estimate is essential only in the tracking mode, not in the PVT and acquisition modes). Within a linear range of operation, the DCO gain can also be expressed as

$$K_{\text{DCO}}(f_V) = \frac{\Delta f_V}{\Delta(\text{OTW})} \tag{3.2}$$

Within a limited range, $K_{\text{DCO}}$ should be fairly linear with respect to the input, although it is possible to generalize the DCO gain as also being a function $K_{\text{DCO}}(\text{OTW})$ of the specific input:

$$K_{\text{DCO}}(f_V, \text{OTW}) = \frac{\Delta f_V}{\Delta(\text{OTW})} \tag{3.3}$$

## 3.2 DCO GAIN ESTIMATION

Due to its analog nature, the $K_{\text{DCO}}$ gain is subject to process and environmental factors that cannot be known precisely. It belongs to one of a few unknown system parameters whose estimate, $\widehat{K}_{\text{DCO}}$, must be determined. As described later, the $\widehat{K}_{\text{DCO}}$ estimate can be calculated entirely in the digital domain by observing the phase-error responses to the past DCO phase-error corrections. The actual DCO gain estimation involves arithmetic operations, such as multiplication or even division, and averaging, and could be performed by dedicated hardware or a *digital signal processor* (DSP).

The frequency deviation $\Delta f_V$ of Eq. 3.2 cannot be measured directly except perhaps in a lab or factory setting. Due to the digital nature of the synthesizer, $\Delta f_V$ can, however, be measured indirectly on the fly by harnessing the power of the existing phase-detection circuitry. Discussion of this method is deferred until Section 4.15.

The oscillator gain dependence on PVT and frequency makes it necessary to estimate it on an as-needed basis within the actual operating environment. This is especially important in battery-operated cellular phones with frequency-hopping operation, where, for example, the supply voltage can vary rapidly with the adaptive transmit power.

## 3.3 DCO GAIN NORMALIZATION

At a higher level of abstraction, a DCO oscillator, together with a DCO gain normalization $f_R/\widehat{K}_{DCO}$ multiplier, logically comprise a *normalized DCO* (nDCO), as illustrated in Fig. 3.1. The DCO gain normalization decouples the phase and frequency information throughout the system from the process, voltage, and temperature variations that normally affect $K_{DCO}$. This phase information is subsequently normalized to the clock period of the oscillator $T_V$, whereas the frequency information is normalized to the value of the external *reference frequency* $f_R$ (see Section 4.9). Digital input to the nDCO is a fixed-point *normalized tuning word* (NTW), whose integer-part LSB corresponds to $f_R$. The reference frequency is chosen as the normalization factor because it is the master basis for the frequency synthesis as defined in Section 1.1. In addition, the clock rate and update operation of this discrete-time system are established by the reference frequency.

The quantity $K_{DCO}$ should be contrasted with the process–temperature–voltage-independent oscillator gain $K_{nDCO}$, which is defined as the frequency deviation (in hertz) of the DCO in response to a 1-LSB change in the integer part of the NTW input. If the DCO gain estimate is exact, $K_{nDCO} = f_R$; otherwise,

$$K_{nDCO} = f_R \frac{K_{DCO}}{\widehat{K}_{DCO}} = f_R r \qquad (3.4)$$

Dimensionless ratio $r = K_{DCO}/\widehat{K}_{DCO}$ is a measure of the DCO gain estimation accuracy. Details of DCO gain estimation are presented in Section 4.15.

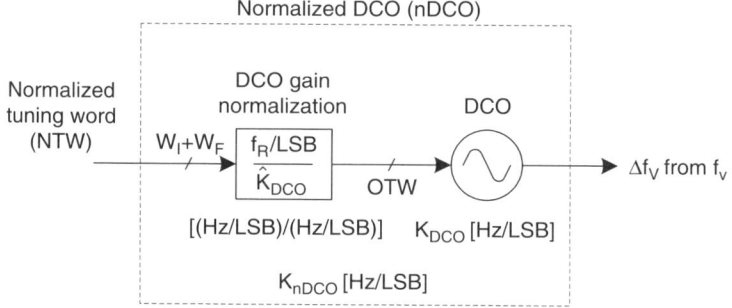

**Figure 3.1** Normalized DCO hardware abstraction layer. (From [58], © 2003 IEEE.)

## 3.4 PRINCIPLE OF SYNCHRONOUSLY OPTIMAL DCO TUNING WORD RETIMING

Figure 3.2 depicts a principle of the synchronously optimal DCO input tuning word retiming method. This idea is based on the observation that changing the tuning control input of an oscillator in order to adjust its phase and frequency in normal PLL operation is quite a disturbing event that reveals itself as jitter or phase noise [59–61]. This is especially noticeable in the case of a sample-mode oscillator such as the DCO, where its oscillating frequency is commanded to change at discrete times. Since the oscillating frequency of an *LC* tank is controlled by a voltage-to-capacitance conversion device (e.g., varactor), the instances when the oscillating energy is fully stored in a capacitor are the worst moments at which to change the capacitance. The total charge must be preserved, so changing the capacitance at those moments causes the electrical potential to exhibit the largest change ($\Delta V = Q/\Delta C$), as shown in Fig. 3.3a. These perturbations are then AM-to-PM-translated by the oscillator circuit into timing jitter. Changing the varactor capacitance at times when it is fully discharged will affect its voltage very little and thus contribute very little to the oscillating jitter (Fig. 3.3b).

The solution is then to control the timing moments when the varactor capacitance change is allowed to occur, thus minimizing jitter due to the tuning word update. This is implemented by feeding the delayed oscillator edge transitions back as clock input to the *synchronous* register retiming stage, as shown in Fig. 3.2. The retiming stage ensures that the input control data, as seen by the oscillator, is allowed to change at a precise and *optimal* time after the oscillator zero crossings. The feedback loop delay is set algorithmically to minimize oscillator jitter. The algorithm to control the delay line value for an optimal timing adjustment takes advantage of the fact that the phase error, which is related to the DCO jitter metric to be minimized, is now in digital form (see Section 4.3) and readily available for processing. Various statistics of this signal, such as mean-squared error, could thus be optimized by utilizing digital signal-processing hardware. The tightly

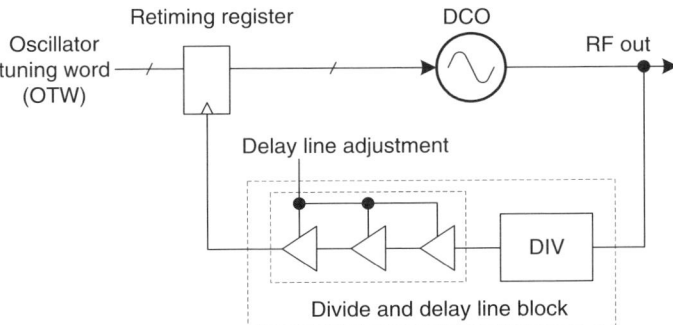

**Figure 3.2** Synchronously optimal sampling and timing adjustment of a DCO input. (From [56], © 2003 IEEE.)

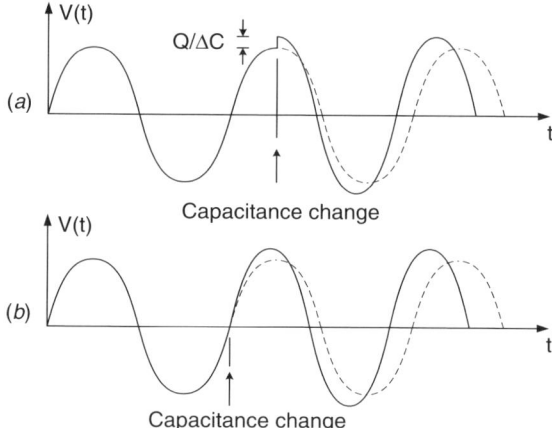

**Figure 3.3** Waveforms for capacitance change of an *LC* tank oscillator. (From [56], © 2003 IEEE.)

integrated companion DSP engine is capable of transferring digital phase-error samples to its own memory at the reference frequency and then postprocess them.

The actual delay could be accomplished by a voltage-controlled delay line (e.g., a long string of inverters with externally controlled supply voltage, $V_{dd}$). The range of delay line $V_{dd}$ voltage allowed is quite limited (from 1 V to about 1.8 V), so the delay change contribution per inverter is not very significant. However, the delay multiplied by the total number of inverters can exceed the DCO clock cycle, thus guaranteeing full 360° coverage. Note that the order of the edge divider and the delay line is reversible, but operating the delay line at a lower clock rate is more advantageous from a power consumption standpoint.

## 3.5 TIME DITHERING OF DCO TUNING INPUT

Phase error is calculated and used to correct the DCO frequency at regular intervals as normally determined by the frequency reference. These regularly spaced events are likely to introduce sharp spurs at the output. The nature of these spurs is similar to those frequently encountered in the fractional-*N* clock division method described in Section 1.3.2, where Fig. 1.26 depicts the deterministic and $\Sigma\Delta$ randomized spectra. The digital control of the oscillator makes it possible to randomize these events somewhat, in the manner described below. It should be noted that it is generally difficult to perform accurate time dithering in the continuous-time domain.

### 3.5.1 Oscillator Tune Time Dithering Principle

Figure 3.4 shows the principle behind the oscillator time dithering scheme. Instead of calculating and applying the tuning word input to the oscillator at evenly spaced

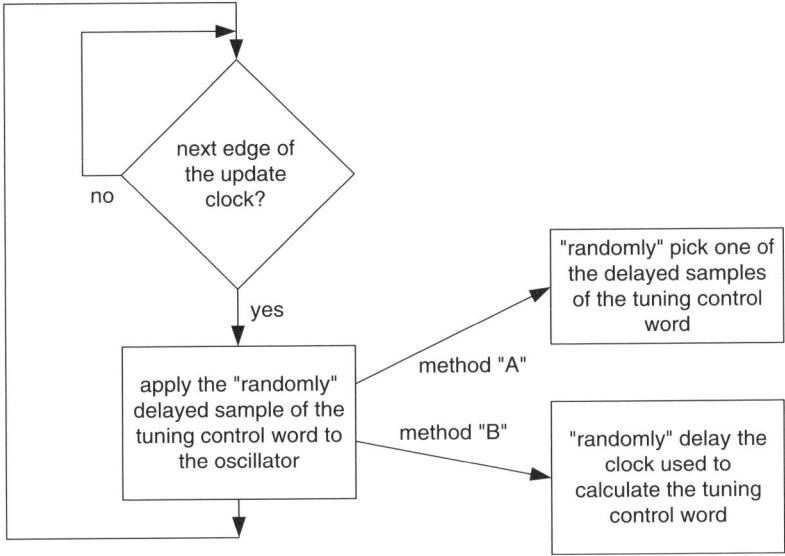

**Figure 3.4** Flowchart of oscillator tune time dithering.

and deterministic time intervals, as conventionally defined by the frequency reference clock, "random" timestamp deviations at each update are exercised. The statistical properties of these time-shift deviations will determine how much of the spectral spur energy gets spread into the background. The time dithering of the oscillator tuning input could be implemented fundamentally in one of two ways: time dithering of the OTW itself or time dithering of the actual time the OTW gets calculated and applied.

### 3.5.2 Direct Time Dithering of Tuning Input

Figure 3.5 shows the main idea behind the tuning input time dithering scheme. The *oscillator tuning word* (OTW) is a digital signal and is synchronous to the comparison events of the phase detection operation (described in Chapter 4). The oscillator tuning word would normally be connected to the DCO input after a gain stage (with possible low-level signal conditioning) if a loop filter is not used. In this scheme, time-shifted replicas of the OTW signal are pseudorandomly selected to randomize the exact timestamps of regular DCO frequency updates.

Figure 3.6 reveals a simple time-causal method to obtain the time-shifted replicas of the OTW with delay stages. An accurate discrete-time dithering of the OTW signal is obtained by reclocking it by the high-frequency oversampling clock and passing it through a delay shift register. A multibit input multiplexer synchronously selects the appropriate output of the delay register chain. This method provides a

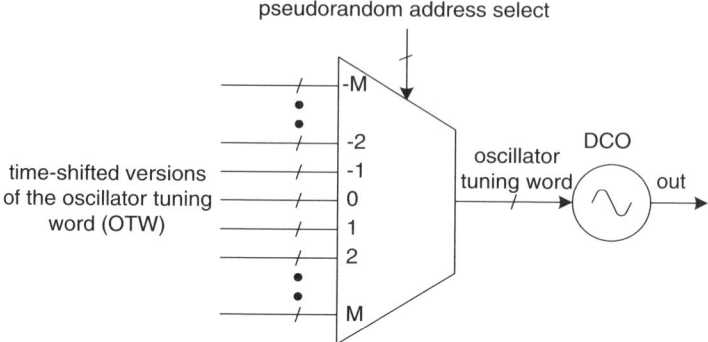

**Figure 3.5** Basic idea of discrete-time dithering of the DCO tuning input.

means of dynamically offsetting actual DCO update timing, which is done discretely by the oversampling clock at the frequency reference rate. The discrete delay control stream is further constrained to promote favorable properties of the resulting OTW signal. Consequently, it is advantageous to use $\Sigma\Delta$-modulated time-shift control to move the time-dither quantization energy into higher frequencies, where it is easy to filter it out by the $Q$ of the oscillator, power amplifier, and antenna band filter. Properties of the modulator should be selected based on the quantization noise characteristics desired.

The digitized RF output of the synthesizer would be used as the high-frequency oversampling clock directly or after appropriate frequency division by the

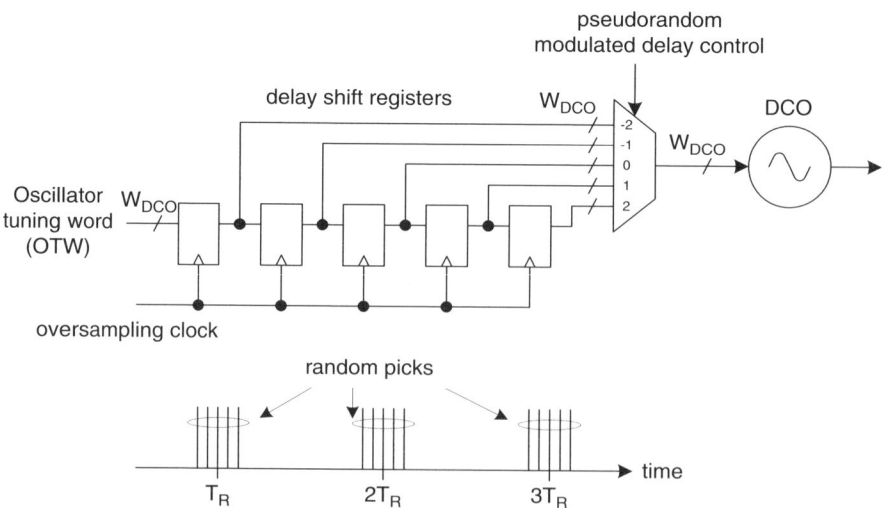

**Figure 3.6** Time dithering method of DCO tuning input through a multiplexer.

## 3.5 TIME DITHERING OF DCO TUNING INPUT

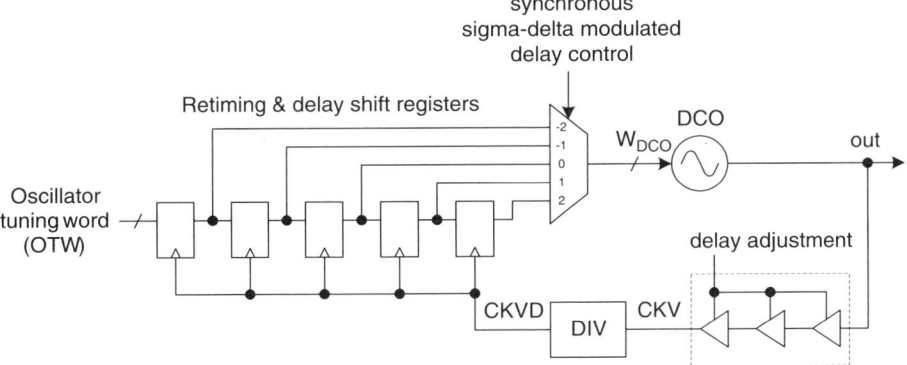

**Figure 3.7** Time dithering with DCO synchronous tuning input retiming.

edge-divider[1] block DIV, shown in Fig. 3.7. In this embodiment, the oscillator is implemented as an ASIC cell with truly digital I/Os, even at RF (shown in Fig. 2.14). Consequently, this circuit, and even the entire normalized DCO, are implemented in a digital fashion. This figure also reveals the synchronously optimal DCO input tuning word retiming method (described in Section 3.4), which could be used optionally in conjunction with the discrete-time dithering method.

The order of fine time delay (covering at least one RF clock period) and edge division could be reversed, but less power is usually consumed to delay the lower-repetition-rate clock edges.

### 3.5.3 Update Clock Dithering Scheme

Figure 3.8 depicts an improved time dithering method (method B in Fig. 3.4), which randomizes sampling edges of the *update clock* (CKU) instead of the oscillator tuning word input, as in the preceding method. The update clock is then used to trigger the generation and sampling of the DCO tuning input. The operational order of calculating the tuning word and time dithering is thus reversed. This leads to substantial hardware savings since delaying the clock, which takes a single bit, is preferred to delaying a multibit tuning word. Another clear benefit in a digitally intensive system is that the complex OTW calculation operation is more randomly spread in time and exhibits less temporal correlation. Consequently, this leads to further reduction of frequency spurs.

If the silicon chip die contains a microprocessor and a DSP on the same substrate, which is often the case with modern RF transceivers, it is advantageous to clock these processors *synchronously* to the time-dithered update clock CKU. Two significant benefits could thus be obtained: First, randomly modulating the clock period prevents substrate noise with strong periodical correlation from coupling

---

[1] *Frequency divider*, *edge divider*, and *clock divider* are interchangeable terms in the frequency synthesizer design field.

# 60  NORMALIZED DCO

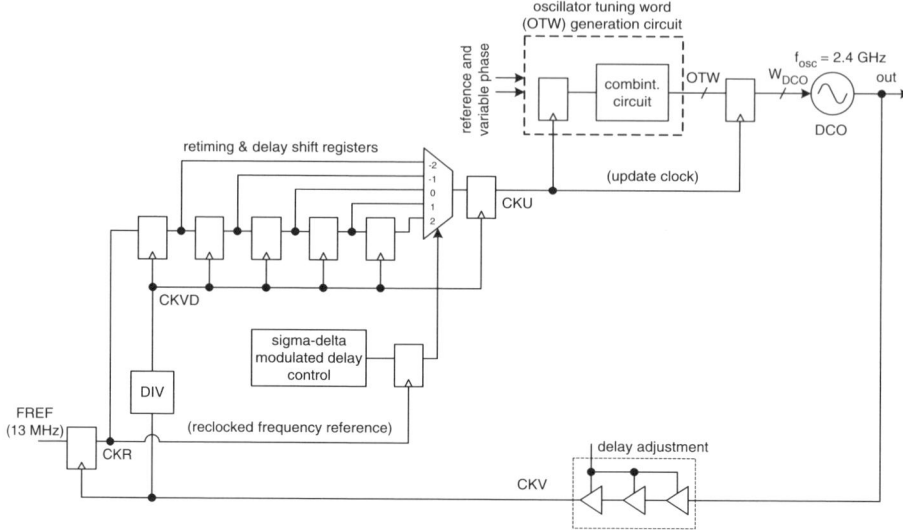

**Figure 3.8** Time dithering implementation of an update clock.

from the digital baseband to the RF section. Second, if the processor clock exhibits enough delay from the synthesizer update clock, the phase detection and tuning word adjustment operations occur during the "quiet" periods of the DSP.

## 3.6  IMPLEMENTATION OF PVT AND ACQUISITION DCO BITS

Figure 3.9 shows an implementational block diagram of the three separate DCO loop filter gain paths for the three modes of operation: PVT, acquisition, and tracking, as defined originally in Fig. 2.9. The blocks on the left side of the diagram (phase detector and loop gain section) are introduced formally in Chapter 4. The tracking path additionally splits into integer and fractional parts, due mainly to their significantly different clock rates. Each of the switched capacitor array banks (introduced with Fig. 2.10) is controlled individually by an oscillator interface circuit.

Also shown in Fig. 3.9 is the phase detector output signal PHE being fed into the three gain circuits (GP, GA, and GT, for the PVT, acquisition, and tracking modes, respectively). Due to their vastly different gain ranges, each gain circuit could use a different subset of the full range of phase error. The gain circuits multiply the phase error by the associated factors, which are split into two parts: the loop normalizing gain (MEM_ALPHA[2] set to $\alpha$[3]) and the DCO normalization gain (MEM_GAIN set to $f_R/\widehat{K}_{\mathrm{DCO}}$). Only the second normalizing multipliers belong formally to

---
[2]The prefix MEM_ indicates that these variables and other calibration data are stored in controller registers.
[3]Loop normalizing gain $\alpha$ is defined in Chapter 4.

## 3.6 IMPLEMENTATION OF PVT AND ACQUISITION DCO BITS 61

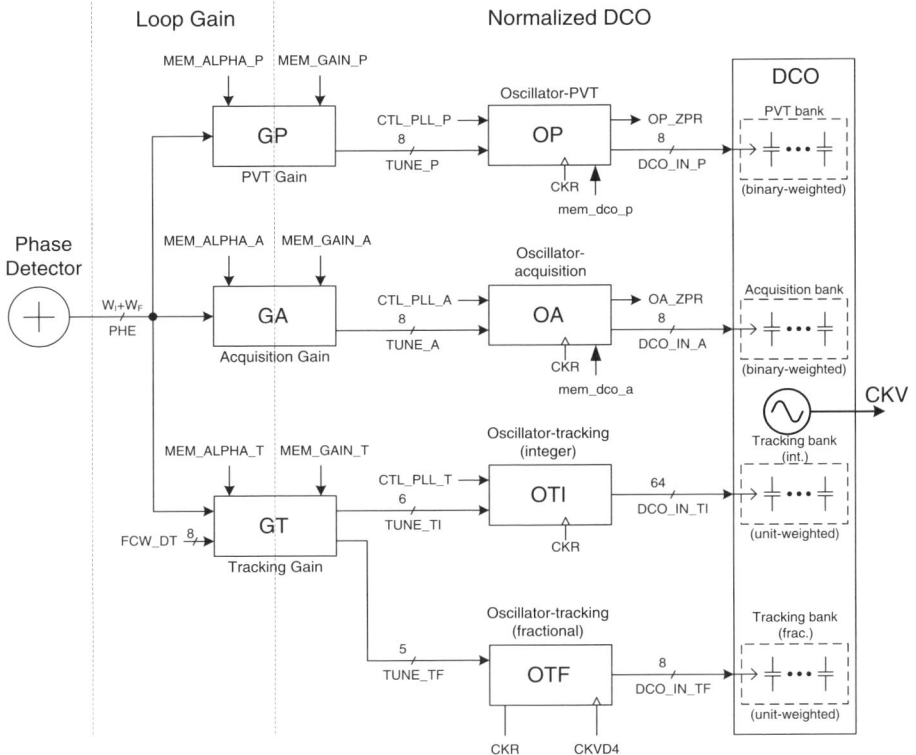

**Figure 3.9** DCO gain paths: implementational block diagram.

the nDCO layer, but are combined physically with the loop gain multipliers for implementational reasons. The outputs of the gain circuits are the oscillator tuning words (OTWs) that control the oscillator control circuits OP, OA, OTI, and OTF.

The PVT and acquisition oscillator interface is shown in Fig. 3.10. Both capacitor banks are built as 8-bit binary-weighted arrangements. Their direct digital control is expressed arithmetically as *unsigned* number arguments into the total contributive capacitances $C^P$ and $C^A$ of Eqs. 2.10 and 2.11, respectively. Digital PLL control, however, natively operates at an offset frequency with respect to a certain center or "natural" frequency. Consequently, the control logic inherently uses an arithmetic encoding representation that is in a *signed 2's-complement* notation. A conversion mechanism, simply inverting the MSB bit, is thus required only at the interface point. In this scheme, $-2^7 \cdots 0 \cdots (2^7 - 1)$ maps to $0 \cdots 2^7 \cdots (2^8 - 1)$, so the MSB bit inversion could be thought of as an addition of $+2^7$ to the 8-bit 2's-complement signed number with the carry-outs disregarded. It should be noted that the intrinsic offset of $2^7$ on the DCO side is part of establishing the natural oscillating frequency.

## 62  NORMALIZED DCO

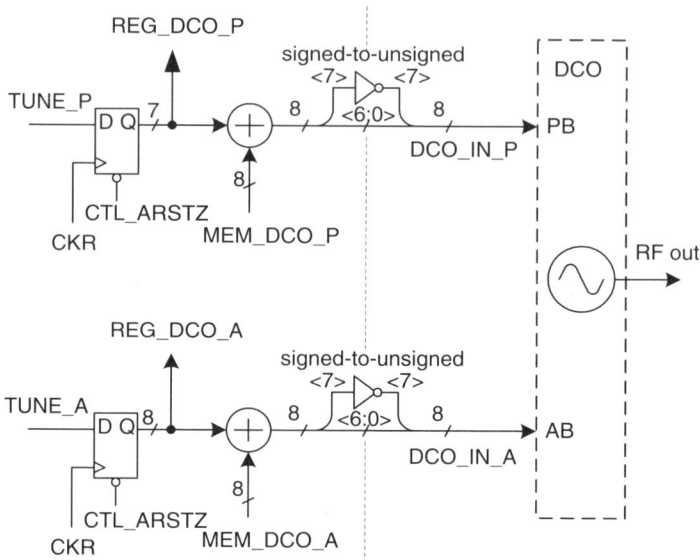

**Figure 3.10** Oscillator interface with PVT and acquisition bits.

The observation above raises a very important point. The center or natural frequency of a digitally controlled oscillator (DCO) must be handled differently than that of a voltage-controlled oscillator (VCO). In a VCO, the natural frequency is defined at the zero or ground level (or halfway between the positive and negative supply rails) of the tuning voltage input. This results in a maximum tuning range in both frequency directions. With a DCO, however, the situation is not as straightforward, especially with multiple tuning words. The issue being faced could be considered a generalization of the natural oscillator tuning frequency, which so far had not seen practical consequences with VCO-based designs, due to a difficulty in freezing analog tuning voltages.

Figures 3.11 and 3.12 illustrate almost identical interface control structures (OP and OA blocks of Fig. 3.9) for both PVT and acquisition bits, respectively, whereas Fig. 3.10 showed only interface details. There are two sets of register memory interface words. MEM_DCO_P and MEM_DCO_A could be the last frequency estimate from the controller's look-up table, in order to speed up the loop operation. REG_DCO_P and REG_DCO_A are the frequency offset status words reported back to the controller.

At reset, the DCO is placed at the center of the operational frequency range (possibly redefined by MEM_DCO_P and MEM_DCO_A) through asynchronous clear CTL_ARSTZ of the driving registers. This is an important mechanism that prevents the oscillator from failing to oscillate if the random power-up values of tuning word registers set it above the oscillating range, which might happen at the slow process corner.

3.6 IMPLEMENTATION OF PVT AND ACQUISITION DCO BITS  63

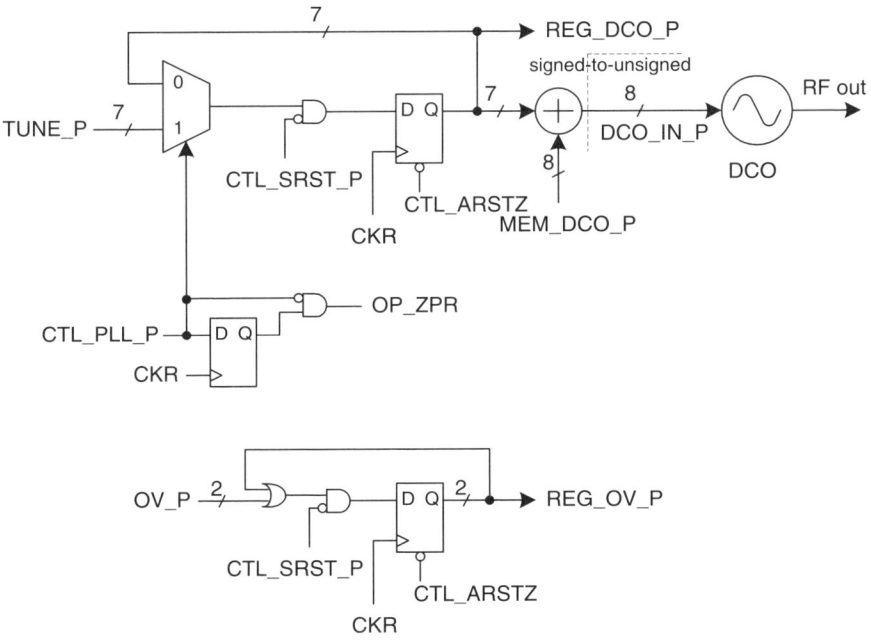

**Figure 3.11** Control circuit of oscillator PVT bits (OP).

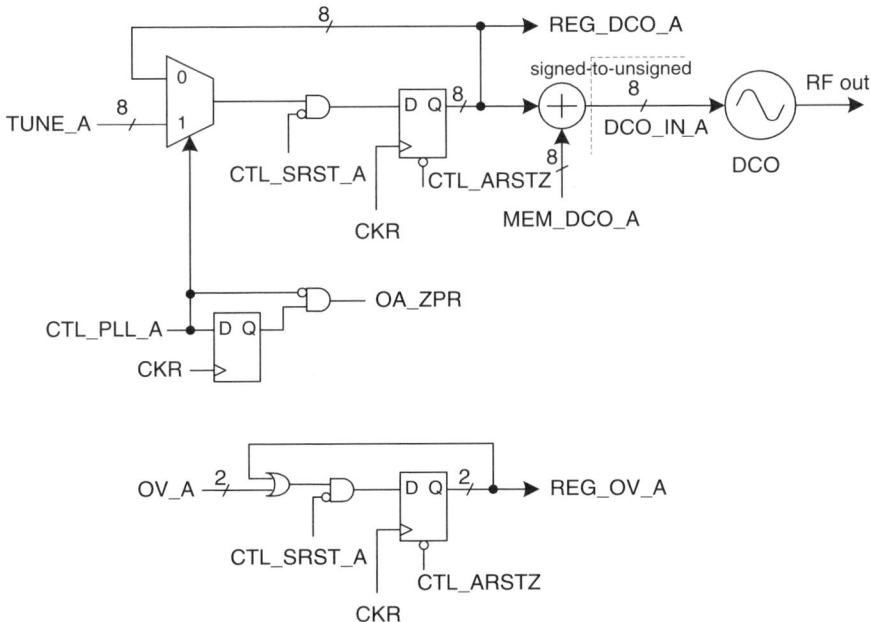

**Figure 3.12** Control circuit of oscillator acquisition bits (OA).

During the active mode of operation, the new tuning word is latched by the register with every clock cycle. Upon DCO operational mode changeover (see Fig. 2.9), the last value of the tuning word stored is maintained by the register. Consequently, during regular operation, only one interface path of Fig. 3.9 can be active at a given time, whereas modes executed previously maintain their final DCO control states. The *zero-phase restart* (ZPR) is used to zero out the phase detector output to avoid any discontinuities in the oscillator tuning word during mode switchover. A short explanation of the ZPR principle is as follows: At mode switchover, the tuning word of the last mode corresponds to a certain value of the phase error. This tuning word is now frozen, so the phase-error value that maintains it is no longer needed. However, the new mode is always referenced to the new center frequency established by the last mode. Consequently, it operates on *excess*, rather than absolute, phase error. Therefore, the old value of the phase error that corresponds to the frozen tuning word of the last mode would constantly have to be subtracted from the new phase error. A better solution is to use the method of zero-phase restarting. In this way, a *hitless* (i.e., free of perturbation) progression through the three DCO operational modes is accomplished. This method is described in more detail in Section 4.16.

## 3.7 IMPLEMENTATION OF TRACKING DCO BITS

The PVT and acquisition bits described above are used in the preparatory steps to quickly establish the center of the operating frequency. They are inactive during subsequent normal operation when the settled synthesized frequency is used. The tracking bits of a DCO oscillator, on the other hand, need much greater care and attention to detail since any phase noise or spurious tone contribution of the tracking bits will degrade the synthesizer performance.

### 3.7.1 High-Speed Dithering of Fractional Varactors

Figure 3.13 describes the notion of increasing the frequency resolution of the DCO. The tracking part of the *oscillator tuning word* (OTW) is split into two components: $W_{TI} = 6$ integer bits and $W_{TF} = 5$ fractional bits. The LSB of the integer part corresponds to the basic frequency resolution of the DCO oscillator. The integer part is thermometer-encoded to control the same-size DCO varactors of an *LC* tank oscillator. In this scheme all the varactors are unit weighted, but their switching order is predetermined. This guarantees monotonicity and helps to achieve excellent linearity, especially if their switching order agrees with the physical layout. The transients are minimized since the number of switching varactors is no greater than the code change. This compares very favorably with binary-weighted control, where a single LSB code change can cause all the varactors to toggle. In addition, due to equal load throughout for all bits, the switching time is equalized in response to code changes. As described below, a slightly more general unit-weighted capacitance

## 3.7 IMPLEMENTATION OF TRACKING DCO BITS

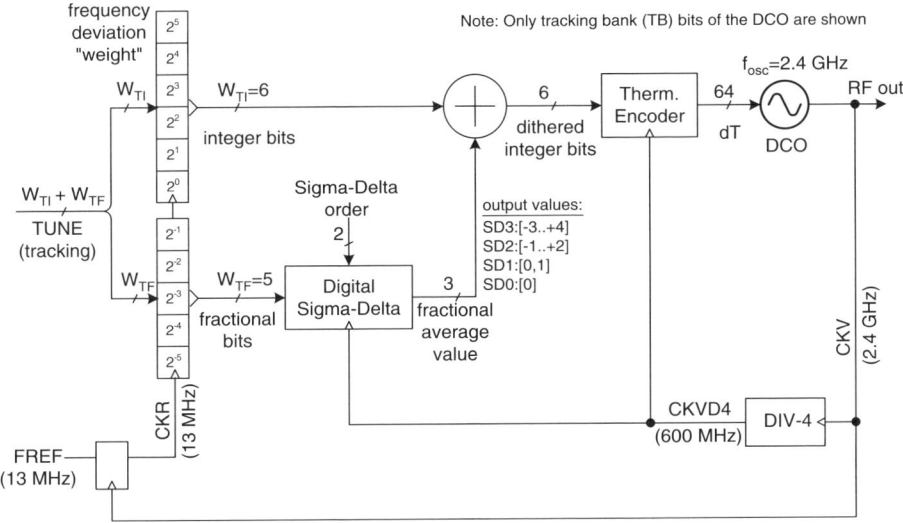

**Figure 3.13** Improving frequency resolution with $\Sigma\Delta$ dither of DCO tracking bits. (From [56], © 2003 IEEE.)

control can be used to add some extra coding redundancy, which lends itself to various algorithmic improvements of system operation.

The fractional part, on the other hand, employs a time-averaged dithering mechanism to further increase frequency resolution. The dithering is performed by a digital $\Sigma\Delta$ modulator that produces a high-rate integer stream whose average value equals the lower-rate fractional input. The digital $\Sigma\Delta$ modulator is considered an essential part of the DCO solution described for wireless applications. $\Sigma\Delta$ techniques have been used successfully for over two decades in the field of analog data converters. This has developed a rich body of knowledge for other applications to draw on [62].

Appendix A proves that the DCO spurs introduced by the switching of a varactor can be made vanishingly small if performed at a fast enough rate. Below is a summary with implications. A simple first-order $\Sigma\Delta$ modulator pattern [26] is not random at all and is likely to create spurious tones. If the LSB varactor has a frequency resolution of $\Delta f$ and is dithered at a rate of $f_m$, it will produce two spurs $f_m$ away on both sides of the oscillating frequency, with the power level of $20\log(\beta/2)$ relative to the carrier, where $\beta = (2/\pi)(\Delta f/f_m)$. In this implementation, the step size of LSB varactors is $\Delta f = 23$ kHz, and if the dithering clock is 600 MHz such that the maximum dithering rate $f_m = 300$ MHz (for a fractional input of 0.5), the spur level generated is only $-92$ dBc. It could be shown through Fourier series decomposition of the modulating wave that for the slowest nonzero dithering rate of $f_m = 18.75$ MHz (corresponding to a fractional input of $1/32$) and beyond, the spur generated rises only 4 dB, to $-88.3$ dBc. Even though this level is sufficiently low for most wireless applications, it is far from representing the worst case. First, the

DCO input, and consequently the $f_m$ dithering rate, is not constant but is subject to continual change during normal closed-loop PLL operation, thus spreading the spur energy widely. Second, the $\Sigma\Delta$ modulator is normally selected to operate in the second or third order to further randomize the dithering pattern. With such an insignificant amount of $\Sigma\Delta$ quantization energy, no phase-noise degradation should be observed in the system during normal operation.

The integer part of the tuning word is then added to the integer-valued high-rate-dithered fractional part. The resulting binary signal is thermometer encoded to drive the 64 tracking bank varactors. In this simplest embodiment, the high-rate fractional part is added arithmetically to the low-rate integer part, thus making its output, as well as the entire signal path terminating at the varactors inside the DCO, high rate. The separate DCO fractional bits $d^{TF}$ of Fig. 2.10 are not used here. A preferred solution to implement this approach is presented below. It should be noted here that a small delay mismatch between the integer and fractional parts due to the $\Sigma\Delta$ modulator group delay is intentionally left uncorrected. The precise alignment may not be necessary since the resulting degradation may be insignificant due to the high-rate dithering and small amount of quantization energy.

As explained in Section 2.2, the dithering method trades the sampling rate for the frequency granularity. As an example of the design implemented, if the frequency resolution of the 2.4-GHz DCO is $\Delta f^T = 23$ kHz with a 13-MHz update rate, the effective time-averaged frequency resolution within one reference or update cycle after 600-MHz $\Sigma\Delta$ dither with five sub-LSBs would be $\Delta f^{T-\Sigma\Delta} = 23$ kHz$/2^5 = 718$ Hz. The frequency resolution improvement achieved here is $2^5 = 32$. This corresponds roughly to the sampling rate speed-up of 600 MHz$/13$ MHz $= 46.2$.

The point above requires further elaboration. As mentioned in Section 2.2, the finest resolution improvement per reference cycle is upper bounded by the sampling rate speed-up. In the example, this condition is not quite satisfied. In fact, the *operational* frequency resolution achieved by the frequency synthesizer is much finer. It is determined by the fractional wordlength $W_F = 15$ of the reference phase accumulator (introduced in Chapter 4) and equals 13 MHz$/2^{15} = 396.7$ Hz. The difference between the two frequency resolution numbers is that the former assumes a single reference cycle during which the loop does not make corrections. The latter, on the other hand, involves multiple FREF cycles and harnesses the PLL averaging power over the longer observation period. Closed-loop operation is deferred until Chapter 4.

The structure of the digital $\Sigma\Delta$ modulator is depicted in Fig. 3.14. It is implemented as a third-order MASH-type architecture [28] that could, conveniently and efficiently, be scaled down to a lower order. It is clocked by a 600-MHz divide-by-4 oscillator clock, CKVD4. Its topology is based on reference [26] and was shown in Fig. 1.25. The original structure is not the best choice for high-speed designs because the critical path spans through all the three accumulator stages and the carry-sum adders. A critical path retiming transformation needs to be performed to shorten the longest timing path to only one accumulator, so that 600-MHz clock operation can be reached. Since the structure is highly modular, the lower-order modulation characteristics (plotted in Fig 1.26) could be set by

## 3.7 IMPLEMENTATION OF TRACKING DCO BITS

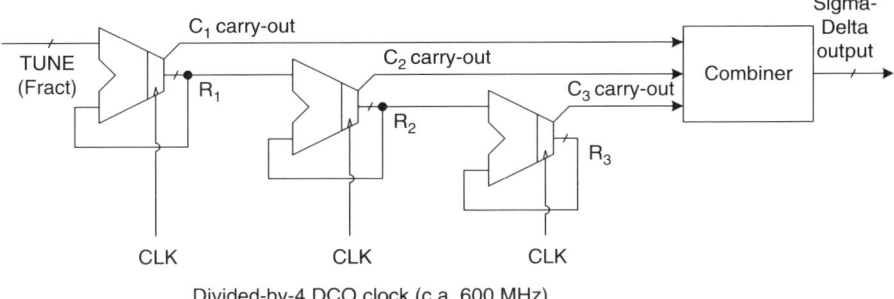

**Figure 3.14** MASH-3 $\Sigma\Delta$ digital modulator structure. (From [56], © 2003 IEEE.)

disabling the tail accumulators by gating off the clock, which is a preferred method from a power-saving standpoint.

A schematic diagram of the first accumulator cell of the $\Sigma\Delta$ digital modulator is shown in Fig. 3.15. "DITHER" is a 1-bit pseudorandom dithering sequence signal that serves to increase the randomness of a $\Sigma\Delta$ digital modulator. Its chief purpose is to randomize the stream of the fractional tuning word when a small static or slowly varying fractional value might produce low-frequency oscillations. Randomizing the LSB of the first accumulator stage would effectively break any long periodic sequences.

The combiner circuit (originally shown in Fig. 1.25) merges the three single-bit carry-out streams such that the resulting multibit output satisfies the third-order $\Sigma\Delta$ spectral property. The $\Sigma\Delta$ stream equation is a result of register retiming of the architecture originally described in reference [26]:

$$\text{out}_{\Sigma\Delta} = C_1 D^3 + C_2(D^2 - D^3) + C_3(D - 2D^2 + D^3) \tag{3.5}$$

where $D \equiv z^{-1}$ is the delay element operation. It must be noted that when realized in digital logic, the unsigned output has a dc offset of 3, which is equal to the number of negation operations. This equation is easily scaled down to the second- or first-order $\Sigma\Delta$ by disregarding the third or third and second terms, respectively.

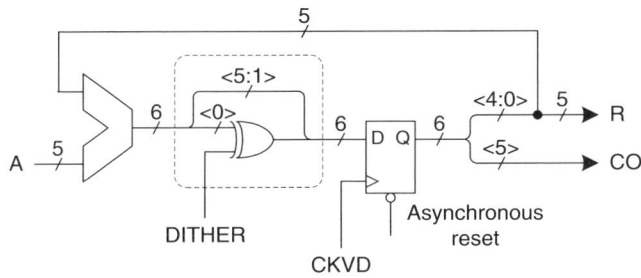

**Figure 3.15** First cell of a $\Sigma\Delta$ modulator.

**68** NORMALIZED DCO

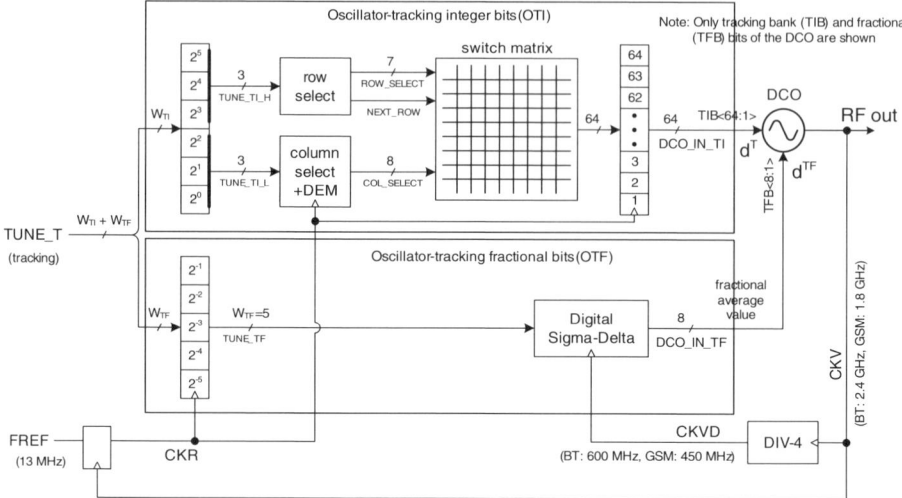

**Figure 3.16** Implementation block diagram of DCO tracking bits (OTI + OTF) with DEM of the integer part and $\Sigma\Delta$ dithering of the fractional part. Critical high-speed arithmetic operations are performed in the "analog" domain through capacitive additions inside the DCO.

Figure 3.16 reveals the preferred method of implementing the integer and fractional oscillator tracking control (OTI and OTF of Fig. 3.9) from a lower-power standpoint [56]. The fractional path of the DCO tracking bits, which undergoes high-rate dithering, is now entirely separated from the lower-rate integer part. It even has a dedicated DCO input just to avoid "contaminating" the rest of the tracking bits with frequent transitions. The switch matrix, together with the row and column select logic, operates as a binary-to-unit-weight encoder in response to the integer part of the tracking tuning word. The $\Sigma\Delta$ modulator is only responsive to the fractional part of the tracking tuning word. The actual merging of both parts is performed inside the oscillator through time-averaged capacitance summation at the $LC$ tank.

Another important benefit of the approach chosen is that the high-speed arithmetic operation of the Eq. 3.5 combiner is now trivial. Figure 3.17 shows the

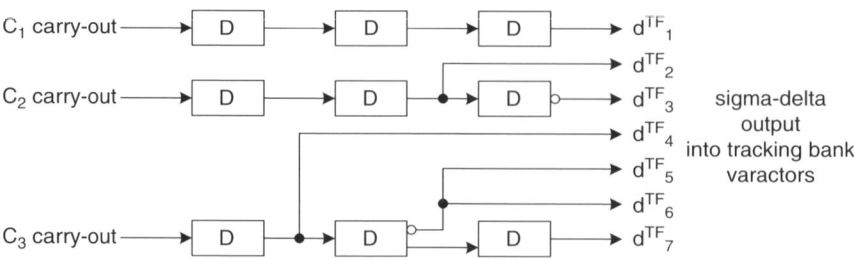

**Figure 3.17** $\Sigma\Delta$ modulator carry-out combiner structure. (From [56], © 2003 IEEE.)

## 3.7 IMPLEMENTATION OF TRACKING DCO BITS 69

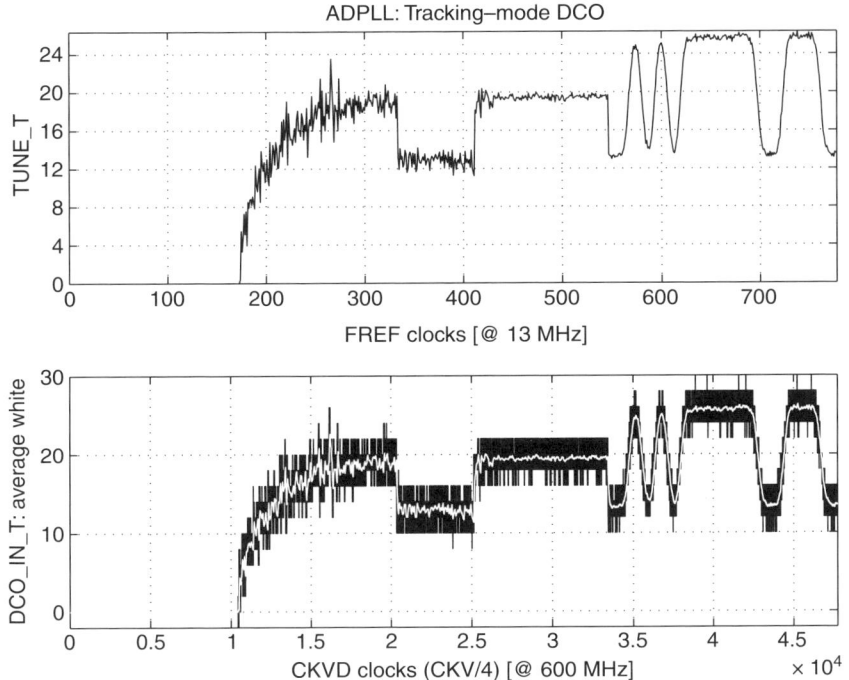

**Figure 3.18** Simulation plot using $\Sigma\Delta$ modulation of the fractional part of the tracking tuning word: top, fixed-point tuning word; bottom, decoded merged DCO integer input. (From [56], © 2003 IEEE.)

implementation. All that is required are flip-flop registers (for the delay operation) with complementary outputs (for the negation). The arithmetic addition is performed inside the oscillator through capacitance summation.

Figure 3.18 illustrates second-order MASH-type $\Sigma\Delta$ modulation of the fixed-point tracking DCO tuning control word with 5 fractional bits. The fixed-point tuning word (TUNE_T, upper plot) consists of 6 integer bits and 5 fractional bits and is clocked at a 13-MHz reference frequency. The $\Sigma\Delta$ modulates the 5-bit fractional part at 600-MHz clock rate and outputs the integer stream that controls the DCO frequency. The lower plot shows the $\Sigma\Delta$ output stream (DCO_IN_TF) "merged" with the integer part stream (DCO_IN_TI). For the purposes of visualization only, the DCO_IN_TI integer stream is mathematically decoded into an unsigned number representation and added to the mathematically decoded DCO_IN_TF signed fractional stream. This mathematical operation is performed by a MATLAB software package based on the data files generated by a VHDL MODELSIM simulator.[4] The running average, the solid white curve on the lower plot, faithfully reproduces the fixed-point tuning control input.

---
[4]Simulation methodology is described in Chapter 6.

## 3.7.2 Dynamic Element Matching of Varactors

Ideally, each of the unit-weighted capacitors of the tracking bank has exactly the same capacitative value. Using a real-world fabrication process, however, the capacitative value of each capacitor will vary slightly from the ideal. As capacitors are turned on and off by the integer tracking oscillator controller OTI of Fig. 3.9, nonlinearities will be evident in the output, due to variations in capacitative values, as shown in Fig. 3.19.

A method to improve the digital-to-frequency conversion linearity is also revealed in Fig. 3.16. It shifts the unit-weighted varactors cyclically using the *dynamic element matching* (DEM) method now being employed in DACs [63]. The integer part of the tuning word is split into upper and lower bits. The upper bits are encoded and control the row selection of the switch matrix. The lower bits are also encoded and select the next column of the switch matrix. The cyclic shift of unit-weighted varactors is performed within the row (see Fig. 3.20) but could extend to other rows. However, the number of active switches does not change for the same control input.

In Fig. 3.20 the capacitors associated with an unfilled row ("next row" signal of Fig. 3.16) of the switch matrix are rotated on each FREF clock cycle. Initially, the first three columns of row 3 are enabled; on the next clock cycle, columns 2 through 4, rather than columns 1 through 3, are enabled; on the next clock cycle, columns 3 through 5 are enabled; and so on. Accordingly, on each clock cycle, the set of capacitors used in the 64-element array changes slightly. Over time, the nonlinearities shown in Fig. 3.19 average out, thereby producing a much more accurate output.

With this DEM scheme, the switches enabled for a single row are rotated. This is accomplished by modulo-incrementing the starting column of the switches enabled on each clock cycle. This method could be varied slightly by including two (or more) rows in the rotation. As a result, a larger frequency range would be subject to

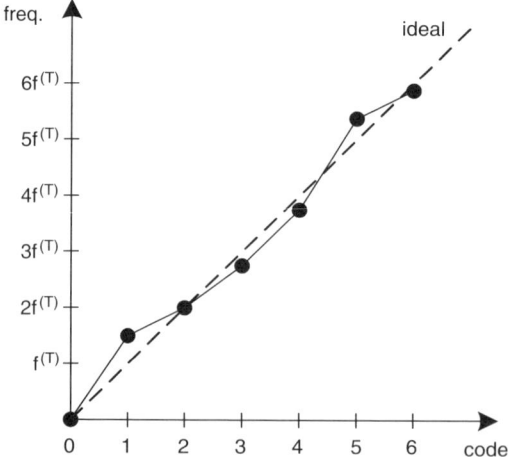

**Figure 3.19** Cumulative nonlinearity of DCO tracking bits.

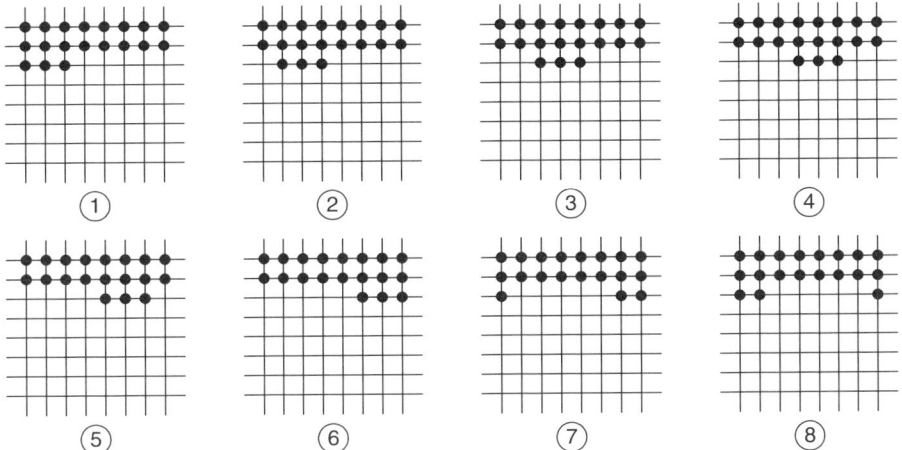

**Figure 3.20** Dynamic element matching through cyclic shift within a matrix row. (From [56], © 2003 IEEE.)

beneficial time averaging by including a greater number of capacitors in the rotation but at the cost of a longer repetition cycle. An alternative method of increasing the DEM frequency span would be to lengthen the number of columns per row, thus creating a nonsquare matrix.

The output bits of the switch matrix are coupled individually to the bank of 64 resampling drivers, which are implemented as flip-flop registers. Each driver controls a single unit-weighted varactor of the *LC*-tank. Using resampling by the CKR clock eliminates delay mismatches due to path differences, such that the timing points of varactor transitions coincide. This helps with spurious noise control. It should be noted that although the switch matrix is shown (from an algorithmic standpoint) in a row/column configuration, the actual implementation is not a precise grid. In fact, a group of rows could be combined physically into a single line.

The principal difference in DFC vs. DAC specification requirements is that the full dynamic range is not required for the number of controlled units available. In DFC applications, frequency headroom is required because it is not expected that the oscillator will operate at the precisely specified frequency before entering the tracking mode.

### 3.7.3 DCO Varactor Rearrangement

As illustrated in Fig. 3.21, the 64 integer tracking-bit varactors of an *LC* tank have a physical layout of two long columns, and the fractional tracking-bit varactors are arranged separately. However, the controlling circuitry is located on only one side. This creates an unbalanced structure in which routing to one varactor column is significantly shorter, with easier access than to the other, and therefore their transient response is different. Moreover, the spatially separated devices are likely to be

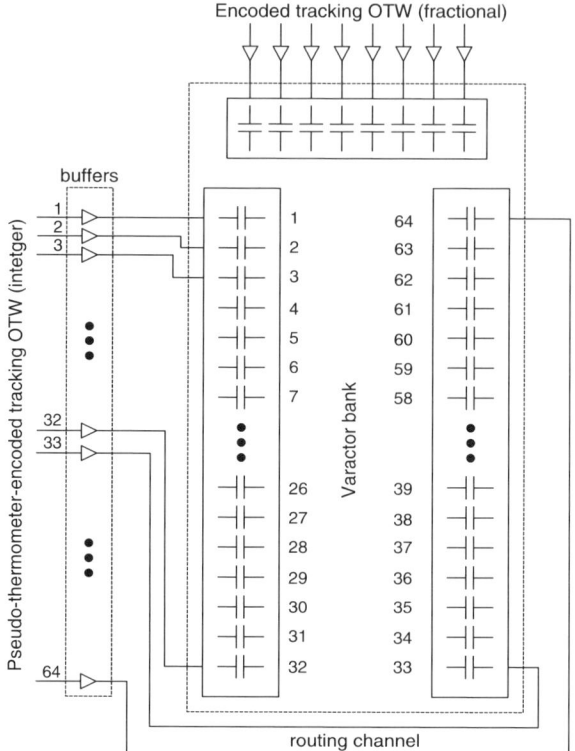

**Figure 3.21** Layout diagram of tracking capacitors. (From [56], © 2003 IEEE.)

more mismatched, due to the process gradient. If, during the course of operation, the varactor selection transitions through the column boundary, it is likely to create larger switching perturbations. Hence before entering the finest tracking mode, a rearrangement of DCO varactors might be performed such that "lower-quality" (varactors farther away from the buffers 1 to 64) capacitors be filled in order to maximize the frequency dynamic range of the most preferred capacitor section. In this architecture, this could be done upon switchover from fast tracking to tracking. It should be noted that whereas in other designs, the tracking capacitors could be arranged differently depending on various layout issues, a certain set of capacitors will always be favored based on the proximity metric to the control logic.

Figure 3.22 illustrates a method of improving the quality of DFC conversion. In the initial state, half of the capacitors in each column are turned on (designated by a "+") and half are turned off (designated by a "−"). During fast tracking, the capacitors of the less desirable right-hand column are, to the extent possible, enabled or disabled in order to fine-tune the oscillator to the channel selected. If additional capacitors need to be enabled or disabled, the capacitors from the left-hand column may be used, preferably those capacitors at the edges of the column. After channel tuning, the capacitors in the left-hand column are used for modulation

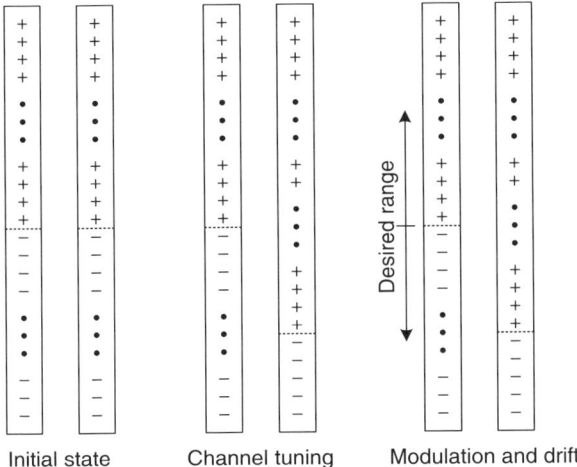

**Figure 3.22** Tracking capacitors rearrangement after initial settling. (From [56], © 2003 IEEE.)

and drift control. In this way, the most desirable capacitors are used for maintaining lock and for generating the signal once data are being transmitted. It should be noted that while the preferred center point was shown as the middle of the left-hand column, the preferred center point could be set at any location that is far from layout and routing discontinuities.

## 3.8 TIME-DOMAIN MODEL

The time-domain nDCO model is presented in Fig. 3.23. It is build on the DCO time-domain model presented in Fig. 2.18. Provided that the DCO gain is estimated correctly, the normalized tuning word NTW (denoted as $\tilde{d}[i]$) at the nDCO input will change its operating frequency by $\Delta f_V = \tilde{d}[i] f_R$. On every rising DCO edge event, $\Delta f_V$ multiplied by a "constant" $1/f_V^2$ will be accumulated (see Eq. 2.22). The accumulation interval is established by the inverse of the reference frequency

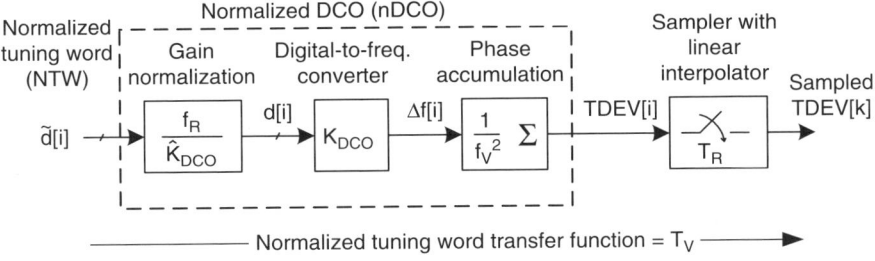

**Figure 3.23** nDCO time-domain model. (From [56], © 2003 IEEE.)

# 74  NORMALIZED DCO

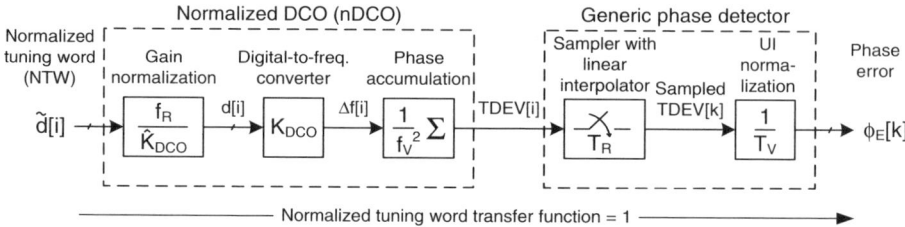

**Figure 3.24** nDCO time-domain model with generic phase detection. The transfer function is unity.

$T_R = 1/f_R$. It is related to the nominal oscillation frequency $f_V$ by $T_R = NT_V$, where $N$ is the traditionally-defined (possibly fractional) PLL frequency-division ratio. During the $T_R$ interval, the $\tilde{d}[i]$ frequency tuning input is assumed constant. It is justified by the fact that between the reference events there are no updates to the tuning word. At the end of $N$ cycles of $f_V$, the *sampled* difference of the accumulated timing deviation TDEV will be

$$\text{TDEV}[k] - \text{TDEV}[k-1] = N\Delta T = N\frac{\Delta f_V}{f_V^2} = \tilde{d}[i]\frac{Nf_R}{f_V^2} = \tilde{d}[i]T_V \quad (3.6)$$

where $k = \lceil i/N \rceil$ (since $N$ can be fractional). The samples out of the nDCO are at the DCO rate, while the generic phase detector operates at the reference clock. The change of data rate is accomplished using a sampler that employs a first-order interpolation [64].

As stated in Section 2.7, the DCO phase accumulation is not tied to any hardware. It simply reflects the workings of a progression of time. However, the sampling mechanism requires explicit hardware that would make periodic snapshots of the evolving TDEV.

A glimpse into Chapter 4 is given in Fig. 3.24. A generic hardware circuitry is added to the nDCO that would detect TDEV and convert it to digital bit format. At the same time, it deals with the troublesome unit of time by performing normalization to the clock period ($T_V$) of the DCO oscillation, defined as the *unit interval* (UI). The diagram does not suggest any particular mechanism; it merely indicates and describes mathematically a timing deviation detector that would determine any frequency and phase deviations of the oscillator, which would then be fed back as loop corrections. The transfer function from the normalized tuning word input to the detector output is 1 (bits/bit) within one frequency reference clock cycle. An appropriate digital scaler/filter between $\phi_E$ and $\tilde{d}$ will give rise to a phase-locked loop, which is covered in Chapter 4.

## 3.9  SUMMARY

In this chapter we presented the first hierarchical layer of arithmetic abstraction over a raw digitally controlled oscillator, to make it easier to operate it algorithmically.

The main task of this overlay block is to perform DCO calibration and normalization such that the normalized DCO transfer function is largely independent of process and environmental factors. Other improvements, such as increasing frequency resolution through $\Sigma\Delta$ dithering and dynamic element matching, were also introduced. This layer serves the purpose of concealing the implementational details to the layers above in order to make their algorithms and implementation simpler.

A block diagram of the datapath portion of the normalized DCO layer was presented on the right-hand side of Fig. 3.9. It consists of the oscillator interface (OP, OA, OTI, and OTF) and normalizing gain blocks (second part of GP, GA, and GT) for each of the operational modes. Non-datapath operation of $K_{DCO}$ estimation is performed in software. It utilizes full synthesizer features and is described in subsequent chapters.

# CHAPTER 4

# ALL-DIGITAL PHASE-LOCKED LOOP

*Digital-to-frequency conversion* (DFC) of the normalized digitally controlled oscillator described in Chapter 3 operates in an open-loop manner. Consequently, its stability is quite poor, due to drift or wander of the self-generated phase and frequency. In this chapter we describe a phase correction mechanism by which the output phase, and hence frequency, is corrected periodically by comparison with a stable reference phase as established by the *frequency reference* (FREF) input of Fig. 1.1. In this way, the long-term frequency stability of the synthesizer matches that of the reference. The phase correction mechanism is performed entirely in the digital domain by phase-locking to the reference input the DCO clock generated. Design and building of its constituent blocks also follow the digital methodology.

Following the PLL classification in Best's widely cited book [33], the frequency synthesizer described in this chapter is not a classical *digital PLL* (DPLL), which actually is considered a semianalog circuit, but an *all-digital PLL* (ADPLL), with all building blocks defined as digital at the input/output level. It uses digital design and circuit techniques from the ground up. At the heart lies a digitally controlled oscillator (DCO), which deliberately avoids analog tuning voltage controls. The DCO is analogous to a flip-flop—the cornerstone of digital circuits—whose internals are analog, but the analog nature does not propagate beyond its boundaries. This allows its loop control circuitry to be implemented in a fully digital manner, as illustrated in Fig. 4.1.

---

*All-Digital Frequency Synthesizer in Deep-Submicron CMOS*, by Robert Bogdan Staszewski and Poras T. Balsara
Copyright © 2006 John Wiley & Sons, Inc.

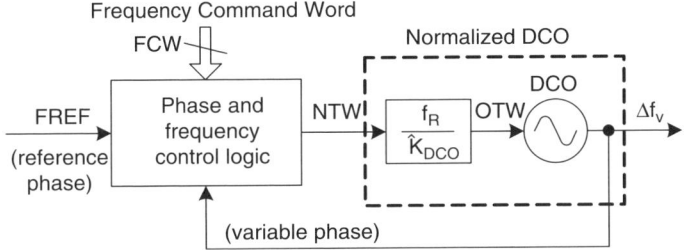

**Figure 4.1** DCO within all-digital PLL architecture.

In this chapter we also introduce gear-shift and zero-phase restart techniques, which in cooperation with synthesizer loop operation, control the progressive refinement of resolution as the operational frequency approaches the desired frequency. These techniques implement the algorithm described in Fig. 2.9.

## 4.1 PHASE-DOMAIN OPERATION

Let's define the actual clock period of the variable oscillator (DCO or VCO in general) output CKV as $T_V$ and the clock period of the frequency reference FREF as $T_R$. Let's assume that the oscillator runs appreciably faster than the available reference clock, $T_V \ll T_R$, which is the case in RF synthesizers, where the multi-GHz RF carrier frequency generated is orders of magnitude higher than that of a typical 10- to 40-MHz crystal reference. Let's assume further, to simplify the initial analysis, that the actual clock periods are constant or time invariant.

The CKV and FREF clock transition timestamps (or events measured in units of a second) $t_V$ and $t_R$, respectively, are governed by the following equations:

$$t_V = iT_V \tag{4.1}$$

$$t_R = kT_R + t_0 \tag{4.2}$$

where $i = 1, 2, \ldots$ and $k = 1, 2, \ldots$ are the CKV and FREF clock transition index numbers, respectively, and $t_0$ is an initial time offset between the two clocks, which without loss of generality is absorbed into the FREF clock.

It is convenient in practice to normalize the transition timestamps in terms of actual $T_V$ units [referred to as *unit intervals* (UIs)] since it is easy to observe and operate on actual CKV clock events. Let's define two dimensionless entities: variable clock "phase" and reference clock "phase":

$$\theta_V \equiv \frac{t_V}{T_V} \tag{4.3}$$

$$\theta_R \equiv \frac{t_R}{T_V} \tag{4.4}$$

The term $\theta_V$ is defined only at CKV transitions and indexed by $i$. Similarly, $\theta_R$ is defined only at FREF transitions and indexed by $k$. This results in

$$\theta_V[i] = i \tag{4.5}$$

$$\theta_R[k] = k\frac{T_R}{T_V} + \frac{t_0}{T_V} = kN + \theta_0 \tag{4.6}$$

The normalized transition timestamps $\theta_V[i]$ of the variable clock, CKV, could be estimated as $R_V$ by accumulating the number of significant (rising or falling) edge transitions of the actual clock:

$$R_V(iT_V) \equiv R_V[i] = \sum_{l=1}^{i} 1 \tag{4.7}$$

Without the frequency reference retiming (described in Section 4.2), the normalized transition timestamps $\theta_R[k]$ of the frequency reference clock, FREF, could be obtained by accumulating the *frequency command word* (FCW) on every significant (rising or falling) edge of the frequency reference clock:

$$R_R(kT_R) \equiv R_R[k] = \sum_{l=1}^{k} \text{FCW} \tag{4.8}$$

The initial values of the estimates, $R_V[0]$ and $R_R[0]$, are not critical, but for simplicity are assumed to be zero.

FCW is formally defined as the frequency-division ratio of the *expected* variable frequency to the reference frequency:

$$\text{FCW} \equiv \frac{\mathcal{E}(f_V)}{f_R} \tag{4.9}$$

The reference frequency is usually of excellent long-term accuracy, at least as compared to the variable oscillator. For this reason, we do not use the expectation operator on $f_R$. FCW control is generally expressed as being comprised of integer ($N_i$) and fractional ($N_f$) parts:

$$\text{FCW} = N = N_i + N_f \tag{4.10}$$

Alternatively, FCW could be defined in terms of the division of the two clock periods in the mean sense:

$$\text{FCW} \equiv \frac{T_R}{\mathcal{E}(T_V)} \tag{4.11}$$

where $\mathcal{E}(T_V) \equiv \overline{T_V}$ is the average clock period of the oscillator. Equation 4.11 gives another interpretation of phase-domain operation. The FCW value establishes how many high-frequency CKV clocks are to be contained within one lower-frequency FREF clock. It suggests counting the number of CKV clocks and dividing it by the number of FREF cycles in order to get the estimate. It should also be noted here that the instantaneous clock period ratio might be slightly off, due to the phase noise effects of the DCO oscillator. However, the long-term value should be very precise and approach FCW in the limit.

In steady-state conditions, the PLL operation achieves a zero or constant averaged phase difference between the variable $\theta_V[i]$ and the reference $\theta_R[k]$ phases. Attempt to formulate the phase error as this unitless phase difference $\phi_E = \theta_R - \theta_V$ would be unsuccessful due to nonalignment of the time samples. This is addressed in Section 4.2.

An additional benefit of operating a PLL with phase-domain signals is to alleviate the need for the frequency detection function within the phase detector. This allows us to operate the PLL as type I (only one integrating pole, due to the DCO frequency-to-phase conversion), where it is possible to eliminate a low-pass loop filter between the phase detector and the oscillator input, resulting in a high bandwidth and fast response of the PLL. It should be noted that conventional phase-locked loops, such as a charge-pump-based PLL (Fig. 1.19), do not truly operate in the phase domain. There, the phase modeling is only a small-signal approximation under the locked condition. Their reference and feedback signals are edge based, and their closest distance is measured as a proxy for the phase error. Gardner describes this as "converting the timed logic levels into analog quantities" [21]. False frequency locking is a deficiency that results directly from not truly operating in the phase domain, and it requires some extra measures, such as use of a phase/frequency detector.

## 4.2 REFERENCE CLOCK RETIMING

It must be recognized that the two clock domains, as described in Section 4.1, are not entirely synchronous, and it is difficult to compare the two digital phase values physically at different time instances $t_V$ and $t_R$ without having to face metastability problems.[1] During frequency acquisition, their edge relationship is not known, and during phase lock, the edges will exhibit rotation if the fractional FCW is nonzero. Therefore, it is imperative that the digital-word phase comparison be performed in the same clock domain. This is achieved by oversampling of the FREF clock by the high-rate DCO clock, CKV (see Fig. 4.2),[2] and using the resulting CKR clock to accumulate the reference phase $\theta_R[k]$ as well as to sample the high-rate DCO phase $\theta_V[k]$ synchronously, mainly to contain the high-rate transitions.

---

[1]Mathematically, $\theta_V[i]$ and $\theta_R[k]$ are discrete-time signals with incompatible sampling times and cannot be compared directly without some sort of interpolation.
[2]The retiming circuit is much more complicated in practice and is discussed in Section 4.7.

# 80   ALL-DIGITAL PHASE-LOCKED LOOP

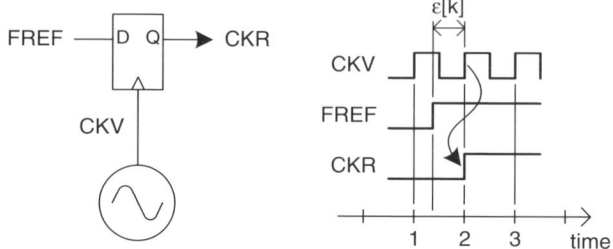

**Figure 4.2** Concept of synchronizing clock domains by retiming the frequency reference.

The CKR clock is thus stripped of the FREF timing information and is used throughout the system. Since the phase comparison is now performed synchronously at the rising edge of CKR, Eqs. 4.5 and 4.6 ought to be rewritten as

$$\theta_V[k] = \theta_V[i]|_{iT_V = \lceil kT_R \rceil} = \lceil kN \rceil \tag{4.12}$$

$$\theta_R[k] = kN + \theta_0 + \varepsilon[k] \tag{4.13}$$

where $\varepsilon[k]$ is the CKV clock edge *quantization error* in the range $\varepsilon \in (0, 1)$.

$\varepsilon[k]$ could be further estimated and corrected by other means, such as the fractional error correction circuit. This operation is illustrated in Fig. 4.3 as an example of integer-domain quantization error for a simple case of the frequency-division ratio of $N = 2\frac{1}{4}$. Unlike $\varepsilon[k]$, which represents rounding to the next DCO edge, a conventional definition of the phase error represents rounding to the closest DCO edge and is shown as $\phi[k]$ in Fig. 1.22. An exact definition of the correction signal is not extremely important as long as it is consistent and provides a proper negative feedback.

The reference retiming operation of Fig. 4.2 can be recognized as a quantization in the DCO clock transitions integer domain, where each CKV clock rising edge is the next integer and each rising edge of FREF is a real-valued number. Since the

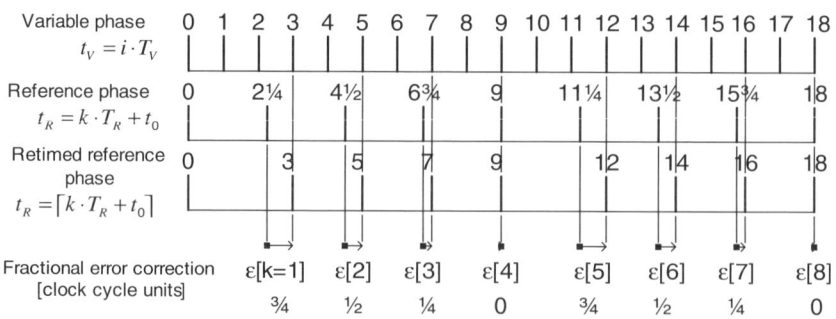

**Figure 4.3** Fractional-$N$ division ratio timing example with $N = 2 + \frac{1}{4}$. (From [65], © 2005 IEEE.)

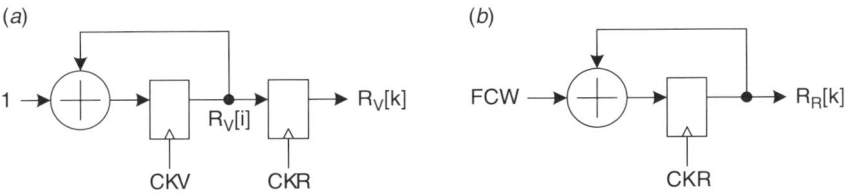

**Figure 4.4** Hardware implementation of (a) the variable phase $R_V[k]$ and (b) the reference phase $R_R[k]$ estimators. (From [65], © 2005 IEEE.)

**Table 4.1  ADPLL Clock Names**

| Notation | Name | Frequency |
|---|---|---|
| CKV | Variable (DCO) clock | $f_V$ |
| FREF | Frequency reference | $f_R$ |
| CKR | Retimed frequency reference clock | $f_R$ |

system must be time-causal, quantization to the next DCO transition (next integer), rather than the closest transition (rounding off to the closest integer), could only be performed realistically.

The set of phase estimate equations (4.7 and 4.8) should be augmented by the sampled variable phase

$$R_V[k] = \sum_{l=1}^{i} 1 \bigg|_{iT_V = \lceil kT_R \rceil} \qquad (4.14)$$

The index $k$ is now the $k$th transition of the retimed reference clock CKR, not the $k$th transition of the reference clock FREF. By constraint, it contains an integer number of CKV clock transitions. As shown in Fig. 4.4, $R_V[k]$ is implemented as an incrementer followed by a flip-flop register, and $R_R[k]$ is implemented as an accumulator.

Table 4.1 summarizes major clocks used in the architecture described. Because of the clock edge displacement as a result of retiming, the CKR clock is likely to have an instantaneous frequency different from its *average* frequency.

## 4.3  PHASE DETECTION

The phase error $\phi_E[k]$ is the difference between the reference phase $\theta_R[k]$ and the variable phase $\theta_V[k]$:

$$\phi_E[k] = \theta_R[k] - \theta_V[k] \qquad (4.15)$$

Additionally, the radian is not very useful here as a unit of measure because the loop operates on the whole and fractional parts of the variable period, and true unitless variables are more appropriate.

The initial temporary assumption made in Section 4.1 about the actual clock periods to be constant or time invariant could be relaxed at this point. Instead of producing a constant ramp of the phase error detected, the phase detector will now produce an output according to real-time clock timestamps.

The phase error can be estimated in hardware by the phase detector operation, defined by

$$\widehat{\phi}_E[k] = R_R[k] - R_V[k] + \varepsilon[k] \tag{4.16}$$

As suggested by Fig. 4.3, the conventional definition of the phase error, expressed as a difference between the reference and variable phases, needs to be augmented here to account for the quantization correction $\varepsilon$.

It is possible to rewrite Eq. 4.16 in terms of independent integer and fractional parts such that the integer part of the reference phase $R_{R,i}$ is added to the integer-only $R_V$, and the fractional part of the reference phase $R_{R,f}$ is added to the fractional-only $\varepsilon$:

$$\widehat{\phi}_E[k] = \left(R_{R,i}[k] - R_V[k]\right) + \left(R_{R,f}[k] + \varepsilon[k]\right) \tag{4.17}$$

In light of this equation, the fractional error correction $\varepsilon$ is to track the fractional part of the reference phase $R_{R,f}$, which is similar in operation to the variable phase $R_V$ tracking the integer part of the reference phase $R_{R,i}$. Therefore, the three-term phase detection mechanism performs dual-phase error tracking, with separate paths for the integer and fractional parts. The fractional-term tracking should be contrasted with the integer-term tracking, due to the apparently different arithmetic operations. In the integer part, both terms should ideally subtract to $-1$, whereas in the fractional part, both terms should ideally add to 1. This operation is a result of the $\varepsilon$ definition and has no implications for circuit complexity. Even the resulting bias of 1 is easily absorbed by the variable-phase accumulator.

Table 4.2 summarizes the major phase-domain signals and cross-references them with the implementational notation used later. For the architecture implementation described in this book, $W_I = 8$ and $W_F = 15$.

Figure 4.5 illustrates a general block diagram of the phase detection mechanism of Eq. 4.16. It consists of the phase detector itself, which operates on the three phase sources: reference phase $R_R[k]$, variable phase $R_V[k]$, and fractional error correction $\varepsilon[k]$. The actual variable phase $R_V[i]$ is clocked by the CKV clock of index $i$ and must be resampled by the CKR clock of index $k$. After the PHV resampling, all three phase sources are synchronous to the CKR clock,[3] which guarantees that the resulting phase error $\phi_E[k]$ will also be synchronous. The blocks are summarized

---

[3] For implementation reasons, the TDC output could optionally be resampled by CKR.

**Table 4.2 Phase Detection Signal Name Cross-Reference**

| Math Notation | Implementation Notation | Name | Bus Width |
|---|---|---|---|
| $N$ | FCW | Frequency command word | $W_I + W_F$ |
| $\theta_R[k]$ | — | Reference phase | — |
| $R_R[k]$ | PHR | Reference phase (estimated) | $W_I + W_F$ |
| $\theta_V[i]$ | — | Variable phase | — |
| $R_V[i]$ | PHV | Variable phase (estimated) | $W_I$ |
| $\theta_V[k]$ | — | Sampled variable phase | — |
| $R_V[k]$ | PHV_SMP | Sampled variable phase (estimated) | $W_I$ |
| $\varepsilon[k]$ | PHF_F | Fractional error correction | $W_F$ |
| — | PHF_I | TDC edge skip | 1 |
| $\phi_E[k]$ | PHE | Phase error | $W_I + W_F$ |

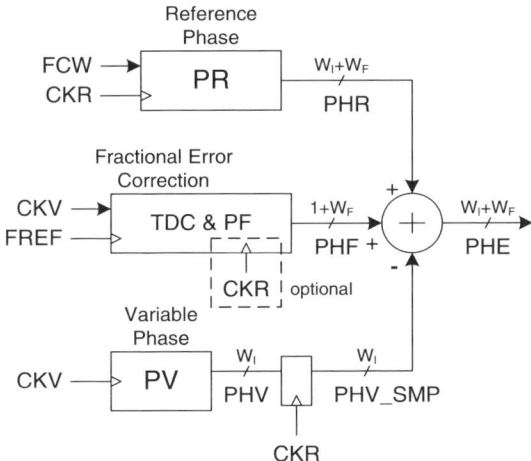

**Figure 4.5** General block diagram of phase detection.

in Table 4.3. An extra output bit from the fractional error correction PF due to metastability avoidance is explained in Section 4.6.

Figure 4.6 reveals the internal structure of the phase detector circuit. All inputs are synchronous. The integer and fractional parts of the fixed-point phase signals

**Table 4.3 Phase Detection Block Names**

| | |
|---|---|
| PV | Variable-phase accumulator (incrementer) |
| PR | Phase reference accumulator |
| PF | Fractional error correction |
| TDC | Time-to-digital converter (part of PF) |
| PD | Phase detector |

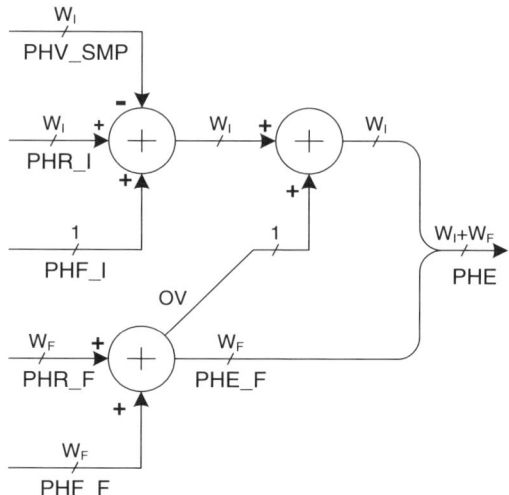

**Figure 4.6** Phase detector structure.

are split and processed independently with proper bit alignment. The integer portion uses modulo arithmetic in which $W_I$-width rollovers are expected as a normal occurrence.

Table 4.4 is a numerical version of the example shown in Fig. 4.3. It shows an ideal scenario: when all the clock edges come at the right time, due to which $\widehat{\phi}_E[k]$ turns out to be zero. However, if the actual periods of CKV or FREF clocks were different, the error would be nonzero.

**Table 4.4 Numerical Example Corresponding to Fig. 4.3 ($N = 2.25$ and $\theta_0 = 0$)**

| $i$ | $R_V[i]$ | $k$ | $R_V[k]$ | $kN + \theta_0$ | $\varepsilon[k]$ | $R_R[k]$ | $\widehat{\phi}_E[k]$ |
|---|---|---|---|---|---|---|---|
| 0 | 0 | 0 | 0 | 0 | 0 | 0 | 0 |
| 1 | 1 | 0 | 0 | 0 | 0 | 0 | 0 |
| 2 | 2 | 0 | 0 | 0 | 0 | 0 | 0 |
| 3 | 3 | 1 | 3 | 2.25 | 0.75 | 2.25 | 0 |
| 4 | 4 | 1 | 3 | 2.25 | 0.75 | 2.25 | 0 |
| 5 | 5 | 2 | 5 | 4.50 | 0.50 | 4.50 | 0 |
| 6 | 6 | 2 | 5 | 4.50 | 0.50 | 4.50 | 0 |
| 7 | 7 | 3 | 7 | 6.75 | 0.25 | 6.75 | 0 |
| 8 | 8 | 3 | 7 | 6.75 | 0.25 | 6.75 | 0 |
| 9 | 9 | 4 | 9 | 9.0 | 0 | 9.0 | 0 |
| 10 | 10 | 4 | 9 | 9.0 | 0 | 9.0 | 0 |
| 11 | 11 | 4 | 9 | 9.0 | 0 | 9.0 | 0 |
| 12 | 12 | 5 | 12 | 11.25 | 0.75 | 11.25 | 0 |

### 4.3.1 Difference Mode of ADPLL Operation

Equation 4.16, describing phase-error hardware estimation, is repeated below:

$$\widehat{\phi}_E[k] = R_R[k] - R_V[k] + \varepsilon[k] \tag{4.18}$$

The previous phase-error sample at $k-1$ is simply expressed as

$$\widehat{\phi}_E[k-1] = R_R[k-1] - R_V[k-1] + \varepsilon[k-1] \tag{4.19}$$

Equation 4.8 for the reference phase is rewritten here in an accumulative form:

$$R_R[k] = R_R[k-1] + \text{FCW} \tag{4.20}$$

Similarly, the variable phase of Eq. 4.14 is rewritten as

$$R_V[k] = R_V[k-1] + \Delta R_V[k] \tag{4.21}$$

where $\Delta R_V[k]$ is a number of whole CKV clock cycles between the two consecutive edges of the retimed FREF clock (CKR). Plugging Eqs. 4.20 and 4.21 into Eq. 4.18 results in

$$\widehat{\phi}_E[k] = (R_R[k-1] - R_V[k-1]) + (\text{FCW} - \Delta R_V[k]) + \varepsilon[k] \tag{4.22}$$

and further substitution of Eq. 4.19 simplifies the phase-error equation to

$$\widehat{\phi}_E[k] = \widehat{\phi}_E[k-1] + (\text{FCW} - \Delta R_V[k]) + (\varepsilon[k] - \varepsilon[k-1]) \tag{4.23}$$

This equation, shown graphically in Fig. 4.7, is the difference form of the ADPLL phase detector output, which is used in future designs. Despite the mathematical equivalency, the new architecture simplifies handling of various exceptions, such as by means of freezing the DCO updates.

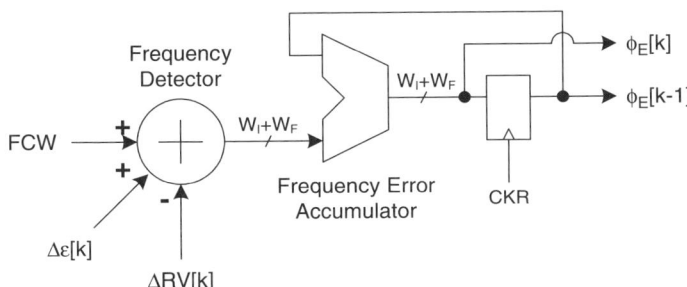

**Figure 4.7** Phase detector using the difference mode of ADPLL operation.

### 4.3.2 Integer-Domain Operation

If $\varepsilon[k]$ could not be estimated, this operation might be compensated in the phase domain by the ceiling operation of the reference phase:

$$\tilde{R}_R[k] = \lceil R_R[k] \rceil \tag{4.24}$$

The reference phase $R_R[k]$ is generally a fixed-point arithmetic signal with a sufficiently large fractional part to achieve the required frequency resolution as governed by Eq. 4.10. The ceiling operation of Eq. 4.24 could easily be implemented in hardware by discarding the fractional bits and incrementing the integer bits.[4] In fact, the statement above agrees quite well with Eq. 4.17, where $\varepsilon$ tracks the fractional part of the reference phase by perfectly complementing it to 1 such that $R_{R,f}[k] + \varepsilon[k] = 1$:

$$\tilde{\phi}_E[k] = R_{R,i}[k] - R_V[k] + 1 \tag{4.25}$$

It should be emphasized that even though the integer-domain quantization error $\varepsilon[k]$ due to reference clock retiming in Eq. 4.13 is compensated by next-integer rounding operation (ceiling) of the reference phase in Eq. 4.24, the phase resolution still cannot be better than $\pm\frac{1}{2}$ of the DCO clock cycle. It means that the CKV clock drift within its full clock cycle cannot be detected, and consequently, compensated by the loop. In other words, the transfer function of the variable phase is quantized, and slight phase-error variations of less than CKV clock cycle are uncorrected. This could also be illustrated by the timing diagram of Fig. 4.2. If the FREF edge wanders between the first and second CKV edges, this movement cannot be detected and the CKR always transitions on the second CKV edge. The consequence of operating in the integer domain without the $\varepsilon$ correction is a larger phase noise at lower frequencies, which for many nonwireless applications might be quite adequate.

The integer-phase detector in the synchronous digital phase environment can now be realized as a simple arithmetic subtraction of the DCO phase $R_V$ from the reference phase. This operation is performed on every rising edge of the CKR clock.

$$\tilde{\phi}_E[k] = \tilde{R}_R[k] - R_V[k] = \lceil R_R[k] \rceil - R_V[k] = R_{R,i}[k] - R_V[k] + 1 \tag{4.26}$$

## 4.4 MODULO ARITHMETIC OF THE REFERENCE AND VARIABLE PHASES

The variable and reference accumulators $R_V[i]$ and $R_R[k]$, respectively, are implemented in modulo arithmetic in order to practically limit the wordlength of the arithmetic components. Table 4.2 reveals that the integer part of the accumulators is $W_I$ ($= 8$ in this book) and the fractional part of the reference

---

[4]This method improperly handles the case when the fractional part is zero, but this has no practical consequences.

## 4.4 MODULO ARITHMETIC OF THE REFERENCE AND VARIABLE PHASES

accumulator is $W_F$ (=15 in this book). These accumulators represent the variable and reference phases, $\theta_V$ and $\theta_R$, respectively, which are linear and grow without bound with the development of time. The registers, on the other hand, cannot hold arbitrarily large numbers, so they must be restricted to a small stretch of line from zero to infinity, which repeats itself indefinitely such that any such stretch is an alias of the fundamental stretch from 0 to $2^{W_I}$.

A good example of such an approximation is modulo arithmetic. Any accumulator of length $W_I$, whose carry-out bits are simply disregarded and which does not perform saturation, is a modulo-$2^{W_I}$ accumulator. In fact, modulo arithmetic is very natural in the digital logic. Both the variable and reference modulo-$2^{W_I}$ accumulators are not absolutely linear in the strictest sense of the word. They are, however, linear in the local sense (for a constant frequency, of course). The phase equations 4.7 and 4.8 are now rewritten to acknowledge the implicit modulo operation:

$$R_V[i+1] = \mathrm{mod}(R_V[i] + 1, 2^{W_I}) \qquad (4.27)$$

$$R_R[k+1] = \mathrm{mod}(R_R[k] + \mathrm{FCW}, 2^{W_I}) \qquad (4.28)$$

Figure 4.8 shows a modulo-16 (4-bit) operation with a frequency-division ratio of $N = 10$ and perfect alignment between the two phases. If the system is in a settled state, both phases follow a sawtooth trajectory with the same speed. The variable phase $R_V[i]$ (top plot) traverses all integers at a very fast rate. The reference phase

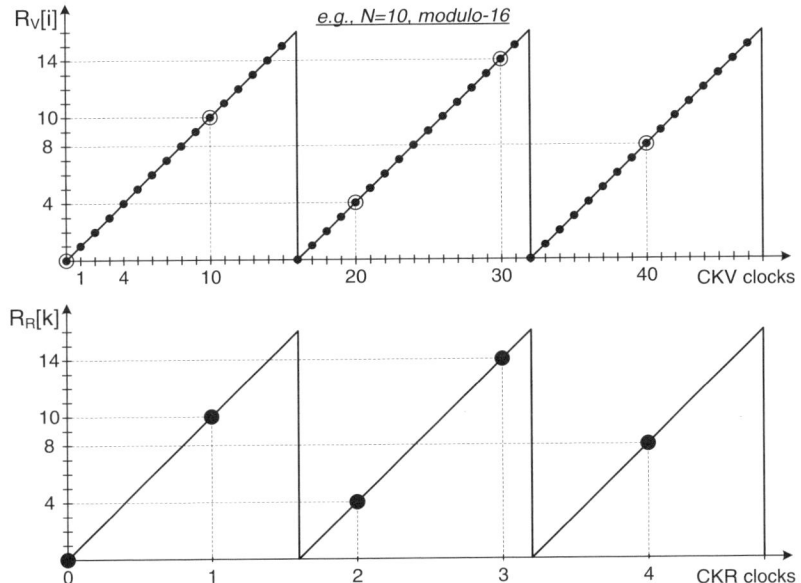

**Figure 4.8** Modulo arithmetic of reference and variable-phase registers with zero-phase alignment.

$R_R[k]$ (bottom plot), on the other hand, moves infrequently (10 times less often) but with large steps (10 times larger). As a result, their traversal velocity is the same, and their respective locations will correspond to the same register readout (0, 10, 4, 14, 8,...), which happens at the same time on the CKR clock, thus producing zero-phase error, as expected in a system with zero-phase and frequency errors. Also shown every 10 CKV clock cycles is the sampling process of $R_V[k] = R_V[i]$ during activity at $R_R[k]$.

A situation similar to Fig. 4.8 but exhibiting a small phase offset of 3 is depicted in Fig. 4.9. For this case, the difference between $R_R[k]$ (3, 13, 7, 1, 11, ...) and $R_V[k]$ (0, 10, 4, 14, 8, ...) readout sequences should always be 3, and it is, except for the fourth comparison. Here, the error of $-13$ lies beyond the $[-8, 7]$ linear range, and performing modulo-16 arithmetic will get it to 3, which is in line with the others.

The modulo arithmetic on $R_V$ and $R_R$ could be visualized as two rotating vectors, with the smaller angle between them constituting the phase error (Fig. 4.10). $R_V$ and $R_R$ are positive numbers, and their maximum possible value without rollover depends on the counter width or integer part of the FCW, and equals $2^{W_I}$. The phase error has the same range but is symmetric around zero (i.e., it is a 2's-complement number). This figure also helps to demonstrate that the phase detector is not only the arithmetic subtractor of two numbers but also performs a cyclic adjustment, so that under no circumstances would the larger angle between the two vectors be decided. This could happen if, for example, the larger vector appears just before the smaller vector, which is already on the other side of the "zero" radius line. Because the PD output is a $W_I$-bit constrained signed number, the conversion is

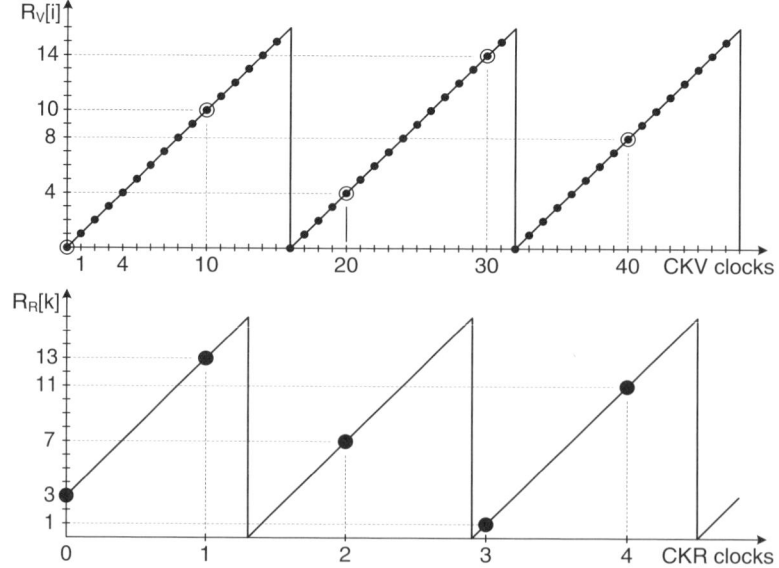

**Figure 4.9** Modulo arithmetic for phase offset $\phi_E$ of 3. (From [65], © 2005 IEEE.)

## 4.4 MODULO ARITHMETIC OF THE REFERENCE AND VARIABLE PHASES

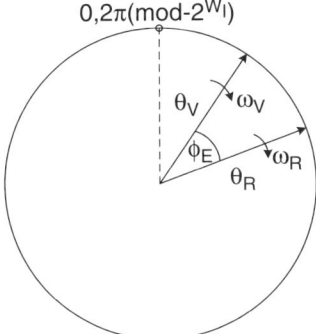

**Figure 4.10** Rotating vector interpretation of the reference and variable phases. (From [65], © 2005 IEEE.)

always made implicitly and the output lies within $[-2^{W_I-1}, 2^{W_I-1} - 1]$. Consequently, no extra hardware is required and none is shown in Fig. 4.6.

Due to the modulo arithmetic, there might be a possibility of aliasing for large values of FCW and large variable frequencies. For example, FCW $= 2^{W_I} + x$ will alias to $x$. Similarly, $f_V = f_R(2^{W_I} + x)$ will alias to $f_V = f_R x$. The "Nyquist frequency" could be increased by making the integer width $W_I$ sufficiently large. In the implementation presented in this book, however, both variable and reference frequencies are tightly controlled and there is no possibility of getting into the foldover region.

### 4.4.1 Variable-Phase Accumulator (PV Block)

As shown in Fig. 4.4, the variable-phase accumulator implements the DCO clock count incrementing as defined in Eq. 4.7, with the rollover effect as described above.

The current CMOS processes are fast enough to perform an 8-bit binary incrementer at a 2.4-GHz clock in one cycle using a simple carry-ripple structure. Critical timing of this operation would comprise a chain of seven half-adders and an inverter. However, for an actual commercial application, it might be necessary to add an extra timing margin to guarantee robust operation with acceptable yield for all the process and environmental conditions, as well as anticipated clock distribution skew statistics. This extra margin can be obtained by increasing the maximum operational speed through topological means. The carry-ripple binary incrementer can be transformed into two separate smaller incrementers, as shown in Fig. 4.11. The first incrementer operates on the two lower-order bits and triggers the higher-order increment whenever its count reaches "11". The second incrementer operates on the same CKV clock, but the 6-bit increment operation is now allowed to take four clock cycles. The long critical path of the 8-bit carry-ripple incrementer has thus been split into smaller parts, allowing for the necessary timing margin.

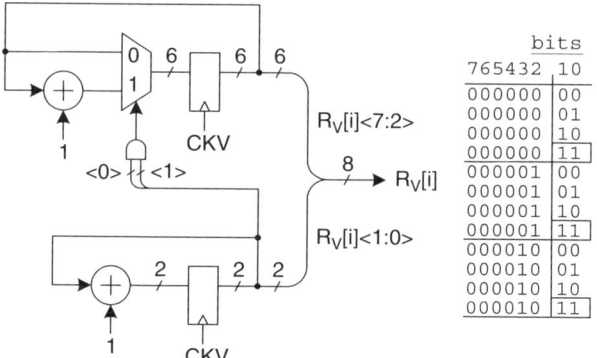

**Figure 4.11** Variable-phase incrementer with separate calculation between lower- and higher-order bits.

It can be seen that the critical path of the partitioning above was actually the increment trigger path from the lower-order registers, through the AND gate and the six single-bit multiplex select lines, and terminating on the higher-order registers. The fanout of 6 was taking a disproportional toll on the delay, so a slight modification can be made by retiming that control path. The modified version of the variable-phase block is shown in Fig. 4.12. The triggering state is now one count earlier, at "10". The chief improvement of this solution is that the register output Q_INC is now capable of driving six multiplex select lines.

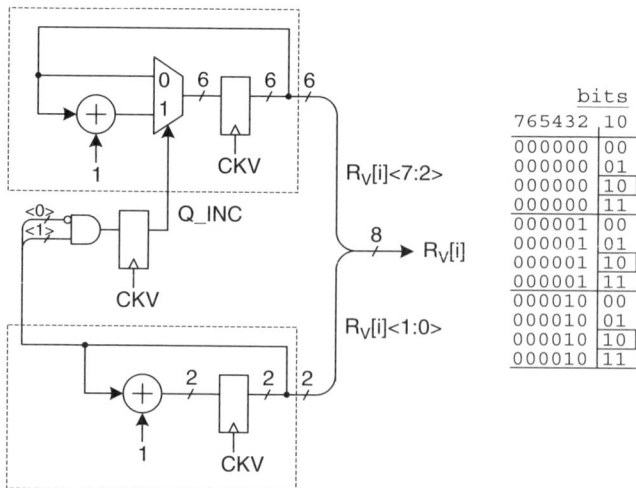

**Figure 4.12** Implementation of a variable-phase incrementer with higher-order increment retiming.

## 4.5 TIME-TO-DIGITAL CONVERTER

Due to the DCO edge counting nature of a PLL, the phase quantization resolution of the integer precision configuration, described in Section 4.3.2, cannot be better than $\pm\frac{1}{2}$ of the DCO clock cycle. For wireless applications, finer phase resolution might be required. This must be achieved *without* forsaking the digital signal processing capabilities. Figure 4.5 shows the method by which the integer-domain quantization error $\varepsilon[k]$ gets corrected by means of a *fractional-period error estimator* (PF).

The fractional (sub-$T_V$) delay difference $\varepsilon$ between the reference clock (FREF) and the next significant edge of the DCO clock (CKV) is measured using a *time-to-digital converter* (TDC) with a time quantization resolution $\Delta t_{res}$ of an inverter delay $t_{inv}$, and the time difference is expressed as a fixed-point digital word. This operation is shown in Fig. 4.13. The raw TDC output could not be used in the system in its integer form since the time resolution is a varying physical parameter; therefore, it has to be normalized by the oscillator clock period. Only the fractional error correction $\varepsilon$ is used by the phase detector.

The smallest time interval that is readily resolved in the digital fractional-phase detector is a TDC inverter delay $t_{inv}$ of the given technology, which is about 30 ps for a typical 0.13-$\mu$m CMOS process. The string of inverters forms the backbone of the simplest possible implementation of time-to-digital conversion. In a digital

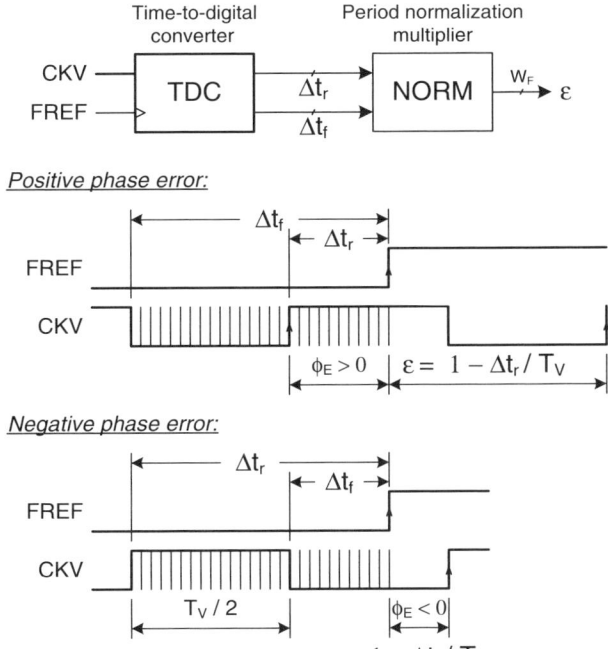

**Figure 4.13** Fractional (sub-$T_V$) phase-error estimation. (From [67], © 2004 IEEE.)

deep-submicron CMOS process, the inverter could be considered a basic precision time-delay cell which has full digital-level regenerative properties.

It should be noted that it is possible to achieve substantially better resolution than the inverter delay for the TDC function. An example is given in [66] that utilizes a Vernier delay line with two nonidentical strings of buffers. The slower string of buffers is stabilized by negative feedback through a delay line. The buffer time propagation difference establishes the resolution. The disadvantage of this approach is higher power consumption and extra analog circuitry.

The digital fractional phase is determined by passing the DCO clock through a chain of inverters (see Fig. 4.14), such that each inverter output would produce a clock slightly delayed from that of the previous inverter [67]. The staggered clock phases are then sampled by the same reference clock. This is accomplished by an array of registers whose Q outputs form a pseudo-thermometer code. In this arrangement there will be a series of 1's and 0's. In this example, a series of four 1's (half period: $T_V = 8$ inverters) start at position 3 and extend to position 6. The series of four 0's follow, starting at index 7. The position of the transition detected from 1 to 0 would indicate a quantized time delay $\Delta t_r$ between the FREF sampling edge and the rising edge of the DCO clock, CKV, in $t_{\text{inv}}$ multiples. Similarly, the position of the transition detected from 0 to 1 would indicate a quantized time delay $\Delta t_f$ between the FREF sampling edge and the falling edge of the DCO clock, CKV (in Fig. 4.14, $\Delta t_r = 6\, t_{\text{inv}}$ and $\Delta t_f = 2\, t_{\text{inv}}$). Because of the time-causal nature of this operation (one cannot process the future edges), both delay values must be interpreted as positive. This is fine if $\Delta t_r$ is smaller than $\Delta t_f$. This corresponds to the positive phase error ($\phi_E > 0$) of the classical PLL in which the reference lags the closest DCO edge and therefore the phase sign has to be negated. However, it is not as straightforward if $\Delta t_r$ is greater than $\Delta t_f$. This corresponds to the negative phase error ($\phi_E < 0$) of a classical PLL. The time lag between the reference edge and the *following* rising edge of CKV must be calculated based on the available information of the delay between the *preceding* rising edge of CKV and the reference edge and

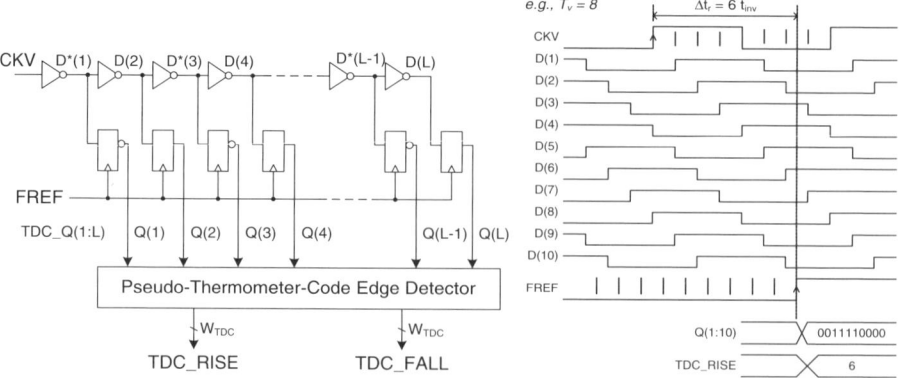

**Figure 4.14** Time-to-digital converter. (From [67], © 2004 IEEE.)

the clock half-period, which is the difference $T_V/2 = \Delta t_r - \Delta t_f$ (see Fig. 4.13). In general,

$$T_V/2 = \begin{cases} \Delta t_r - \Delta t_f & \Delta t_r \geq \Delta t_f \\ \Delta t_f - \Delta t_r & \text{otherwise} \end{cases} \quad (4.29)$$

The number of taps $L$ required for the TDC of Fig. 4.14 is determined by how many inverters are needed to cover the full DCO period:

$$L \geq \frac{\max(T_V)}{\min(t_{\text{inv}})} \quad (4.30)$$

If too many inverters are used, the circuit is more complex and consumes more power than necessary. For example, in Fig. 4.14, inverters 9 and 10 are beyond the first full cycle of eight inverters and are not needed since the pseudo-thermometer decoder is based on a priority detection scheme, and earlier bits would always be considered first. It is good engineering practice, however, to keep some margin to guarantee proper system operation at the fast process corner and the lowest DCO operational frequency, even if it is below the operational band.

In this implementation of TDC, we have chosen a symmetric sense-amplifier-based flip-flop (adapted from reference [68]) with differential inputs to guarantee substantially identical delays for rising and falling input data. This is discussed in Section 4.7.1.

### 4.5.1 Frequency Reference Edge Estimation

Power consumption is a very crucial parameter for battery-operated wireless terminals. Even though in this regard the transmitter demonstrated is very competitive with other devices available on the market, dramatically increasing the battery life would open up a new range of applications. Figure 4.15 reduces current

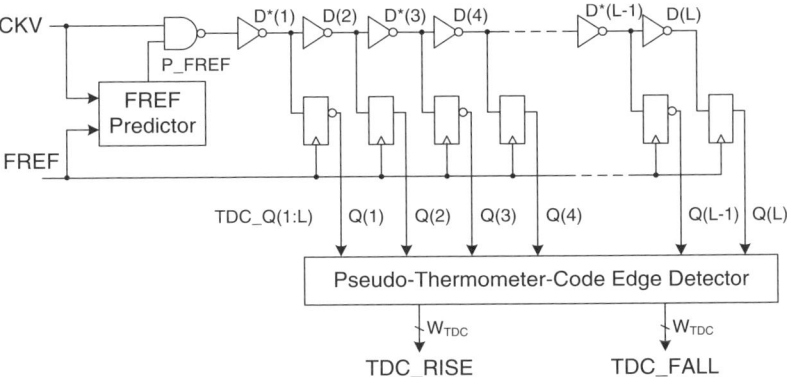

**Figure 4.15** Time-to-digital converter with FREF prediction.

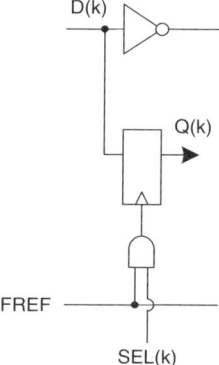

**Figure 4.16** Time-to-digital converter with local control of registers.

consumption through an intelligent power management scheme. It performs periodic gating of the oscillator clock inside the TDC by predicting where the next FREF edge might lie. The timing information lies in the edges of the FREF clock; thus, operating the fractional phase detector in between its edges unnecessarily propagates the gate-switching transitions. Statistics about the past edges and hit/miss ratios could be used to allow the prediction window to be as narrow as practically possible.

After a certain length of operational time, enough information could be gathered about both clock phases and their statistics to reduce power and noise coupling. Instead of clocking all the TDC word registers in Fig. 4.14 with the FREF clock, only a selected few could be clocked, as shown in Fig. 4.16, thus saving the transition power and reducing noise. The region selected should be large enough to ensure good "hit" probability and positioned based on the expected occurrence as demonstrated in Fig. 4.3. It should be noted that a "miss" is not a serious occurrence; it means that no phase information could be used for that particular compare event. In this case, the next compare event could engage *all* the thermometer-coded TDC registers, just to guarantee no consecutive misses.

## 4.6 FRACTIONAL ERROR ESTIMATOR

In this implementation the conventional phase $\phi_E$ is not needed. Instead, $\Delta t_r$ is used for the $\varepsilon[k]$ correction of Eq. 4.13 that is positive and $\varepsilon \in (0, 1)$. It has to be normalized by dividing it by the clock period (unit interval) and complementing to 1, in order to combine it properly with the fractional part of the reference phase output $R_{R,f}$. The fractional correction $\varepsilon[k]$ is represented as a fixed-point digital word:

$$\varepsilon[k] = 1 - \frac{\Delta t_r[k]}{T_V} \qquad (4.31)$$

In practice, it is preferable to obtain the clock period $T_V$ through longer-term averaging in order to ease the calculation burden and linearize the transfer function of $1/T_V$. The averaging time constant could be as slow as the expected drift of the inverter delay, possibly due to temperature and supply voltage variations. The instantaneous value of the clock period is an integer, but averaging it would add significant fractional bits with longer operations:

$$\overline{T}_V = \frac{1}{N_{\text{avg}}} \sum_{k=1}^{N_{\text{avg}}} T_V[k] \tag{4.32}$$

For each cycle $k$, $T_V[k]$ is computed using Eq. 4.29. For the design presented here, it was found that accumulating 128 FREF clock cycles would produce accuracy within 1 ps of the inverter delay. The length of the operation is chosen to be a power of 2 since the division by the number of samples $N_{\text{avg}}$ could now be replaced with a simple right-shift.

The actual fractional output of the $\varepsilon$ error correction needs one extra bit, due to the fact that the entire CKV cycle would have to be skipped if the rising edge of FREF transitions too close to the rising edge of CKV, since the retiming circuit of Fig. 4.2 will not be able to capture it correctly. As a safety precaution, the previous falling CKV edge would have to be used is then always resampled by the following rising edge of CKV. PHF_I is of integer LSB weight. This scenario is illustrated in Fig. 4.17, in which there is a full-cycle skipping if

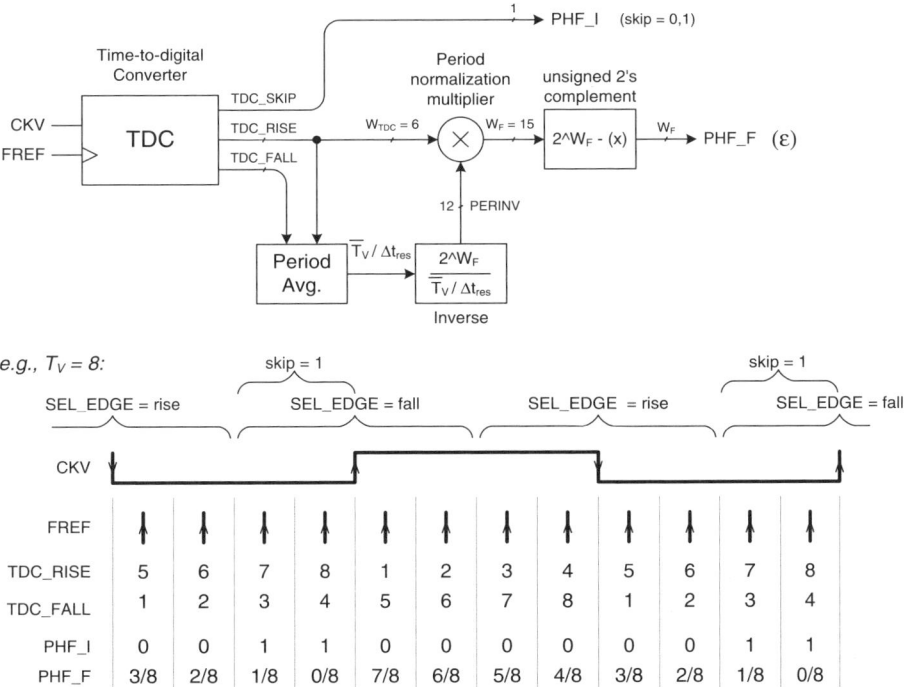

**Figure 4.17** TDC normalization and edge-skipping operation.

**Table 4.5 TDC Signal Names Cross-Reference**

| Math Notation | Implementation Notation | Name | Bus Width |
|---|---|---|---|
| — | TDC_Q | Sampled timing state vector | 48 |
| $\Delta t_r$ | TDC_RISE | CKV rising edge to FREF | $W_{TDC} = 6$ |
| $\Delta t_f$ | TDC_FALL | CKV falling edge to FREF | $W_{TDC} = 6$ |
| — | TDC_SKIP | TDC edge skip | 1 |

FREF happens as close as two inverter delays before the rising edge of CKV. The clock skipping operation is described further in Section 4.7.

This configuration of the ADPLL synthesizer is designed to work when CKV is much faster than FREF. This is typically the case in wireless communications, where FREF (created by an external crystal) is at most a few tens of MHz, and CKV (RF carrier) is in the GHz range. In the test chip implemented, $f_R = 13$ MHz and $f_V = 2.40$ to 2.48 GHz, resulting in a division ratio $N$ of about 180. The large value of $N$ puts more emphasis on the CKV edge counting operation (Eq. 4.7), which is *exact*, and less emphasis on the $\varepsilon$ determination (TDC operation), which is less precise, due to the continuous-time nature of device delays. This architecture would still work correctly if the $N$ ratio is much smaller. The main requirement is that resolution of the fractional error correction be at least an order of magnitude better than the CKV period.

Table 4.5 summarizes the major TDC signals and cross-references them with the implementational notation. TDC_Q is the pseudo-thermometer encoded timing state vector in Fig. 4.14 feeding the priority decoder. TDC_RISE and TDC_FALL are small integer quantizations of the $\Delta t_r$ and $\Delta t_f$ time delays. They are outputs of the priority decoder. The TDC_Q bus width was conservatively chosen to be 48, so 6 bits are required to represent the decoded data.

### 4.6.1 Fractional-Division Ratio Compensation

The $\varepsilon[k]$ samples are roughly constant if the DCO clock period $T_V$ is an integer division of the frequency reference clock period $T_R$. For a more general case where this ratio is fractional, the $\varepsilon[k]$ samples increase linearly within the modulo (0, 1) range (Fig. 4.3 shows an example). This sample pattern could easily be predicted in a digital form that closely corresponds mathematically to the well-known analog fractional phase compensation scheme of fractional-$N$ PLL frequency synthesizers [27]. However, the architecture described here tracks this periodic pattern naturally in a complement-to-1 manner, so no extra processing is needed. In fact, unlike conventional PLL-based frequency synthesizers, this architecture was designed from the ground up to handle a real-valued general frequency multiplication, limited only by the wordlength of the reference phase accumulator.

The complement-to-1 fractional phase tracking works as follows: The fractional rotational speed is determined by the fractional part of the fixed-point FCW control word. Let's assume that FCW_F = 1/4, as in Fig. 4.3. Accumulating it on every FREF cycle would circularly ramp up the fractional part of the fixed-point reference phase register according to the sequence 1/4, 2/4, 3/4, 4/4, 1/4, .... The fractional error correction $\varepsilon$ is always referenced to the next CKV edge, not from the *preceding* CKV edge as before, so it follows the opposite pattern by *ramping down* according to the sequence 3/4, 2/4, 1/4, 0/4, 3/4, .... Hence, $\varepsilon$ tracks the fractional rotation in a complement-to-1 manner. This is also described mathematically by Eq. 4.15.

### 4.6.2 TDC Resolution Effect on Estimated Frequency Resolution

In a conventional PLL, the phase detector is, at least theoretically, a linear device whose output is proportional to the timing difference between the reference and the feedback oscillator clocks. In the all-digital implementation presented, the $\varepsilon$ fractional phase-error correction is also linear but is quantized in $\Delta t_{res}$ time units, where $\Delta t_{res} \approx t_{inv}$. Figure 4.18 shows the quantization effects of the $\varepsilon$ transfer function of Eq. 4.31. The TDC quantum step $\Delta t_{res}$ determines the quantum step of the normalized fractional error correction, which in normalized units is expressed as $\Delta \varepsilon_{res} = \Delta t_{res}/T_V$. The transfer function has a negative bias of $\Delta t_{res}/2$ but is inconsequential since the loop will compensate for it automatically.

As noted in Section 3.8 while referring to Fig. 3.24, the purpose of the phase detection mechanism in the architecture presented is to convert the accumulated timing deviation TDEV, which is a pure time-domain quantity, into digital bit format. At the same time, as the TDC transfer function in Fig. 4.18 confirms, the phase detector is to perform the output normalization such that TDEV = $T_V$ corresponds to unity.

Under these circumstances, the phase detector output $\phi_E$ could be interpreted as a frequency deviation estimator (from a center or natural frequency) of the output

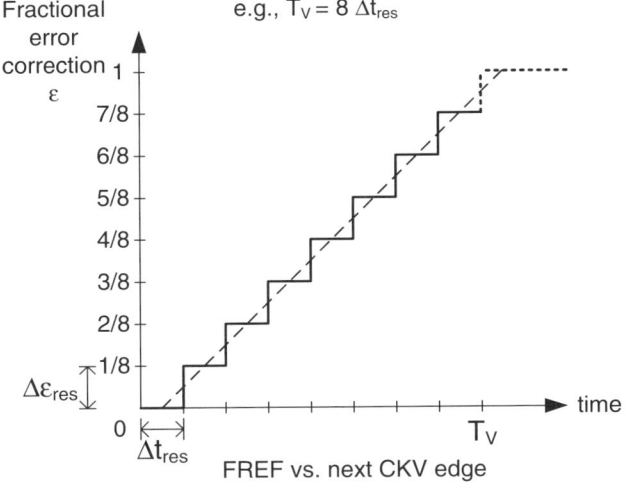

**Figure 4.18** TDC quantized transfer function.

CKV clock and normalized to frequency reference $f_R$. Within one reference clock cycle, $T_R = 1/f_R$,

$$\widehat{\Delta f_V} = -\widehat{\phi_E} f_R \quad (4.33)$$

The estimate above increases linearly with the number of reference cycles.

Resolution of the phase detector is determined directly by the TDC resolution, $\Delta\phi_{E,\text{res}} = \Delta\varepsilon_{\text{res}}$. Adopting the frequency estimation view of the phase detector, the quantum step in the $f_V$ frequency domain, based on one reference cycle, would be

$$\Delta f_{V,\text{res}} = \Delta\varepsilon_{\text{res}} f_R = \left(\frac{\Delta t_{\text{res}}}{T_V}\right) f_R = \left(\frac{\Delta t_{\text{res}}}{T_V}\right) \frac{1}{T_R} \quad (4.34)$$

For example, assuming that $\Delta t_{\text{res}} = 40$ ps, $f_V = 2.4$ GHz, and $f_R = 13$ MHz, the resulting frequency estimate quantization level of a single FREF cycle is $f_{V,\text{res}} = 1248$ kHz. For $\Delta t_{\text{res}} = 30$ ps, $f_V = 1.8$ GHz, and $f_R = 26$ MHz, the resulting $f_{V,\text{res}} = 1404$ kHz. Obviously, this may be unacceptably large for most wireless applications, so advantage can be taken here of the fact that the frequency is a derivative of phase with respect to time, and the frequency resolution could be enhanced with a longer observation interval (i.e., over multiple FREF cycles). In this case, Eq. 4.34 could be modified by multiplying $T_R$ by the number of FREF cycles.

### 4.6.3 Active Removal of Fractional Spurs Through TDC (Optional)

Fractional spurs are defined as those lying at FCW_F $\times f_R$ on both sides of the carrier. Especially problematic in a GSM transmitter are the 400-kHz fractional spurs, which happen when the oscillator operates at 400 kHz away (below and above) from the closest FREF multiple. Consequently, fractional spurs are channel dependent, with one channel per FREF span for a specific spur type.

There are two independent but not entirely orthogonal mechanisms for creation of fractional spurs. The first is the imperfection of the fractional compensation method and is well known in literature for analog-intensive implementations of fractional-$N$ PLLs with analog phase interpolation. In the ADPLL architecture, these imperfections are due to:

1. Estimation error of the TDC period inversion calculation engine
2. TDC quantization
3. TDC nonlinearity

The second mechanism is through coupling of an FREF harmonic into the oscillator. This mechanism is largely architecture independent and is normally addressed by proper layout, matching, and isolation techniques.

It is proposed here that the TDC period normalization mechanism be used additionally to compensate for the spurs resulting from FREF coupling. The spurs are to be identified and detected by performing digital signal processing on fixed-point phase-error samples. It is preferred that the detection and compensation

## 4.6 FRACTIONAL ERROR ESTIMATOR

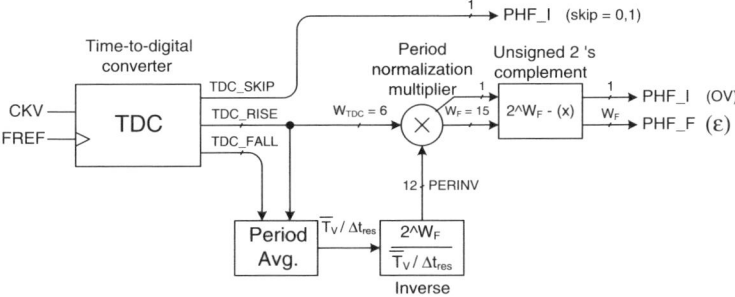

**Figure 4.19** TDC normalization with redefined $\varepsilon$.

process be iterative. Depending on the phase of the coupling signals, the TDC period normalization could be offset in either direction.

The TDC normalization block diagram was presented originally in Fig. 4.17. The raw TDC output cannot be used by the ADPLL in this small positive integer format and therefore had to be normalized to lie within the (0,1) range. Under normal operation, the circuit works as designed. The maximum integer could sometimes exceed the fixed-point averaged value of the quantized DCO period. In this case, the $\varepsilon$ output would be clamped to zero. It should be noted that the process above might present a problem if the PERINV multiplicand is deliberately overestimated and the output has to be saturated. Figure 4.19 reveals a slight modification that allows $\varepsilon$ to formally exceed that range in the negative direction. Figure 4.20 illustrates the normalized TDC output in case of overestimated PERINV with and without saturation. It should be contrasted with Fig. 4.18 when the DCO period is set precisely.

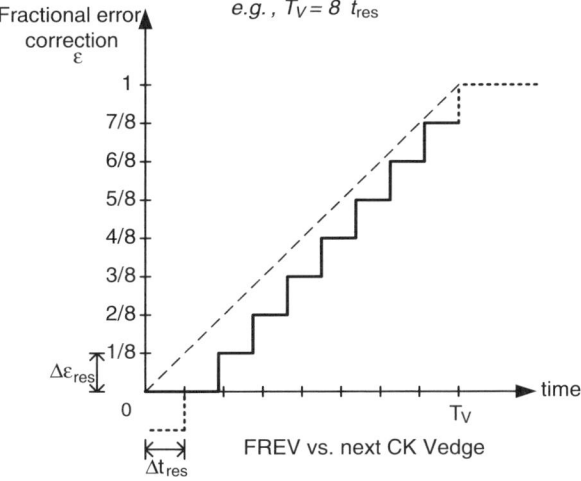

**Figure 4.20** TDC quantized transfer function when PERINV is overestimated.

## 4.7 FREQUENCY REFERENCE RETIMING BY A DCO CLOCK

Frequency reference clock retiming was presented conceptually in Fig. 4.2. Mathematically, the flip-flop register performs the ceiling operation of the real-valued FREF timestamps through retiming it by the integer-valued CKV timestamps. The resulting CKR clock timestamps are integer-valued and, of course, synchronous to the DCO clock. Unfortunately, this simple and elegant mathematical model stumbles in the face of real-world limitation of metastability.

Metastability is a physical phenomenon that limits the performance of comparators and digital sampling elements, such as latches and flip-flops. It recognizes that it takes a nonzero amount of time from the start of a sampling event to determine the input level or state [69,70]. This resolution time gets exponentially larger if the input state change gets close to the sampling event. In the limit, if the input changes at exactly the same time as the sampling event, it might theoretically take an infinite amount of time to resolve. During this time, the output can dwell in an illegal digital state somewhere between 0 and 1. Figure 4.21 shows a clock-to-output (CLK-to-Q) delay vs. input-to-clock (D-to-CLK) skew of a high-performance flip-flop from the GS40 [45]

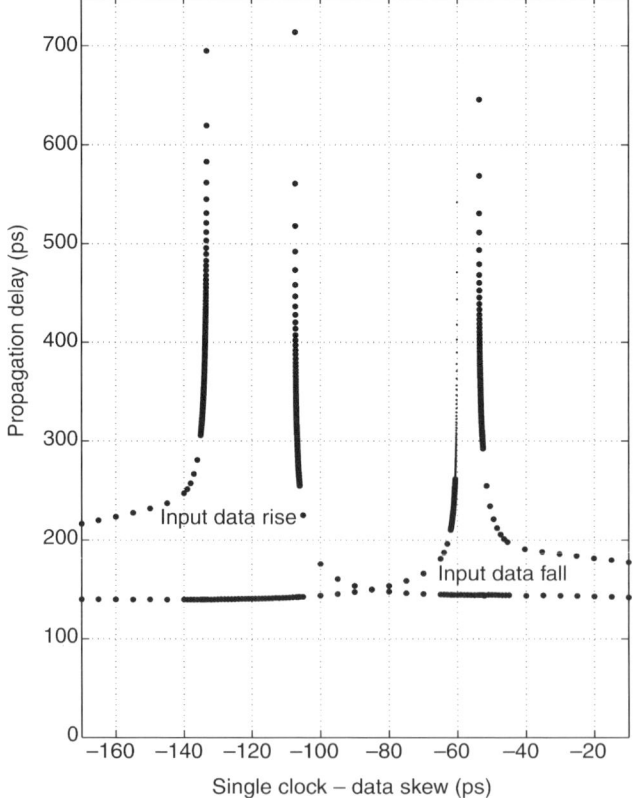

**Figure 4.21** CLK-to-Q delay as a function of a data-clock timing skew relationship: a high performance standard-cell flip-flop from an ASIC library.

digital standard-cell library for a 130-nm CMOS process. It reveals that the fall and rise transitions are not symmetric and that the uncertainty zone differs in location by as much as 65 ps between the edges. This would clearly not be acceptable if the resolution required were below 65 ps. Figure 4.22 shows the same plot for a tactical flip-flop based on [68] and developed specifically for 2.4-GHz digital operations. It enjoys an extremely small metastability window and symmetric response for rise and fall transitions. It is described further in Section 4.7.1.

The metastable condition of the retimed reference clock CKR is not acceptable, primarily for two reasons. The first is general in nature: The metastability of any clock could introduce glitches and double clocking in the digital logic circuitry being driven. The second reason is more specific: A nonbounded relationship between CKV and CKR violates the very principle of the common synchronous plane between them. It is quite likely that within a certain metastability window between FREF and CKV, the CLK-to-Q delay of the flip-flop would have the potential to make CKR span multiple DCO clock periods. This amount of uncertainty is not acceptable for proper system operation. The method presented solves stochastically the metastability problem of the frequency reference retiming of Fig. 4.2.

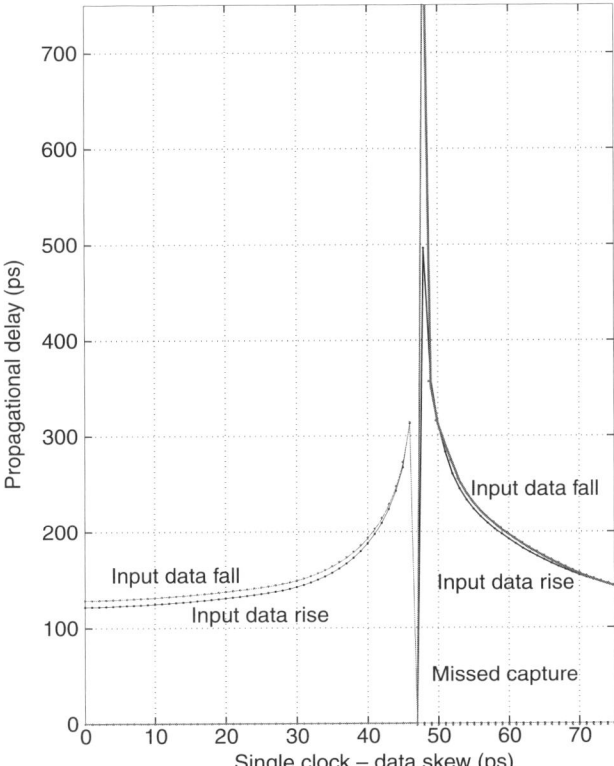

**Figure 4.22** CLK-to-Q delay as a function of a data-clock timing skew relationship: a tactical flip-flop.

### 4.7.1 Sense Amplifier–Based Flip-Flop

Figure 4.23 is a schematic of a tactical sense amplifier–based flip-flop. This topology was compared against other flip-flop architectures for similar current consumption and was found to be the fastest. It was also used by the authors on the previous CMOS technology nodes for an ultrahigh-speed read channel as a sequential element of a *finite-impulse response* (FIR) filter [71,72] and a companion *least-mean-squares* (LMS) adaptation algorithm circuit [73].

The tactical flip-flop consists of two blocks: a pulse generator *sense amplifier* (SA) (top of the figure) and a symmetric slave *set–reset* (SR) latch (bottom of the figure). It is different from a conventional master–slave latch combination in that the pulse generator is not level sensitive but generates a pulse of sufficient duration on either Sb or Rb outputs as a result of changes in clock and data values. The SR latch captures the transition and holds the state until the next rising edge of the clock.

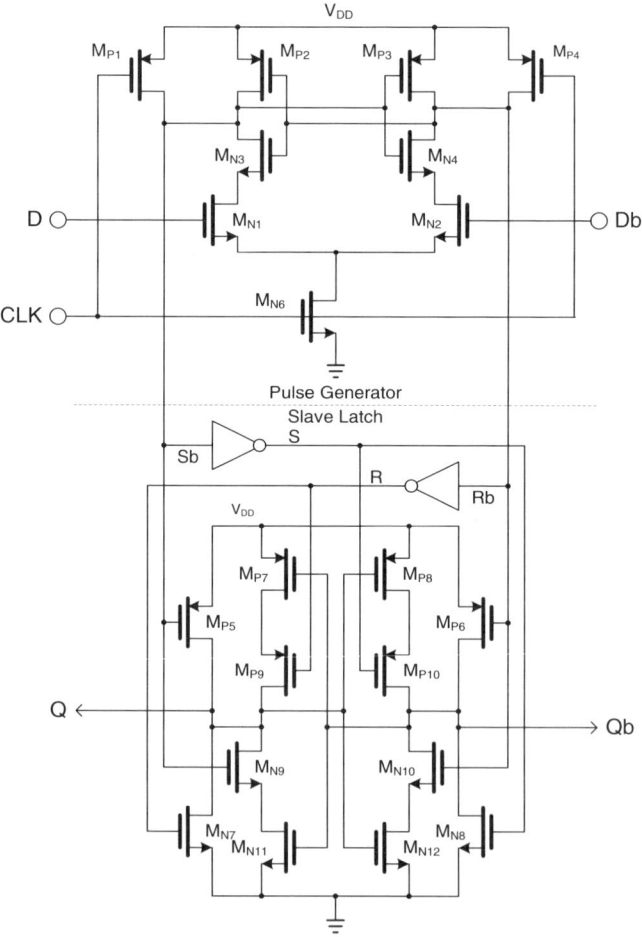

**Figure 4.23** Schematic of the tactical flip-flop. (Adapted from [68].)

After the clock returns to its inactive state, the Sb and Rb outputs of the SA stage both assume a logic-high value.

### 4.7.2 General Idea of Clock Retiming

The frequency reference retiming by the oscillator clock is the key idea of the all-digital synthesizer architecture that allows it to work in a clock-synchronous manner. This method is a general solution to the problem of retiming a lower-frequency timing signal (or clock) by a higher-frequency clock when the signals are asynchronous to each other. Normally, this problem is solved by passing the lower-frequency signal through a series of flip-flops (registers) that are clocked by the higher-frequency clock. There is a certain probability of a metastability condition per register stage, and the overall probability of metastability at the system output decreases exponentially with each register stage. The number of register stages is established such that the *mean time between failures* (MTBF) rate is acceptably large. Unfortunately, the metastability condition keeps the timing uncertainty to one high-speed clock cycle (or possibly higher) since, during metastability, the output could be resolved at a given clock cycle or at the next. Even though the output levels are defined, this timing error is unacceptable in some applications.

In this architecture, the timing signal is the frequency reference FREF, and the oversampling clock is the digitally controlled oscillator output. Obviously, these signals are completely asynchronous with respect to each other. In an ADPLL, we need to represent the reference phase, variable (DCO) phase, phase error, and all other phase signals as fixed-point digital word signals that are synchronous to each other and which cannot be corrupted by noise. If this is accomplished, the phase error could be simply an output of a synchronous arithmetic subtractor used as a phase detector. Thus, the retimed reference (CKR) is used as a synchronous system clock.

The solution could be summarized as follows: Use the *time-to-digital converter* (TDC) circuit described previously (Section 4.6) to determine which edge of the higher-frequency clock (oversampling clock) is farther away from the edge of the lower-frequency timing signal. At the same time, the oversampling clock performs sampling of the timing signal by two registers: one on the rising edge and the other on the falling edge. Then the register of "better quality" retiming (i.e., sufficiently away from the metastable region), as determined by the fractional phase detector decision, is selected to provide the retimed output.

The asynchronous retiming mechanism can be summarized as follows:

1. Sample the frequency reference clock (FREF) using both edges of an oversampling clock (CKV) that is derived from a controllable oscillator, such as a DCO. The sampling is performed by a pair of clocked memory elements, such as flip-flops or registers, one operating on the positive or rising transition of the CKV clock, and the other operating on the negative or falling transition of the CKV clock. The effect of sampling the FREF clock by the CKV clock is to retime FREF to either the rising or falling edge of the CKV clock.

**104**  ALL-DIGITAL PHASE-LOCKED LOOP

2. Delay both retimed FREF clocks by a controllable amount.
   (a) The delay operation could be accomplished by inserting shift register stages clocked by CKV or its derived clock.
3. Retime the falling retimed clock with the rising edge of the CKV clock. If this path is chosen, the CKR clock will always be synchronous with the rising edge of the CKV clock and the $\varepsilon$ definition will be consistent.
4. Feed the outputs of both rising and falling retimed paths into a selection element such as a multiplexer.
   (a) The output of the multiplexer could be resampled further by the CKV clock or a clock derived from it.
5. Use a midedge detector to select one of the two retimed clock paths that would be sufficiently far away from the metastability region.
   (a) The midedge detector could be a TDC clocked by FREF that would determine which of the two CKV edges lies father away (or just far enough away: e.g., a few inverter delays) from the FREF edge. Use the selection choice to amend the TDC output if used as part of the phase-domain all-digital PLL (ADPLL).
   (b) The midedge detector could be a register sampling the delayed CKV clock by the FREF clock such that the retimed clock selected would be sufficiently far away from the metastability condition.

### 4.7.3 Implementation

Figure 4.24 shows an implementation of the key idea that allows an ADPLL to synchronize reference and variable clock planes. Its sole purpose is to resample the reference clock with a variable DCO clock as required in Fig. 4.2 in a manner that would avoid metastability stochastically. It does so by resampling FREF by both edges of CKV and choosing the one that has a larger (or sufficiently large) data-to-clock separation. The select decision is based on the existing TDC output (a TDC portion of Fig. 4.14 is replicated here for clarity); thus, little additional cost is incurred. In the case of FREF resampling with the falling edge of CKV, additional rising-edge CKV retiming is required to satisfy the consistent definition of $\varepsilon$ and for the CKR to have a fixed timing relationship with the rising edge of CKV. This figure also shows two examples of potential metastability and how the correct path selection is made in each case.

The following description of Fig. 4.24 is based on the asynchronous retiming mechanism steps of Section 4.7.2. On the rising edge of the FREF clock, the CKV delay state is being sampled (step 5a). A fraction of the DCO cycle afterward, the FREF clock is sampled by both the falling and rising edges of the CKV clock using a pair of flip-flops (step 1), producing QN and QP outputs, one of which might have a potential for metastability. The QN and QP signals are delayed by several reclocking stages (step 2), and the end of the QN path is further resampled by the rising edge (step 3). Output of the midedge selector, SEL_EDGE, selects one of the two paths that is determined to be far enough

## 4.7 FREQUENCY REFERENCE RETIMING BY A DCO CLOCK

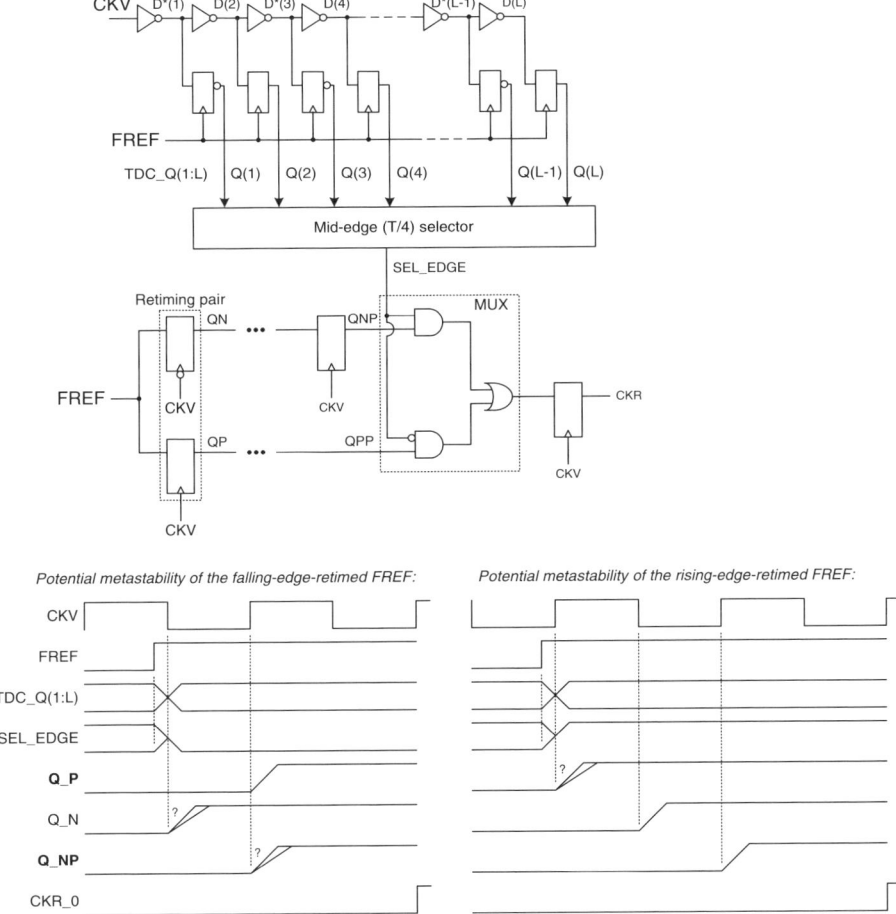

**Figure 4.24** FREF retiming by a DCO clock.

from the metastability condition (step 4). The multiplexer output selected is resampled further (step 4a).

In this implementation, five additional pairs of rising and falling reclocking stages are inserted to allow extra time (in multiples of CKV clock cycles) for the SEL_EDGE selection signal to resolve its metastability. The number of reclocking stages is calculated based on the criteria of metastability such that the selection signal must be valid and free of metastability by the time the two reclocked signals arrive at the multiplexer input. It should be emphasized that not resolving the selection signals on time does not mean that the final CKR clock will be metastable. Since the two reclocked signals pass through multiple flip-flop stages, they will be essentially free of metastability at the MUX inputs (MTBF increases exponentially with the number of stages). If the selection signal has a valid digital level but does not come on time, there is a 50% probability that the wrong decision might be made and a one-time $\pm 1.0$ perturbation to $\varepsilon$ injected into the PLL loop, from which the system can recover. Satisfying the condition of stable flip-flop output described above is quite easy with a sufficiently high gain on the selection signal path. This ensures that the output does not chatter but stays at a legal level until the internal bistable state gets resolved, even though the latter may dwell in an illegal zone until the final moment.

The real metastability can be injected into the CKR output if the reclocked signals arrive at the MUX input at the same time as the selection control. The probability of this happening is much smaller than in the case described above, but the effects can be devastating if the metastable clock finds its path into the digital baseband controller and causes false clocking. There, the system-wide reset might be necessary to recover. For this reason, the output CKR clock gets retimed further by a lower-frequency CKV-derived clock that makes the final MTBF exceedingly large. It should be noted that because the ADPLL state machines in wireless systems get synchronously reset at every packet (up to 3 ms of duration), any effect due to glitches or illegal states there has a very limited lifespan.

Figure 4.25 shows the "raw" TDC output (TDC_Q) as well as the *extended* normalized fractional error correction $\varepsilon$ (PHF_I and PHF_F) for all possible discrete

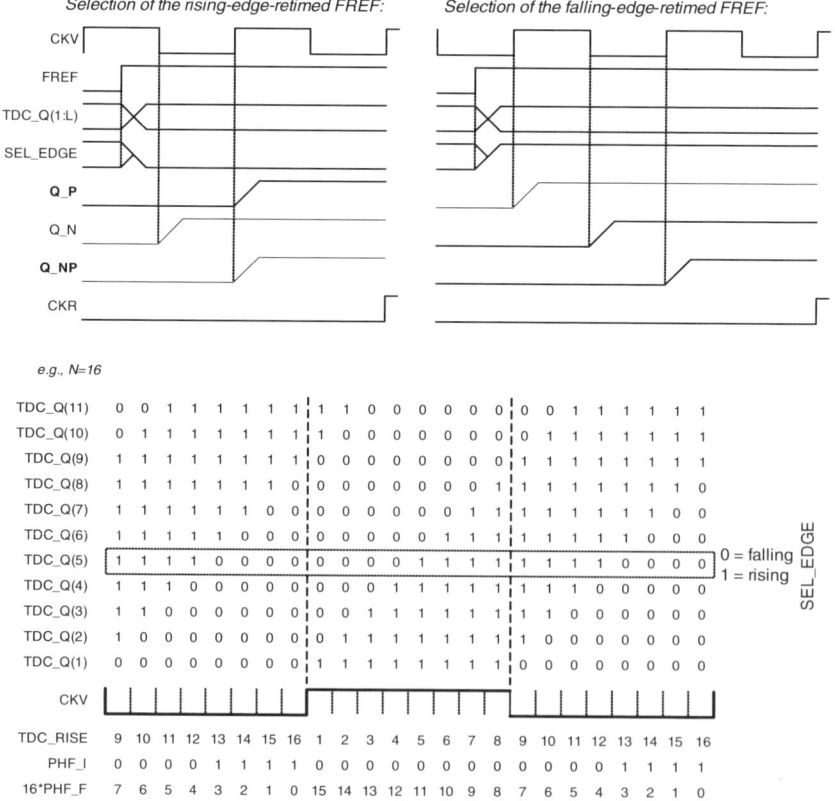

**Figure 4.25** DCO clock retiming details.

timing between CKV and FREF when $T_V = 16 t_{\text{inv}}$. The timing diagram at the bottom is similar to the timing diagram in Fig. 4.17. The number of distinct bins into which the rising edge of FREF can fall relative to CKV has been increased from 8 to 16 simply because now there are not 8 but 16 inverters each CKV clock cycle.

A high degree of redundancy in the pseudo-thermometer-encoded TDC output vector could be exploited to obtain extra error correction and metastability resolution. A majority-voting scheme of at least three neighboring outputs was considered. In the end, however, in this application a simple extraction of the selection control (SEL_EDGE signal) is chosen as the best solution. This timing diagram also shows how a certain single TDC register output [TDC_Q(5) in this example] could be used for edge selection. Inspecting various TDC_Q outputs, we can see that the fifth bit is best centered around the rising edge of CKV: It has four zeros before and four zeros after. In this case, the value of zero on that bit must be assigned to select the other edge (i.e., falling). Inspecting the fifth bit around the falling CKV edge, we can see that it is centered in a group of eight. In this case, a value of 1 must be assigned to the fifth bit to select the other edge (i.e., the rising edge). From the analysis above we can see that selection of the proper CKV edge to sample the FREF signal *without* metastability could be accomplished simply by inspecting a single TDC_Q(5) bit of the sampling state vector. A question might arise here as to what would happen if the TDC_Q(5) signal itself is metastable and cannot be resolved in a timely manner. Inspecting the region around the $0 \rightarrow 1$ and $1 \rightarrow 0$ transitions on that bit, we can see that it is free of any activity on the CKV clock line. The closest CKV transitions happen during about one-fourth of the CKV clock cycle, which is a sufficiently long time to be considered "free of metastability." In this region either of the two decisions could be made, so the SEL_EDGE selection value is immaterial.

It is also shown in Figs. 4.17, 4.24, and 4.25 that it is possible to skip the entire CKV clock cycle if the FREF rising edge is close to the CKV rising edge. If the FREF edge happens to be in any of the four bins before the rising CKV edge, TDC_Q(5) will always select falling-edge sampling followed by rising-edge resampling. Now the rising FREF edge is more than a full clock cycle away from the *final* rising CKV edge, implying that $\varepsilon > 1$. Consequently, an extra bit of information is sent to the phase detector. This could be considered as redefining the range of the fractional error correction to be $\varepsilon \in (0, 2)$.

### 4.7.4 Time-Deferred Calculation of the Variable Phase (Optional)

FREF retiming implementation was described in Fig. 4.24. The chief disadvantage of that method is the need to delay both resampled candidates until the select edge, SEL_EDGE, signal is ready with a sufficiently low probability of metastability. This requires several delay stages operating at the RF rate.

An improved method, implemented in later versions, is revealed in Fig. 4.26. The circuit combines a FREF retiming circuit, variable-phase accumulator, and variable-phase sampler. It also generates a down-divided CKV clock, CKVD8,

**108**   ALL-DIGITAL PHASE-LOCKED LOOP

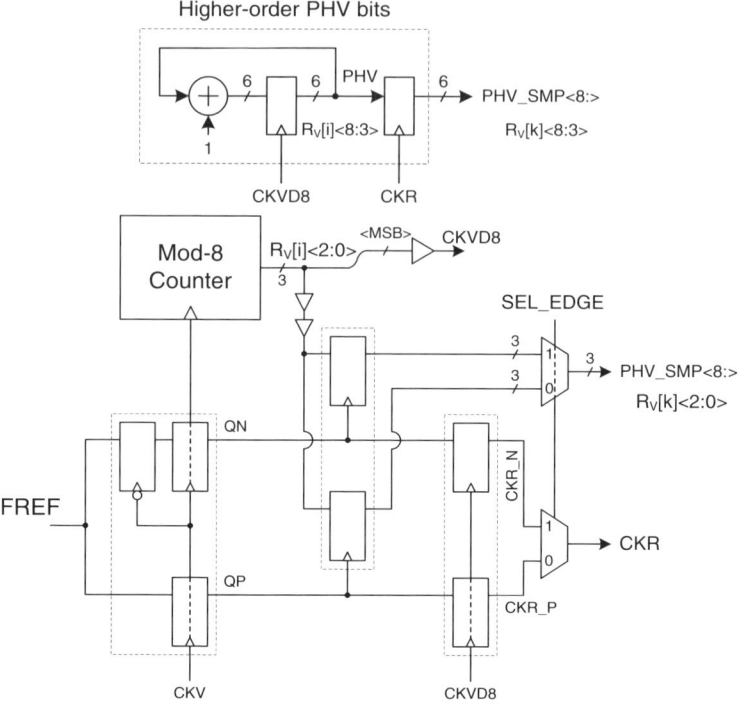

**Figure 4.26**   Variable-phase higher-order (top) and time-deferred lower-order bits (bottom).

as the third-lowest bit of the $RV[i]$ signal. Advantage is taken here of the fact that the retimed reference clock CKR is absolutely necessary only for precise sampling of the variable-phase PHV. All other circuits, which require only an approximate location of the CKR edges, could use a "time-quantized" version of CKR. Consequently, arithmetic increment of lower-order PHV bits and FREF retiming is coupled together tightly into one block. The higher-order bits of PHV are calculated separately and latched with a quantized version of the CKR clock. The concept of separating the calculation of lower- and higher-order PHV bits was introduced in Section 4.41 in conjunction with Fig. 4.11.

In this method, the FREF reference is sampled simultaneously by the rising and falling edges of the CKV clock. Both versions of the retimed reference are used to sample the lower order of the variable phase. The selection signal, SEL_EDGE, is now used to pick one of the two sampled PHV candidates that is far enough from metastability. There is still the need to delay both retimed FREF candidates, but this could now be done with a down-divided CKV clock, thus saving power and area. Operational details of the lower-order bit are shown in Fig. 4.27.

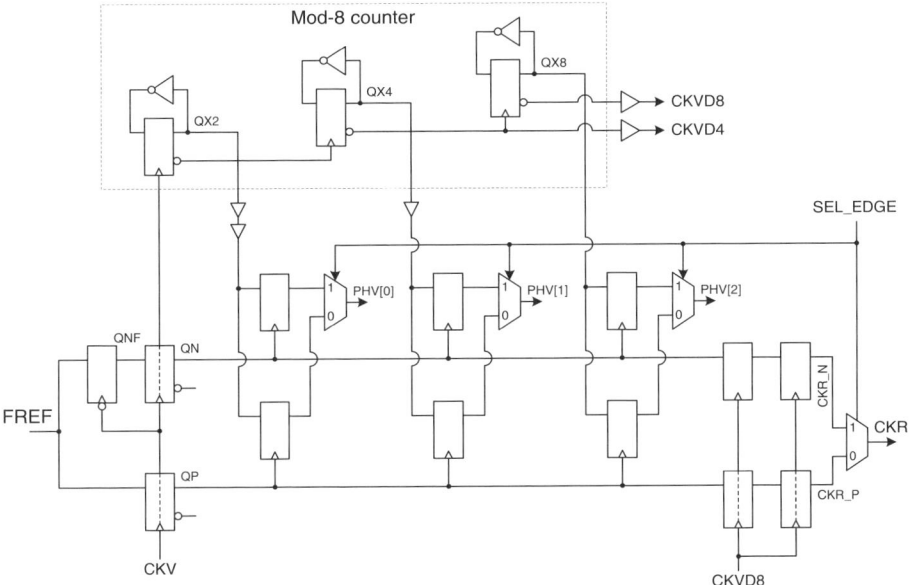

**Figure 4.27** Implementation details of the time-deferred calculation of the lower bits of the variable phase.

## 4.8 LOOP GAIN FACTOR

Figure 4.28 assembles various phase-domain blocks introduced thus far. $f_R/\widehat{K}_{DCO}$ normalization of the DCO gain was covered in Chapter 3. The variable-phase accumulator $R_V[i]$ counts the number of rising edges of the DCO clock. The variable-phase $R_V[i]$ value is sampled by FREF and adjusted through linear interpolation

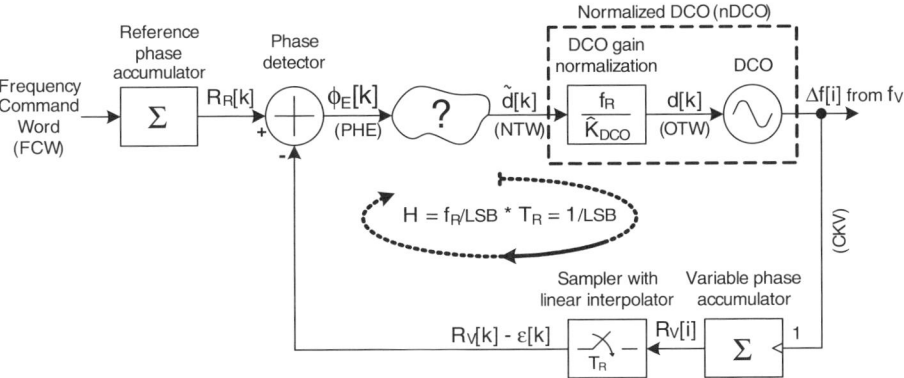

**Figure 4.28** Phase-domain all-digital PLL (ADPLL) architecture. Still missing is the proportional loop gain factor $\alpha$.

(see Section 3.8) by means of a TDC system. The reference phase accumulator $R_R[k]$ addresses Eq. 4.8 by accumulating FCW on each rising edge of the retimed frequency reference. The phase detector implements Eq. 4.16 directly. The architecture is multi-rate with two clocks, FREF and CKV, with respective indices of $k$ and $i$. What is still missing at this point is closing of the loop such that the phase error $\phi_E[k]$ would be used to correct for the frequency and phase drift of the oscillator.

As stated earlier, the phase error $\phi_E[k]$ is expressed in units of the reference frequency $f_R$. Similarly, the normalized DCO (nDCO) control is also normalized to $f_R$. As also revealed in Fig. 4.28, the frequency transfer function from the nDCO input to the phase detector output is unity, which means that a frequency perturbation $\Delta f_V$ will be estimated correctly within the quantization resolution $\Delta f_{V,\text{res}}$ in one FREF cycle $T_R$. Unfortunately, as shown in Section 4.6.2, the quantization of the frequency estimator is excessive. For stability reasons, the output of the phase detector cannot be connected directly to the nDCO input and it must be attenuated.

Figure 4.29 introduces a scaling factor $\alpha$ that controls how much attenuation the phase error must undergo before affecting the nDCO frequency. In the frequency domain it controls the fraction of the frequency detected in response to the frequency changed at the nDCO input. In the time domain it controls how much timing attenuation within a reference clock cycle one should see at the nDCO input in response to a certain change in the nDCO input in the preceding clock cycle. For an overdamped system, $\alpha < 1$; for a critically damped system, $\alpha = 1$; and for an underdamped system, $\alpha > 1$. This also establishes the loop stability.

$$\Delta f_{V,\text{res}} = \alpha \Delta \varepsilon_{\text{res}} f_R = \alpha \frac{\Delta t_{\text{res}}}{T_V} f_R = \alpha \frac{\Delta t_{\text{res}}}{T_V} \frac{1}{T_R} \qquad (4.35)$$

Table 4.6 relates (based on Eq. 4.35) the DCO frequency quantization to the TDC resolution. One can see a trade-off between the quantized loop frequency resolution and the loop dynamics. The frequency quantization of the DCO could be made finer at the cost of making the loop slower or of lower bandwidth.

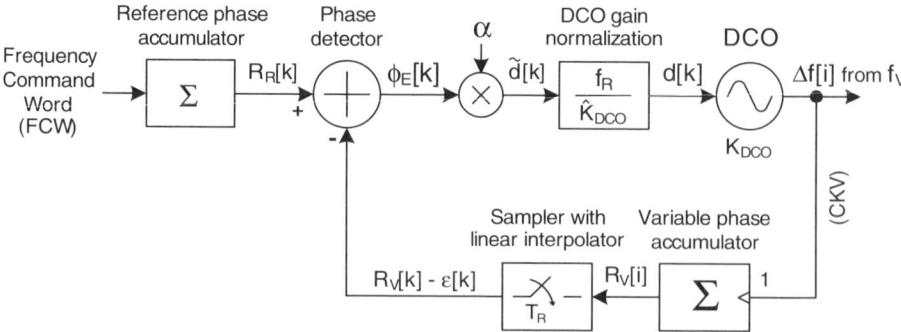

**Figure 4.29** ADPLL from a signal-processing perspective.

Table 4.6  DCO Frequency Quantization for $f_R = 13\,\text{MHz}$

| TDC Resolution, $\Delta t_{\text{res}}$ (ps) | Center Frequency, $f_V$ (MHz) | Loop Gain, $\alpha$ | Frequency Quantization, $\Delta f_V$ (kHz) |
|---|---|---|---|
| 30 | 2450 | 1 | 955.5 |
| 30 | 2450 | 1/8 | 119.4 |
| 30 | 2450 | 1/32 | 29.86 |
| 30 | 2450 | 1/256 | 3.732 |
| 30 | 2450 | 1/1024 | 0.933 |

### 4.8.1 Phase-Error Dynamic Range

A steady-state phase-error signal in this architecture also indicates the frequency offset from the center DCO frequency. To demonstrate it, one should note that the tuning word sets the DCO operating frequency directly, and there exists a proportionality factor $\alpha$ between the normalized tuning word and the phase error, as shown in Fig. 4.29. Consequently, the steady-state frequency offset could be expressed as

$$\Delta f_V = -\phi_E \alpha f_R \tag{4.36}$$

Equation 4.36 should be contrasted with Eq. 4.33, which is only a single reference cycle estimate that is a part of the detection process. Equation 4.36 could also be explained from another perspective. Assume a settled PLL with a phase error of zero, indicating no frequency deviation from the center frequency. If there is a sudden step frequency deviation $\Delta f$ from the oscillator center frequency, in one FREF cycle the phase detector will estimate it as $\widehat{\Delta f_V} = -\Delta f$, following Eq. 4.33. This will correct the DCO frequency by $-\Delta f \alpha$, so the oscillator frequency deviation will get reduced to $\Delta f (1 - \alpha)$. In the second reference cycle, the phase detector output will be $-\Delta f - \Delta f(1 - \alpha) = -\Delta f(2 - \alpha)$, leading to an oscillator correction of $-\Delta f(2 - \alpha)\alpha$. The oscillator frequency deviation will get reduced to $\Delta f - \Delta f(2 - \alpha)\alpha = \Delta f(1 - \alpha)^2$. Similarly, in the third reference cycle, the phase detector output will be $-\Delta f(2 - \alpha) - \Delta f(1 - \alpha)^2 = -\Delta f(3 - 3\alpha + \alpha^2)$, leading to an oscillator correction of $-\Delta f(3 - 3\alpha + \alpha^2)\alpha$. The oscillator frequency deviation will get reduced to $\Delta f - \Delta f(3 - 3\alpha + \alpha^2)\alpha = \Delta f(1 - \alpha)^3$. A clear trend is observable: In the $k$th reference cycle, the oscillator frequency deviation will be $\Delta f(1 - \alpha)^k$ and will decay to zero as $k \to \infty$. Similarly, the phase-error output will be $-\Delta f \sum_{i=0}^{k}(1 - \alpha)^i$ and will approach $-\Delta f / \alpha$ as $k \to \infty$. Consequently, as this process of geometric sequence continues, the DCO frequency gets fully corrected, and the phase detector develops an $-\Delta f / \alpha$ offset normalized to $f_R$.

The loop gain factor $\alpha$ also controls the dynamic range of the phase error. It was mentioned in Section 4.4 that both the variable phase $R_V$ and reference phase $R_R$ are wordlength limited to $W_I$ bits in the integer part. This limitation carries over to the phase error $\phi_E$. Consequently, the phase range of the $\phi_E$ signal is $(-2^{W_I-1}, 2^{W_I-1})$ times $2\pi$ rad of the CKV clock cycle and limits the dynamic range of the DCO to

$$\Delta f_{V,\text{range}} = 2^{W_I} \alpha f_R \tag{4.37}$$

## 4.9 PHASE-DOMAIN ADPLL ARCHITECTURE

At this point, all the major pieces comprising the phase-domain *all-digital PLL* (ADPLL) frequency synthesizer have been introduced. The block diagram is presented in Fig. 4.30. The underlying frequency stability of the system is derived from a frequency reference (FREF) crystal oscillator such as a 13-MHz temperature-compensated crystal oscillator (TCXO) for the host GSM system. The frequency command word (FCW) is defined as the desired frequency-division ratio $f_V/f_R$ and is expressed in a fixed-point format such that an LSB of its integer part corresponds to the $f_R$ reference frequency. It is input to the reference phase accumulator to establish the operating frequency of the channel selected. The central element of the ADPLL is a digitally controlled oscillator (DCO), operating at 2.4 GHz in the BLUETOOTH example, and the PLL build around it is fully digital.

The ADPLL is of type I (i.e., only one integrating pole, due to the DCO frequency-to-phase conversion). Type I loops generally feature faster dynamics and are used where fast frequency/phase acquisition is required or direct transmit modulation is used. The loop dynamics are improved further through avoiding using a loop filter. The issue of the reference feedthrough that affects classical charge-pump PLL loops and shows itself as spurious tones at the RF output is irrelevant here because, as discussed before, a linear, not a correlative, phase detector is

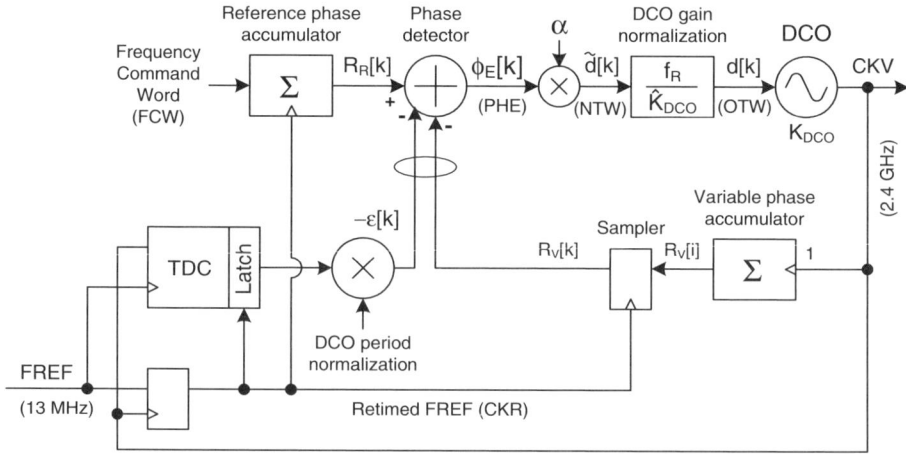

**Figure 4.30** Phase-domain synchronous all-digital PLL synthesizer. (From [65], © 2005 IEEE.)

used. In addition, unlike in type II PLLs, where the steady-state phase error goes to zero in face of a constant frequency offset (i.e., frequency deviation between the actual and center DCO frequencies), the phase error in a type I PLL loop is proportional to the frequency offset. However, due to the digital nature of the implementation, this does not limit the dynamic range of the phase detector or the maximum range of the DCO operational frequency.

The ADPLL is a digitally synchronous fixed-point phase-domain architecture. The variable-phase signal $R_V[i]$ is determined by counting the number of rising clock transitions of the DCO oscillator clock. The reference phase signal $R_R[k]$ is obtained by accumulating the frequency command word (FCW) with every rising edge of the retimed FREF clock. The variable-phase $R_V[k]$ sampled, together with the fractional correction $\varepsilon[k]$, is subtracted from the reference phase $R_R[k]$ in a synchronous arithmetic-phase detector. The $\varepsilon[k]$ corrections by means of the TDC system increase the instantaneous phase resolution of the system to below the basic $2\pi$ rad of the variable phase.

The digital phase-error samples $\phi_E[k]$ are conditioned by a proportional loop gain attenuator $\alpha$ and then normalized by the DCO gain, $K_{DCO}$. The loop gain $\alpha$ is a programmable PLL parameter that controls the loop bandwidth. $K_{DCO}$ normalization is needed to establish the loop bandwidth precisely and to perform direct transmit frequency modulation, as described in Chapter 5. The FREF input is resampled by the RF oscillator clock, and the resulting *retimed clock* (CKR) is used throughout the system. This ensures that the massive digital logic is clocked after the quiet interval of phase-error detection by the TDC.

The chief advantage of keeping the phase information in fixed-point digital numbers is that after conversion, it cannot be corrupted further by noise.[5] Consequently, the phase detector could simply be realized as an arithmetic subtractor that performs an exact digital operation. Therefore, the number of conversion places is kept at a minimum: A single point where the continuously valued clock edge delay is compared in a time-to-digital converter.

It should be emphasized here that it is very advantageous to operate in the phase domain, for several reasons. First, the phase detector used is not a conventional correlative multiplier (such as shown in Fig. 1.19) generating reference spurs. Here, an arithmetic subtractor is used and it does not introduce spurs into the loop. Second, the dynamic range of the phase error could be made arbitrarily large simply by increasing the wordlength of the phase accumulators. This compares favorably with conventional implementations, which typically are limited to only $\pm 2\pi$ of the comparison rate with a three-state phase/frequency detector [22]. Third, phase-domain operation is a lot more amenable to digital implementations than is the conventional approach. Fourth, the phase domain allows algorithmically higher precision than operation in the frequency domain, since frequency is a time derivative of phase and a certain amount of phase quantization (such as in a TDC) decreases its frequency error over time.

---

[5]Obviously, certain digital-state corruption phenomena, such as due to cosmic radiation, are not considered here.

### 4.9.1 Close-in Spurs Due to Injection Pulling

The ADPLL circuits which are not relevant to the foregoing discussion are merged into the digital part of the frequency synthesizer block in Fig. 4.31. Reference clock retiming by the DCO clock strips FREF of its critical timing information (already extracted by the TDC) and produces a retimed clock CKR that is used subsequently throughout the system. This ensures that the massive digital logic is clocked after the quiet interval of phase-error detection by the TDC. The CKR edge timestamps are now synchronous with the RF oscillator, in which time separation between the closest CKR and CKV edges is time invariant. In this example, it is beneficial for avoiding injection pulling [84], in which the slowly varying timing separation between CKR and CKV causes the oscillator to be pulled, thus creating a frequency beating event that exhibits itself as spurs in the output generated.

The injection-pulling mechanism is revealed in Fig. 4.32a. It should be noted that the two frequencies need not be close to each other. The injection pulling could be caused by a harmonic of the lower-frequency FREF clock that falls in the frequency vicinity of the oscillator. In the example, the interfering clock has a frequency $2\frac{1}{4}$ times lower than the oscillator frequency. Each of its edges pulls every second or third oscillator edge.

For a non-retimed FREF clock scenario, the injection-pulling mechanism depends on the fractional part $N_f$ of the frequency division ratio $N$:

$$N = N_i + N_f = \frac{f_V}{f_R} \tag{4.38}$$

where $N_i$ and $N_f$ are the integer and fractional parts of $N$, respectively, and $f_V$ and $f_R$ are oscillator and interfering FREF clock frequencies, respectively. If $N_f = 0$, there is no injection pulling. If $N_f$ is close to zero, it will give rise to a positive beating frequency $f_{\text{beat}} = N_f f_R$. If $N_f$ is close to 1, it will give rise to a negative beating frequency $f_{\text{beat}} = -(1 - N_f) f_R$. The terms *positive* and *negative* here indicate the direction of change of the clock edge pulling force. The higher values of $f_{\text{beat}}$ are generally not dangerous since they are likely to be too fast to pull the oscillator coherently.

The FREF retiming method presented in Fig. 4.31 eliminates the effect of injection pulling, as shown in Fig. 4.32b. It should be noted that a constant

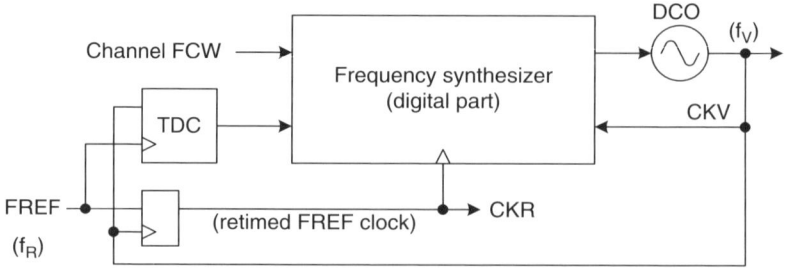

**Figure 4.31** ADPLL highlighting FREF clock resampling to avoid injection pulling. (From [83], © 2005 IEEE.)

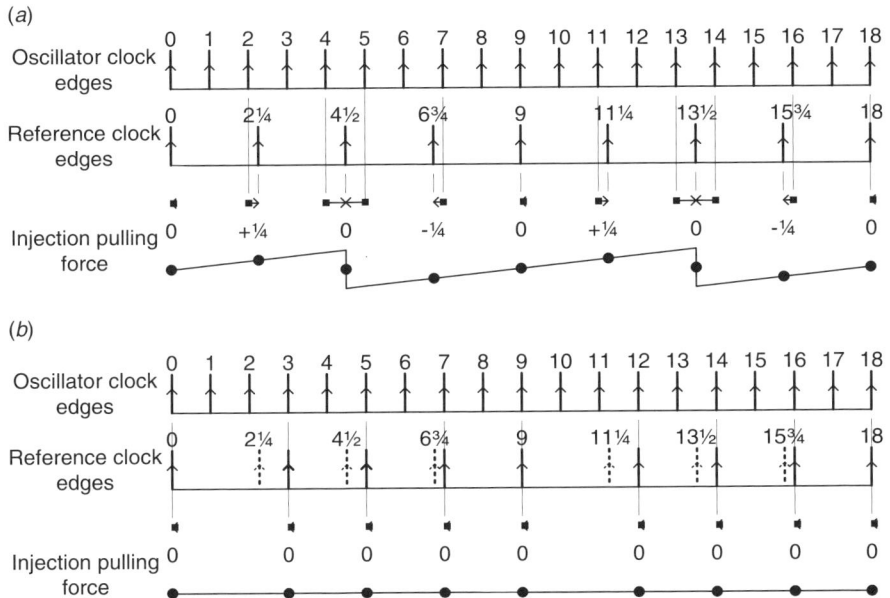

**Figure 4.32** (*a*) Injection pulling of an RF oscillator by a frequency reference clock; (*b*) injection pulling elimination through reference resampling. In this example, $N = 2 + 1/4$. (From [83], © 2005 IEEE.)

nonzero pulling force would present no problem. In fact, due to different propagational delays of various interfering sources through power, ground, and substrate paths, there will be a nonzero equilibrium state of the delays. It is important to realize that as a result, the average frequencies of $f_V$ and $f_R$ no longer not have to have an integer multiple ratio (i.e., $N_f = 0$). The CKR clock has the same average frequency as FREF. The retiming operation only shifts the edges; their expected averaged distances are not affected.

## 4.10 PLL FREQUENCY RESPONSE

Figure 4.33 shows a general *s*-domain model of an ADPLL frequency synthesizer. It is a continuous-time approximation of a discrete-time *z*-domain model [74] and is valid as long as the fluctuation frequencies of interest are much smaller than the sampling rate, which in this case equals $f_R$. (It is commonly accepted that this linear approximation will hold as long as the PLL bandwidth $f_{BW}$ is at least 10 times smaller than the sampling rate [21].) The output is the phase $\phi_V$, not the frequency deviation $\Delta f_V = \Delta \omega_V / 2\pi$, as in Fig. 4.29, which is related by the integration operation: $\phi_V = \Delta \omega_V / s$.

The loop filter is realized here as a normalized gain $\alpha$ stage, thus giving rise to a type I first-order PLL loop, which is defined as having only one integration pole,

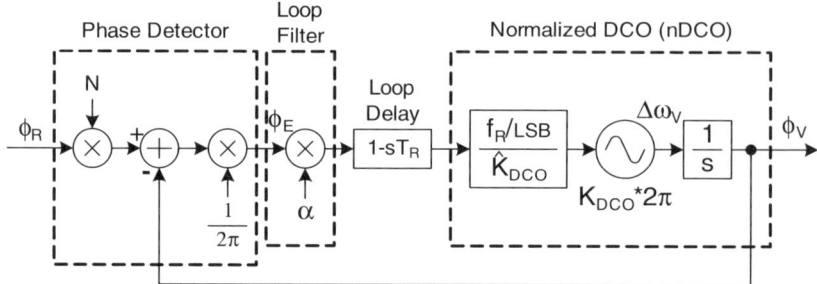

**Figure 4.33** Linearized equivalent $s$-domain model of a type I ADPLL.

due to the DCO frequency-to-phase conversion. For the purposes of this section only, the quantities $\phi_R$ and $\phi_V$ are conventionally defined reference and variable phases, respectively, in radians. $\omega_V$ is the angular frequency in radians per second and is equal to $2\pi f_V$. Introduction of the angular frequency makes it necessary to modify the DCO transfer function by including multiplication by $2\pi$. Strictly speaking, this is viewed as modifying the normalizing gain component $f_R$ (hertz) by $2\pi$ multiplication, which results in $\omega_R$ (rad/s). The quantity $N$ is the frequency-division ratio between the DCO clock and FREF and is equivalent to FCW. Formally, $K_{DCO}$ will be expressed in units of rad/s per LSB, but it is always divided by its estimate, so the radians cancel out, leaving a dimensionless unit, $K_{DCO}/\widehat{K}_{DCO}$. The $e^{-sT_R} \approx 1 - sT_R$ delay operator is the control loop delay, which degrades the phase margin by $2\pi$(cycle delay $\cdot f_{BW}/f_R$). In our case, there is only one cycle control loop delay, and $f_{BW} \ll f_R$; therefore, phase margin degradation is not an issue, and this operator can be disregarded.

The open-loop transfer function $H_{ol}(s)$ is

$$H_{ol}(s) = \frac{1}{2\pi} \alpha \frac{2\pi f_R}{\widehat{K}_{DCO}} \frac{K_{DCO}}{s} = \alpha \frac{f_R}{s} \frac{K_{DCO}}{\widehat{K}_{DCO}} \quad (4.39)$$

Let's assume that the DCO gain is estimated correctly. $H_{ol}(s)$ then reduces to

$$H_{ol}(s) = \alpha \frac{f_R}{s} \quad (4.40)$$

There is one pole at dc, hence type I classification of this ADPLL structure.

The closed-loop transfer function can be expressed as

$$H_{cl}(s) = \frac{NH_{ol}}{1 + H_{ol}} = \frac{N\alpha(f_R/s)}{1 + \alpha(f_R/s)} \quad (4.41)$$

which can be rearranged as

$$H_{cl}(s) = \frac{N}{1 + s/\alpha f_R} \quad (4.42)$$

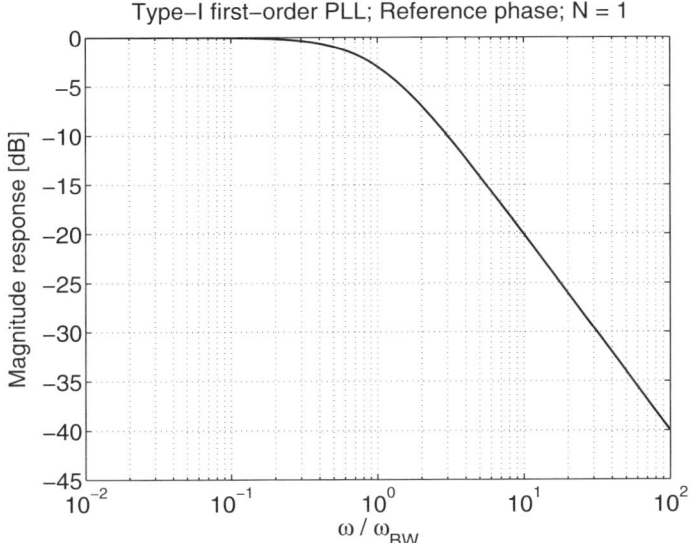

**Figure 4.34** Magnitude response of type I PLL vs. bandwidth-normalized frequency.

Its magnitude response is shown in Fig. 4.34. Substituting $s = j\omega = j2\pi f$ gives

$$H_{cl}(f) = \frac{N}{1 + j(2\pi f/\alpha f_R)} \quad (4.43)$$

From this we obtain the bandwidth or 3-dB cutoff frequency of the low-pass closed-loop PLL (provided that $f_{BW} \ll f_R$, for the s-domain approximation to hold) as

$$f_{BW} = \frac{\alpha}{2\pi} f_R \quad (4.44)$$

Incidentally, the open-loop gain $H_{ol}$ is unity at $f_{BW}$.

Figure 4.35 illustrates the closed-loop transfer function, normalized to the reference frequency, for values of $\alpha$ that are constrained to be negative powers of 2.

It is interesting to note that the ratio divider $N$ is not part of the open-loop transfer function and hence does not affect the loop bandwidth. This is in contrast to conventional PLL-based synthesizers, in which the phase detector uses the divide-by-$N$ oscillator clock (see Fig. 1.8), which is equal to the update rate under the locked condition. In that case, since the phase being compared is expressed in radians (normalized to the update period and multiplied by $2\pi$), the same amount of timing excursion (in seconds) translates into a smaller phase by a factor of $N$. The phase detection mechanism of the digital architecture measures the oscillator timing excursion normalized to the DCO clock cycle, hence no phase division of $\phi_V$ by $N$. Interestingly, this situation is similar to the PLL architecture, which uses a mixer to down-convert the RF oscillator frequency, as opposed to the edge divider.

The frequency reference input phase $\phi_R$, on the other hand, needs to be multiplied by $N$ since it is measured by the same phase detection mechanism normalized to the

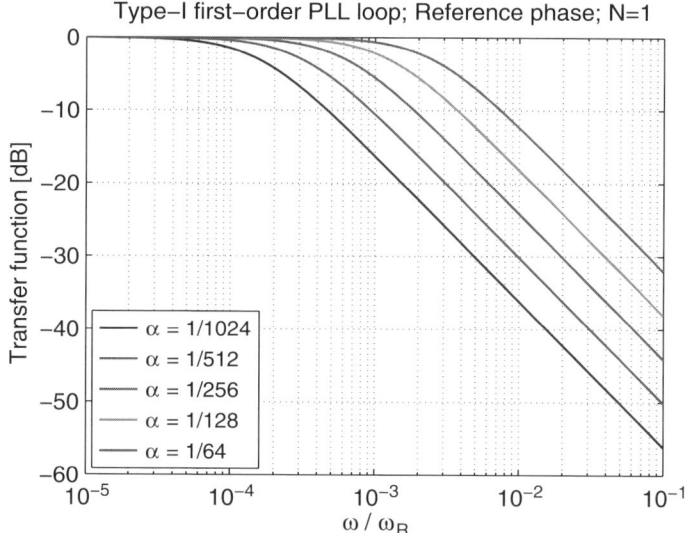

**Figure 4.35** Magnitude response of type I PLL vs. reference-normalized frequency for various values of $\alpha$.

DCO clock cycle. The same amount of timing excursion on the FREF input translates into a larger phase by a factor of $N$ when viewed by the phase detector. There is a multiplication factor of $1/2\pi$ at the output of the phase detector. It is simply due to the fact that the digital-phase error is expressed not in units of radians of the DCO clock, but in the number of its cycles.

There is a small inconsistency in Fig. 4.33 with the way that $s$-domain representations of conventional PLLs are traditionally depicted. It might suggest a phase comparison rate to be that of the DCO oscillator, which is not correct. An alternative but mathematically equivalent $s$-domain representation is shown in Fig. 4.36. The phase comparison rate is now more in line with traditional representations, but the diagram does not reflect the fact that the ADPLL phase signals are normalized to the DCO oscillator.

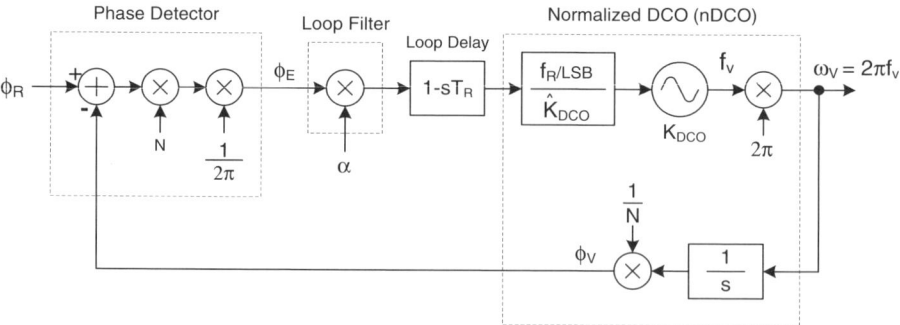

**Figure 4.36** Alternative $s$-domain model of an ADPLL, which indicates the proper comparison rate at the phase detector.

### 4.10.1 Conversion Between the s- and z-Domains

The $z$-operator is defined as $z = e^{j\theta}$, where $\theta = \omega t_0$. $\omega = 2\pi f$ is the angular frequency and $t_0$ is the sampling period. In our case, $t_0 = 1/f_R$, which leads to $z = e^{j\omega/f_R}$. For small values of $\omega$ in comparison with the sampling rate, we can make the following approximation:

$$z = e^{j\theta} \approx 1 + j\theta = 1 + \frac{j\omega}{f_R} = 1 + \frac{s}{f_R} \qquad (4.45)$$

which results in

$$s = f_R(z - 1) \qquad (4.46)$$

Transforming Eqs. 4.40 and 4.42 gives

$$H_{ol}(z) = \frac{\alpha}{z - 1} \qquad (4.47)$$

$$H_{cl}(z) = \frac{N}{1 + (z - 1)/\alpha} \qquad (4.48)$$

The single FREF period delay operator $e^{-sT_R}$ is approximated by $1 - sT_R$ in Fig. 4.33 for the same assumption of small values of $\omega$ frequency.

## 4.11 NOISE AND ERROR SOURCES

An ADPLL linear model, including phase noise sources, is shown in Fig. 4.37. $\phi_{n,R}$ is the phase noise of the reference input that is external to the ADPLL. Its transfer function is expressed by Eq. 4.42. Internal to the system, there are only two places that the noise could be injected. Due to its digital nature, the rest of the system is *completely* immune from time or amplitude-domain perturbations.

The first internal noise source, $\phi_{n,V}$, is the oscillator itself. It undergoes high-pass filtering by the loop. Its closed-loop transfer function is

$$H_{cl,V}(s) = \frac{1}{1 + H_{ol}} = \frac{1}{1 + (\alpha f_R/s)} = \frac{s}{s + \alpha f_R} \qquad (4.49)$$

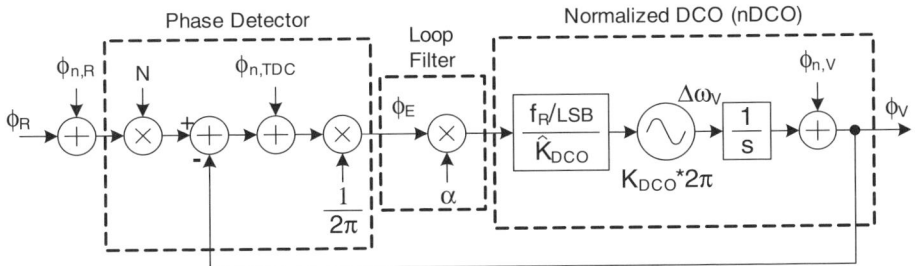

**Figure 4.37** Linear $s$-domain model with noise sources added. (From [67], © 2004 IEEE.)

which can be rewritten as

$$H_{\text{cl},V}(f) = \frac{1}{1 - j(\alpha f_R/2\pi f)} \quad (4.50)$$

The frequency transfer function above indicates that the DCO noise has a high-pass characteristic with a bandwidth or 3-dB cutoff frequency of

$$f_{\text{BW},V} = \frac{\alpha}{2\pi} f_R \quad (4.51)$$

The second internal noise source $\phi_{n,\text{TDC}}$ is the TDC operation of calculating $\varepsilon$. Even though the TDC is a digital circuit, the FREF and CKV inputs are *continuous* in the time domain. The TDC error has several components: quantization, linearity, and randomness due to thermal effects. The TDC quantization noise is governed by Eq. 4.58. It should be noted that the rest of the phase-detection mechanism is digital in nature and *does not* contribute noise. The closed-loop transfer function of the TDC noise can be expressed as

$$H_{\text{cl},\text{TDC}}(s) = \frac{\alpha f_R/s}{1 + H_{\text{ol}}} = \frac{\alpha f_R/s}{1 + \alpha f_R/s} \quad (4.52)$$

which can be rearranged as

$$H_{\text{cl},\text{TDC}}(s) = \frac{1}{1 + s/\alpha f_R} \quad (4.53)$$

or rewritten for easy inspection as

$$H_{\text{cl},\text{TDC}}(f) = \frac{1}{1 + j(2\pi f/\alpha f_R)} \quad (4.54)$$

The TDC-contributed noise has the same transfer function as the reference noise but without the gain of $N$. This is simply due to the fact that the TDC phase signal is normalized to the DCO clock cycle.

### 4.11.1 TDC Resolution Effect on Phase Noise

In closed-loop operation, the TDC quantization $\Delta t_{\text{res}} \approx t_{\text{inv}}$ of timing estimation affects the phase noise at the ADPLL output. Under a large-signal assumption (spanning multiple quantization levels), the variance of the timing uncertainty is

$$\sigma_t^2 = \frac{(\Delta t_{\text{res}})^2}{12} \quad (4.55)$$

The phase noise (rad) is obtained by normalizing the standard deviation of the timing error to the unit interval and multiplying by $2\pi$ radians:

$$\sigma_\phi = 2\pi \frac{\sigma_t}{T_V} \quad (4.56)$$

## 4.11 NOISE AND ERROR SOURCES

The total phase noise power is spread uniformly over the span from dc to the Nyquist frequency. The single-sided spectral density is, therefore, expressed as

$$\mathcal{L} = \frac{\sigma_\phi^2}{f_R} \quad (4.57)$$

Consequently, the phase-noise spectrum at the ADPLL RF output due to the TDC timing quantization is

$$\mathcal{L} = \frac{(2\pi)^2}{12} \left(\frac{\Delta t_{\text{res}}}{T_V}\right)^2 \frac{1}{f_R} \quad (4.58)$$

The TDC noise transfer function is low-pass with unity gain within the loop bandwidth, as governed by Eq. 4.54.

Banerjee's figure of merit (BFM) [75], being a 1-Hz normalized phase noise floor, is defined as

$$\text{BFM}_{\text{dB}} = \mathcal{L}_{\text{dB}} - 10\log_{10} f_R - 20\log_{10} N \quad (4.59)$$

where $f_R$ is a sampling frequency of the phase comparison and $N = f_V/f_R$ is the frequency-division ratio of a PLL. It is used to compare the phase performance of PLLs with different reference frequencies and division ratios. For the TDC-based PLL, BFM is derived as

$$\text{BFM} = \frac{\mathcal{L}}{f_R N^2} = \frac{(2\pi)^2}{12} \Delta t_{\text{res}}^2 \quad (4.60)$$

For the BLUETOOTH example, substituting $\Delta t_{\text{res}} = 40$ ps, $f_V = 2.4$ GHz, $T_V = 417$ ps, $f_R = 13$ MHz, we obtain $\mathcal{L} = 10\log(2.33 \times 10^{-9} \text{ rad}^2/\text{Hz}) = -86.3$ dBc/Hz and BFM $= -203$ dB. For GSM, substituting $\Delta t_{\text{res}} = 30$ ps, $f_V = 1.8$ GHz, $T_V = 556$ ps, and $f_R = 26$ MHz, we obtain $\mathcal{L} = 10\log(3.69 \times 10^{-10} \text{ rad}^2/\text{Hz}) = -94.3$ dBc/Hz and BFM $= -205$ dB. Equation 4.58 reveals that the TDC phase-noise contribution could be minimized by improving the TDC timing resolution and increasing the sampling rate. Even though state-of-the-art conventional PLLs implemented in a SiGe process can outperform the ADPLL presented here in the in-band phase noise, $-213$ dB in reference [75] and $-218$ dB in reference [76], the worst-case BFM of $-205$ dB appears adequate even for GSM applications, since there are no other significant phase-noise contributions as in the conventional PLLs. Later generations of deep-submicron CMOS processes can only bring reductions in $\Delta t_{\text{res}} \approx t_{\text{inv}}$, so the performance gap will narrow in the future.

### 4.11.1.1 Alternative Analysis of the TDC Resolution Effect

The TDC finite resolution effect on ADPLL phase noise could be analyzed alternatively using an open-loop response. Equation 4.35 relates the TDC resolution $\Delta t_{\text{res}}$ to the frequency resolution $\Delta f_{V,\text{res}}$ at the DCO output:

$$\Delta f_{V,\text{res}} = \alpha \frac{\Delta t_{\text{res}}}{T_V} f_R \qquad (4.61)$$

Equation 2.20 establishes the relationship between the DCO frequency deviation and its period deviation. Consequently, the DCO period resolution is

$$\Delta T_{V,\text{res}} = \alpha \frac{\Delta t_{\text{res}}}{T_V} f_R \frac{1}{f_V^2} = \alpha \Delta t_{\text{res}} \frac{1}{N} \qquad (4.62)$$

where $N = f_V/f_R$. If the TDC quantization noise is white, the variance of the DCO period deviation is

$$\sigma_{\Delta T_V}^2 = \frac{1}{12}(\Delta T_{V,\text{res}})^2 = \frac{1}{12}\alpha^2(\Delta t_{\text{res}})^2 \frac{1}{N^2} \qquad (4.63)$$

Equation 1.11 governs the relationship between the oscillator period deviation and the phase noise $\mathcal{L}\{\Delta f\}$ at a certain frequency offset $\Delta f$ from the carrier $f_V$. Since the DCO control word is held constant during $N$ DCO clock cycles (i.e., for the duration of FREF), its spectrum occupies only the $\pm f_R/2$ range instead of the $\pm f_V/2$ in the case of DCO noise. Consequently, the spectrum density must be multiplied further by $N$:

$$\mathcal{L}\{\Delta f\} = \frac{\sigma_{\Delta T_V}^2 f_V^3}{\Delta f^2} N = \frac{1}{12}\left(\frac{\Delta t_{\text{res}}}{T_V}\right)^2 \frac{f_R}{\Delta f^2} \alpha^2 \qquad (4.64)$$

To show equivalence with Eq. 4.58, let's select a convenient frequency fluctuation $\Delta f$,

$$\Delta f = f_{\text{BW}} = \frac{1}{2\pi}\alpha f_R \qquad (4.65)$$

to equal the PLL cutoff frequency $f_{\text{BW}}$ at which the tangential (i.e., straight lines approximation) open-loop response $H_{ol}(f)$ and closed-loop response $H_{cl}(f)$ are both at unity (actually, they are both at $-3$ dB gain):

$$\mathcal{L}\{f_{\text{BW}}\} = \frac{(2\pi)^2}{12}\left(\frac{\Delta t_{\text{res}}}{T_V}\right)^2 \frac{1}{f_R} \qquad (4.66)$$

### 4.11.2 Phase Noise Due to DCO $\Sigma\Delta$ Dithering

#### 4.11.2.1 DCO Quantization Effect on Phase Noise
To gain insight into the quantization effects of the finite DCO frequency resolution $\Delta f_{\text{res}} = \Delta f^T$ on the RF output phase noise, consider its transfer function, shown in Fig. 4.38a.

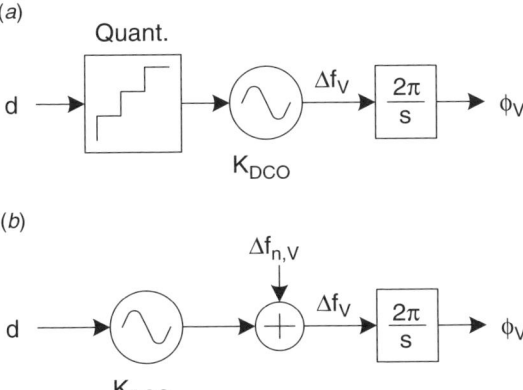

**Figure 4.38** DCO quantization noise model.

The infinite-precision tuning signal $d$ is quantized to a finite-precision tuning word such that it matches the DCO frequency resolution $\Delta f_{\text{res}}$. The actual frequency deviation $\Delta f_V$ will be within $\Delta f_{\text{res}}/2$ of from ideal. The frequency deviation is then converted to phase through the $2\pi/s$ integration. The $2\pi$ multiplication denotes the conversion of a linear frequency (hertz) to an angular frequency (rad/s).

Since the tuning word normally spans multiple quantization levels, the DCO frequency quantization error is modeled in Fig. 4.38b as an additive uniformly distributed random variable $\Delta f_{n,V}$ with white noise spectral characteristics. It is mathematically equivalent to the time-domain DCO model in Fig. 2.19. The quantization error variance is

$$\sigma^2_{\Delta f_V} = \frac{(\Delta f_{\text{res}})^2}{12} \quad (4.67)$$

The total phase-noise power is spread uniformly from dc to the Nyquist frequency (i.e., half of the reference frequency $f_R$). The single-sided spectral density of $\Delta f_{n,V}$ is therefore flat at

$$\frac{1}{2} S_{\Delta f} = \frac{\sigma^2_{\Delta f_V}}{f_R} \quad (4.68)$$

Outside the loop bandwidth, the closed- and open-loop transfer functions from the frequency deviation $\Delta f_{n,V}$ to the phase $\phi_V$ of the RF output are the same:

$$H_{\text{ol},\,\Delta f_V}(s) = \frac{2\pi}{s} \quad (4.69)$$

so the single-sided power spectral density at the output is

$$\mathcal{L}\{\Delta\omega\} = \frac{(\Delta f_{\text{res}})^2}{12 f_R} \left(\frac{2\pi}{\Delta\omega}\right)^2 \quad (4.70)$$

It could be rewritten as

$$\mathcal{L}\{\Delta f\} = \frac{1}{12} \left(\frac{\Delta f_{\text{res}}}{\Delta f}\right)^2 \frac{1}{f_R} \quad (4.71)$$

Actually, the DCO input samples are not impulses to justify the white noise assumption above but are held constant between updates. Consequently, Eq. 4.71 needs to be multiplied by the sinc function corresponding to the Fourier transform of the zero-order hold operation:

$$\mathcal{L}\{\Delta f\} = \frac{1}{12} \left(\frac{\Delta f_{\text{res}}}{\Delta f}\right)^2 \frac{1}{f_R} \left(\text{sinc}\frac{\Delta f}{f_R}\right)^2 \quad (4.72)$$

Equation 4.72 gives rise to the same 20-dB/decade attenuation characteristics as the up-converted thermal noise region of the oscillator phase (see Fig. 1.4), except for the *protective* notches at the DCO input sampling rate and its multiples. Without the dithering, even with the finest varactor resolution of the tracking bank, the resulting phase noise will normally be too high for wireless applications. For example, for a frequency step of $\Delta f_{\text{res}} = \Delta f^T = 23$ kHz and $f_R = 13$ MHz, the resulting phase noise would be $\mathcal{L} = -109$ dBc/Hz at $\Delta f = 500$ kHz offset in the low band. This is to be compared with $\mathcal{L} = -112$ dBc/Hz of the natural DCO phase noise. Adding an extra $W_F = 5$ bits of frequency resolution through dithering would result in $\Delta f_{\text{res}} = \Delta f^T/2^5 = 720$ Hz. This has a potential to bring the phase noise down by 30 dB, to $-139$ dBc/Hz. However, the dithering process itself produces a significant amount of additional phase noise. The design goal is to make the DCO quantization noise contribution significantly lower than the natural DCO noise and the RF standard targeted.

The closed-loop transfer function from the frequency deviation quantization error $\Delta f_{n,V}$ source to the phase $\phi_V$ of the RF output is

$$H_{\text{cl}, \Delta f_V}(s) = \frac{2\pi/s}{1 + H_{\text{ol}}} = \frac{2\pi}{\alpha f_R} \frac{1}{1 + s/\alpha f_R} \quad (4.73)$$

Within the loop bandwidth, the power spectral density of the output phase is flat at

$$\mathcal{L} = \frac{(2\pi)^2}{12} \left(\frac{\Delta f_{\text{res}}}{f_R}\right)^2 \frac{1}{f_R} \frac{1}{\alpha^2} \quad (4.74)$$

and falls off with 20 dB/decade beyond. Similar to the analysis of equivalence of the open- and closed-loop transfer functions in Section 4.11.1.1, Eqs. 4.71 and 4.74 have the same value for $\Delta f = f_{\text{BW}}$.

### 4.11.2.2 Comparison of DCO Dithering and TDC Resolution Effects

The extent of the DCO dithering effect on phase noise can be compared with that of TDC resolution. Equating Eqs. 4.64 and 4.71 yields

$$\Delta f_{V,\text{res}} = f_R \frac{\Delta t_{\text{res}}}{T_V} \alpha \qquad (4.75)$$

For example, for a BLUETOOTH system with $f_R = 13$ MHz, $f_V = 2400$ MHz, and $\alpha = 2^{-8}$, a DCO with a resolution of $\Delta f_{V,\text{res}} = 4.9$ kHz produces the same amount of phase noise as that of a TDC with a resolution of $\Delta t_{\text{res}} = 40$ ps. For a GSM system with $f_R = 26$ MHz, $f_V = 1800$ MHz, and $\alpha = 2^{-6}$, a DCO with a resolution of $\Delta f_{V,\text{res}} = 21.9$ kHz produces the same amount of phase noise as a TDC with a resolution of $\Delta t_{\text{res}} = 30$ ps.

### 4.11.2.3 DCO Dithering Effect on Phase Noise

Equation 4.72 describes the effect of finite DCO frequency resolution. This formula could also be used readily to determine the DCO phase noise due to intentional uniform dithering (i.e., with no noise shaping) of the tracking-bank varactors. The equation is modified for the sampling frequency $f_{\text{dth}}$ being normally much higher than $f_R$:

$$\mathcal{L}\{\Delta f\} = \frac{1}{12}\left(\frac{\Delta f_{\text{res}}}{\Delta f}\right)^2 \frac{1}{f_{\text{dth}}}\left(\text{sinc}\frac{\Delta f}{f_{\text{dth}}}\right)^2 \qquad (4.76)$$

Since its low-frequency content can still be quite high for more demanding wireless standards, noise-shaped DCO dithering is used instead. It should be noted that due to the $W_F$ wordlength limitation of the digital dithering circuit, there will still be a phase-noise contribution due to the finite resolution per Eq. 4.72, with $\Delta f_{\text{res}} = \Delta f^T/2^{W_F}$.

The spectrum of the $\Sigma\Delta$-shaped frequency deviation

$$S_{\Delta f}(\Delta f) = \frac{(\Delta f_{\text{res}})^2}{12}\frac{1}{f_{\text{dth}}}\left(2\sin\frac{\Delta f}{f_{\text{dth}}}\right)^{2n} \qquad (4.77)$$

relates to the phase noise as $S_\phi(\Delta f) = S_f(\Delta f)/\Delta f^2$. Consequently, we obtain the phase-noise spectrum as

$$\mathcal{L}\{\Delta f\} = \frac{1}{12}\left(\frac{\Delta f_{\text{res}}}{\Delta f}\right)^2 \frac{1}{f_{\text{dth}}}\left(2\sin\frac{\pi\Delta f}{f_{\text{dth}}}\right)^{2n} \qquad (4.78)$$

The $\Sigma\Delta$ varactor dithering moves the varactor quantization noise energy to high-frequency offsets at the RF output. The spectrum shows a null at the 600-MHz sampling frequency due to a zero-order hold. There is a spectrum replica in the range 600 to 1200 MHz that is attenuated by the sinc function. The entire $\Sigma\Delta$ dithering process contains two phase-noise components: Eq. 4.72 with $\Delta f_{\text{res}} = \Delta f^T$ and Eq. 4.78 with $\Delta f_{\text{res}} = \Delta f^T/2^{W_F}$.

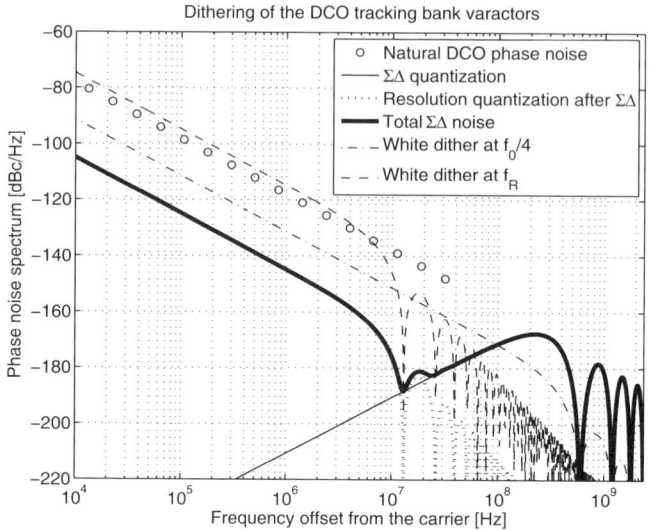

**Figure 4.39** Phase-noise spectrum due to $\Delta f_{res}$ frequency quantization and various dithering schemes. $f_V = 2400$ MHz, $\Delta f_{res} = 23$ kHz, $f_{dth} = f_V/4$, and $W_F = 5$.

Figure 4.39 shows the expected phase-noise spectrum components of the DCO implemented in this design for both white dithering by the $f_R = 13$-MHz clock and second-order $\Sigma\Delta$ dithering by $f_{dth} = 600$ MHz. The white dithering by $f_{dth}$ clock rate is hypothetical only. Figure 4.40 shows similar plots for the GSM

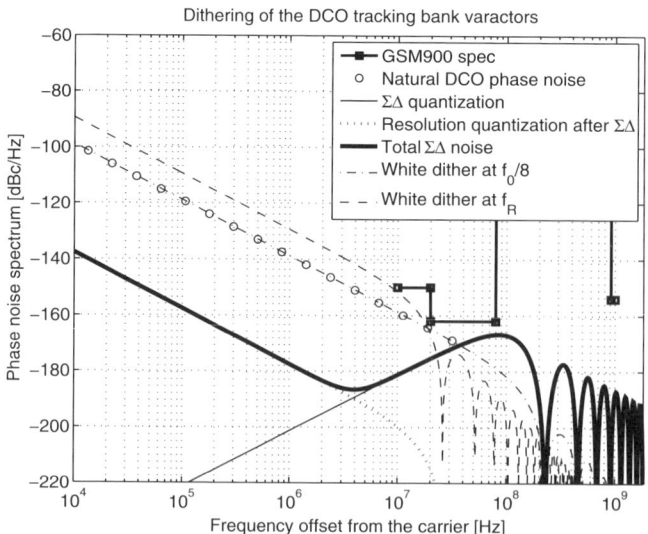

**Figure 4.40** Phase-noise spectrum due to $\Delta f_{res}$ frequency quantization and various dithering schemes. Low-band $f_V = 915$ MHz, $\Delta f_{res} = 12$ kHz at HB, $f_{dth} = f_V/8$, and $W_F = 8$.

example. The second-order MASH $\Sigma\Delta$ dithering by $f_{dth} = f_V/8 = 225$ MHz with a TB step of $\Delta f_{res} = 12$ kHz will produce high-frequency phase noise below $-166$ dBc/Hz at 20- to 80-MHz frequency offsets, which meets the extremely tough GSM specification by a margin of 4 dB.

## 4.12 TYPE II ADPLL

Figure 4.30 is now modified to include a second pole at zero frequency, thus giving rise to a type II ADPLL. This is also known as a *proportional–integral* (PI) *controller* and is used in all-digital PLLs for clock generation and recovery [31–35], and in hard-disk-drive read channels [77]. A body of experience has been developed over the years on this type of structure.

PI control is accomplished in the digital domain by accumulating phase-error samples $\phi_E[k]$ and scaling them by integral loop gain $\varrho$ (Fig. 4.41). The value of $\varrho$ should normally be smaller than $\alpha$. Contributions from the proportional and integral paths are added together.

A chief advantage of type II topology is its better filtering capabilities of oscillator noise, leading to improvements in the overall phase-noise performance. A type I loop can provide only 20-dB/decade filtering of the DCO phase noise, whereas type II can provide up to 40 dB/decade. As a result, the up-converted DCO flicker noise, which exhibits a 30-dB/decade spectral slope (see Fig. 1.4) and could be quite troublesome in deep-submicron CMOS, can now be removed completely.

Another advantage of the type II loop is that it exhibits no steady-state frequency error in face of the reference or variable-frequency ramp. It could be useful in GSM applications, which require low-frequency error under realistic environmental conditions, such as when supply voltages are not settled, the oscillators are drifting and there are external low-frequency perturbations.

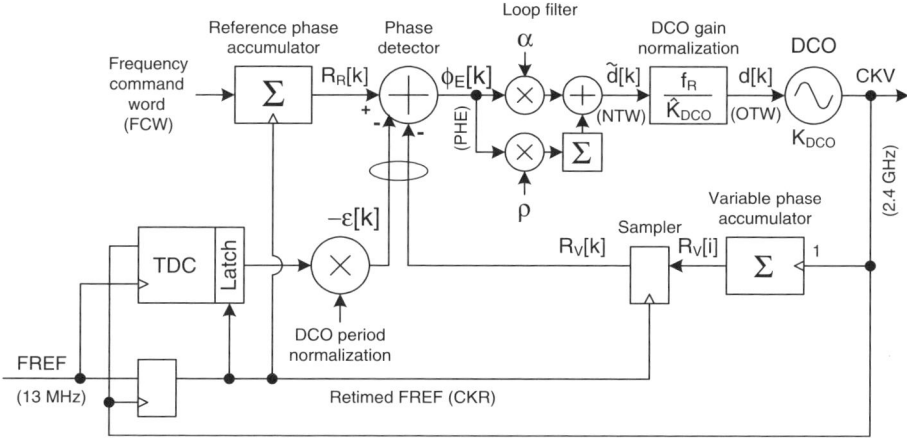

**Figure 4.41** Type II phase-domain all-digital synchronous PLL synthesizer.

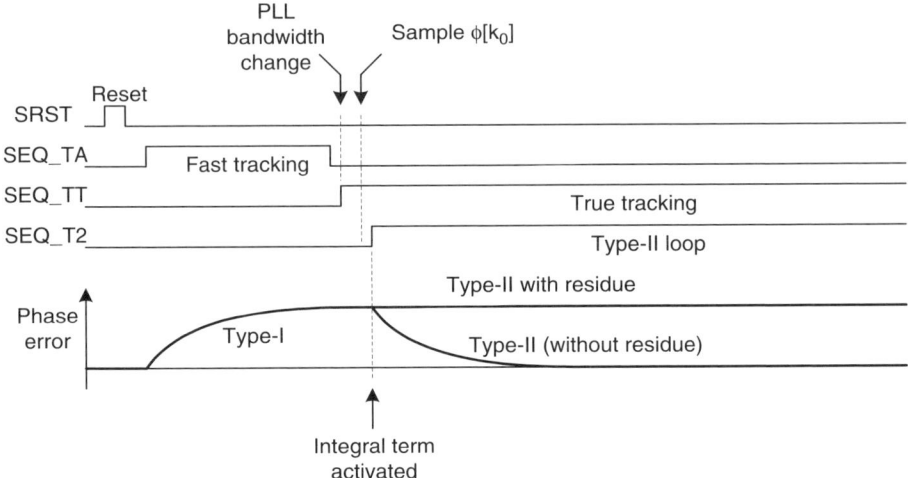

**Figure 4.42** Type II operation with and without phase-error residue.

Another feature of the type II loop is that the phase offset between the reference and variable clocks approaches zero even in the face of constant-frequency offset. This is especially beneficial in clock recovery applications with integer-$N$ frequency-division ratios. In a wireless application such as a local oscillator, however, this feature is not very useful since there is no need to align the RF clock phase.

Due to the longer transients of the PI configuration, its use is beneficial only in the tracking mode. In order not to degrade the switching time, the integral part $\varrho$ should be turned off during acquisition and fast tracking. In this way a fast transient loop characteristic is used during acquisition, and a slower transient, but with a better filtering capability, is used during regular tracking.

To further enhance type II system interoperation, type II loop structure activation should be deferred until the normal tracking mode. During fast acquisition, the tracking-bank varactors are engaged to complete the final frequency settling promptly with a high loop gain setting. At the time of switchover into normal tracking, the proportional loop gain gets attenuated and the integral loop gain sets in, with its internal accumulator set initially to zero. Unfortunately, the residual dc offset of the preceding mode[6] would now be considered an undesired phase-error bias, which might take a long time to work out. A solution is to subtract this bias from the phase error before the accumulation operation.

Figures 4.42 and 4.43 demonstrate type II ADPLL operation. All memory elements (registers) are reset synchronously at the beginning of the operation by asserting the SRST control signal. After the initial frequency is locked roughly using the PVT calibration and acquisition modes (Section 2.5), the ADPLL activates the tracking bank of the DCO varactors. At first, the loop bandwidth is quite high

---

[6] In a type I loop the phase error is proportional to the frequency offset.

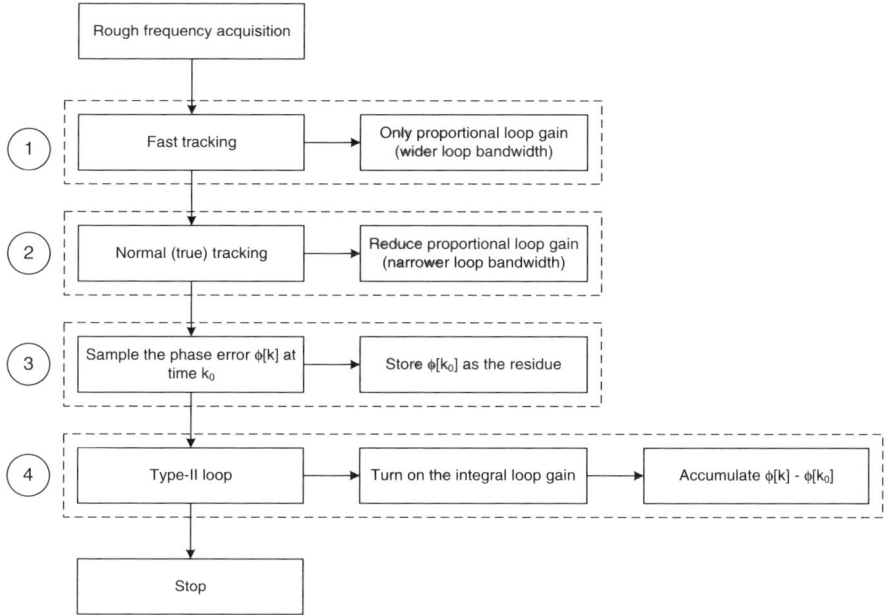

**Figure 4.43** Flowchart of type II operation with phase-error residue.

("fast tracking" when SEQ_TA is asserted) and only a proportional loop gain $\alpha$ is used. This allows us quickly to resolve any frequency quantization error left from the preceding acquisition mode. Then the normal or true tracking mode is entered. The following events happen at about the same time: First, the loop bandwidth is narrowed by scaling down the $\alpha$ proportional gain factor (SEQ_TT asserted). This further filters out the phase error $\phi[k]$, which is then sampled and stored as $\phi[k_0]$. Finally, the ADPLL type I loop turns in to type II (SEQ_T2 asserted) by activating the integral term $\varrho$. The $\phi[k] - \phi[k_0]$ difference is being accumulated, which gives rise to type II ADPLL with phase-error residue. It should be noted that for a type I ADPLL loop, only modes 1 and 2 in Fig. 4.43 are used. In conventionally defined type II operation, mode 3 is not used and $\phi[k_0] = 0$. This forces the built-up quantization frequency error to decay slowly to zero, which creates unnecessary transitioning, as shown by the phase-error plot in Fig. 4.42.

Figure 4.44 reveals hardware realization of the type II loop filter. Both $\alpha$ and $\varrho$ loop gain factors are implemented in an efficient manner as right-bit-shift operations. The residue latch block samples the phase error $\phi[k_0]$ at the beginning of type II loop operation and outputs to the integral accumulator block the adjusted $\phi[k] - \phi[k_0]$ phase-error samples. Conventionally defined type II loop operation ("without residue") could be realized by resetting the SRST signal of the residue latch block, thus forcing $\phi[k_0]$ to zero. Type I loop operation could be realized by additionally resetting the integral accumulator.

**130**  ALL-DIGITAL PHASE-LOCKED LOOP

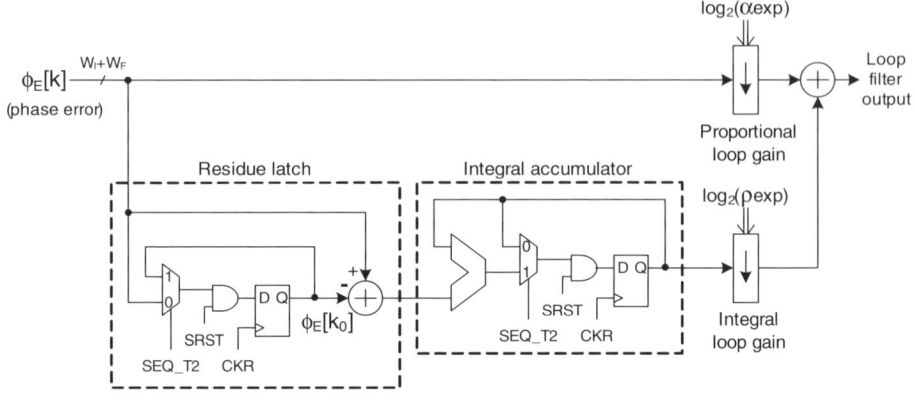

**Figure 4.44** Hardware realization of a type II loop filter.

### 4.12.1 PLL Frequency Response of a Type II Loop

Let's start the analysis by finding the $s$-domain equivalent of the discrete-time accumulator $z^{-1}/(1-z^{-1})$ in the integral path. From Eq. 4.46 we have $(z-1) = s/f_R$, whose inverse is just $z^{-1}/(1-z^{-1})$, which is shown in Fig. 4.45.

Assuming that $K_{\text{DCO}}$ is estimated correctly, the open-loop transfer function, is

$$H_{\text{ol}}(s) = \left(\alpha + \frac{\varrho f_R}{s}\right)\frac{f_R}{s} = \frac{\varrho f_R^2}{s} \cdot \frac{1 + s/(\varrho f_R/\alpha)}{s} \quad (4.79)$$

which shows two poles at the origin and one complex zero at $\omega_z = j(\varrho f_R/\alpha)$. The open-loop unity gain is at

$$\omega_1 = \alpha f_R \left(\frac{1}{2} + \frac{1}{2}\sqrt{1 + \frac{4\varrho}{\alpha^2}}\right) \quad (4.80)$$

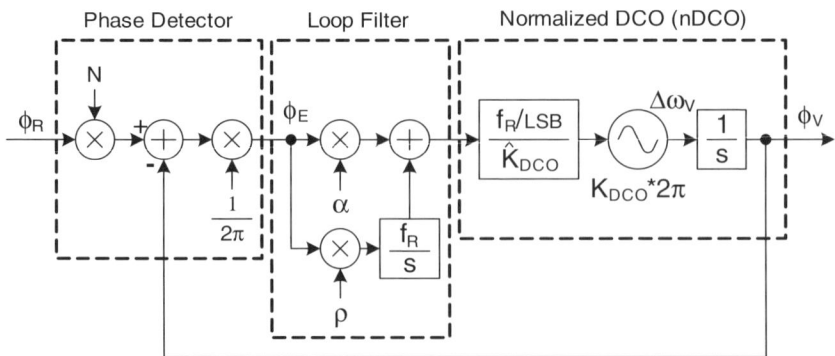

**Figure 4.45** Linearized equivalent $s$-domain model of type II ADPLL. (From [65], © 2005 IEEE.)

The closed-loop transfer function is

$$H_{cl}(s) = N \frac{(\alpha + \varrho f_R/s)(f_R/s)}{1 + (\alpha + \varrho f_R/s)(f_R/s)} = N \frac{\alpha f_R s + \varrho f_R^2}{s^2 + \alpha f_R s + \varrho f_R^2} \quad (4.81)$$

This can be compared to the classical two-pole system transfer function whose magnitude response is shown in Fig. 4.46:

$$H_{cl}(s) = N \frac{2\zeta \omega_n s + \omega_n^2}{s^2 + 2\zeta \omega_n s + \omega_n^2} \quad (4.82)$$

where $\zeta$ is the damping factor and $\omega_n$ is the natural frequency; the zero lies at $\omega_z = -\omega_n/2\zeta$. Fitting the two equations yields

$$\omega_n = \sqrt{\varrho} f_R \quad (4.83)$$

$$\zeta = \frac{\alpha f_R}{2\omega_n} = \frac{1}{2}\left(\frac{\alpha}{\sqrt{\varrho}}\right) \quad (4.84)$$

For example, for GSM, if $\alpha = 2^{-7}$ and $\varrho = 2^{-15}$, then $\zeta = 1/\sqrt{2}$. Figure 4.47 illustrates the closed-loop transfer function for different values of $\alpha$ with constant $\zeta = \frac{1}{4}$.

A type II loop gives one more tuning knob, resulting in more flexibility to adjust the loop noise performance. The $\varrho$ term contributes a square-root effect on the

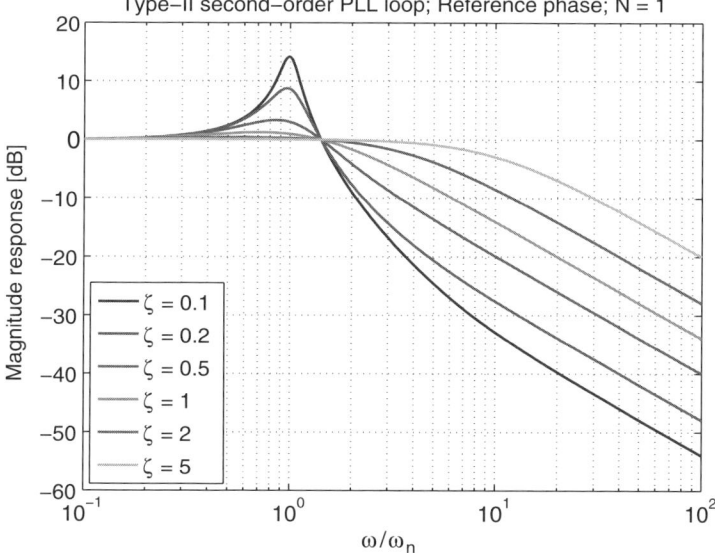

**Figure 4.46** Magnitude response of type II PLL loop vs. $\omega_n$-normalized frequency for various values of $\zeta$.

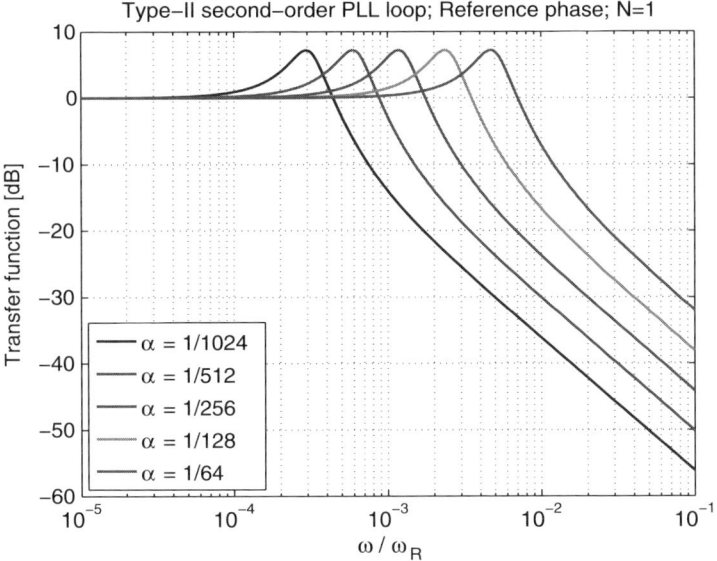

**Figure 4.47** Magnitude response of type II PLL vs. reference-normalized frequency for various values of $\alpha$ with $\zeta = \frac{1}{4}$.

natural frequency and damping factor and thus requires more bits to keep the same dynamic range as $\alpha$.

Transforming Eqs. 4.79 and 4.81 into the $z$-domain gives

$$H_{ol}(z) = \frac{\alpha(z-1) + \varrho}{(z-1)^2} \quad (4.85)$$

$$H_{cl}(z) = N \frac{\alpha(z-1) + \varrho}{(z-1)^2 + \alpha(z-1) + \varrho} \quad (4.86)$$

An analysis similar to that in Section 4.11 can be made to study the effect of the DCO and TDC noise sources. Figure 4.48 is a type II ADPLL version of

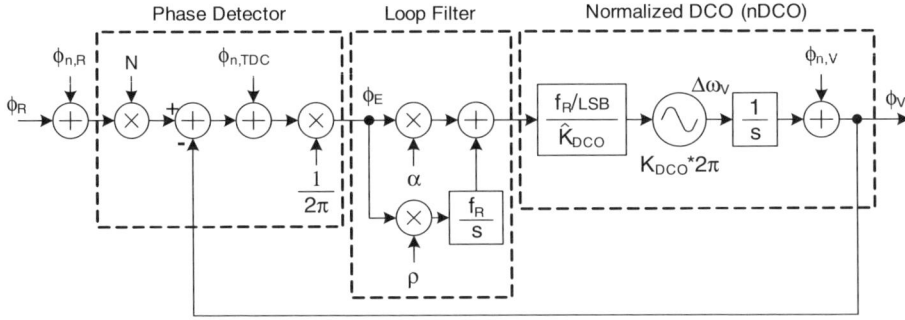

**Figure 4.48** Linear $s$-domain model of a type II ADPLL with noise sources added.

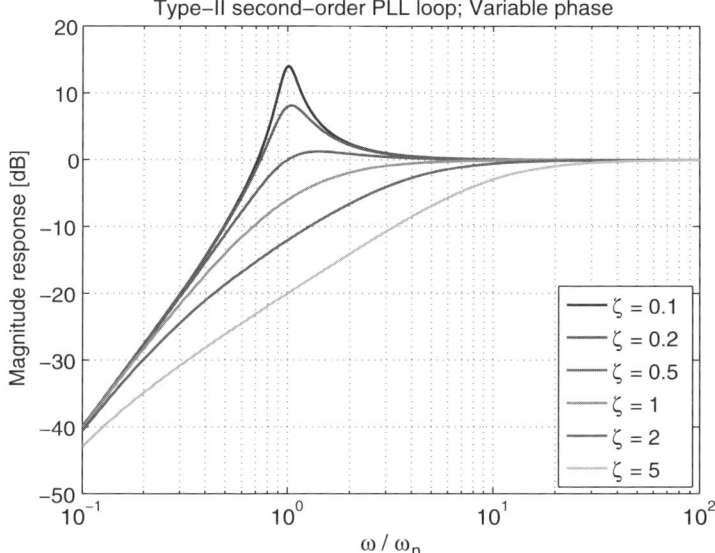

**Figure 4.49** DCO magnitude response of a type II PLL vs. $\omega_n$-normalized frequency for various values of $\zeta$.

Fig. 37. The closed-loop transfer function of the oscillator phase noise is

$$H_{\text{cl}, V}(s) = \frac{1}{1 + (\alpha + \varrho f_R/s)(f_R/s)} = \frac{s^2}{s^2 + \alpha f_R s + \varrho f_R^2} \quad (4.87)$$

Figure 4.49 illustrates the magnitude response of the DCO phase. Compared to type I response, the lower-frequency noise components could be attenuated further at some risk of peaking at the natural frequency $\omega_n$.

The closed-loop transfer function of the TDC phase noise is

$$H_{\text{cl}, \text{TDC}}(s) = \frac{(\alpha + \varrho f_R/s)(f_R/s)}{1 + (\alpha + \varrho f_R/s)(f_R/s)} = \frac{\alpha f_R s + \varrho f_R^2}{s^2 + \alpha f_R s + \varrho f_R^2} \quad (4.88)$$

## 4.13 HIGHER-ORDER ADPLL

Replacing a conventional phase/frequency detector with a TDC, as well as a conventional oscillator with a DCO, now allows us to use a fully digital loop filter. The loop filter could be constructed as a combination of FIR and IIR filters as well as an accumulator (i.e., an IIR filter with a pole at dc). The FIR and IIR filters are usually cascaded with the proportional loop gain $\alpha$, whereas the accumulator, which gives rise to type II PLL configuration, is connected in parallel. It is generally more beneficial to use IIR filters than FIR filters since the former are usually more compact and provide stronger filtering capability. On the other hand, complex

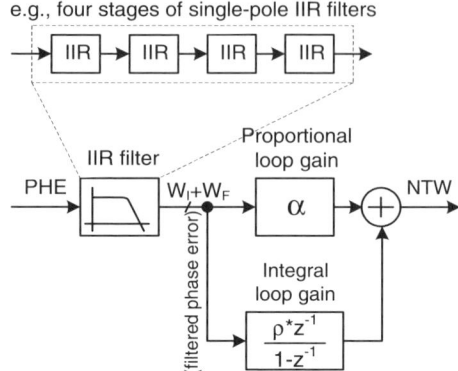

**Figure 4.50** Loop filter with single-pole IIR stages and an accumulator that provides a pole at dc.

IIR filters could easily become unstable. This problem can be solved by using a cascade of single-pole IIR filters, which are unconditionally stable. The phase-delay variation of IIR filters is usually not a problem for fluctuation frequencies of interest that are at least 10 times smaller than the sampling frequency.

Figure 4.50 shows a loop filter with four single-pole IIR stages and an accumulator $z^{-1}/(1-z^{-1})$ that provides a pole at dc. The order of the arrangement is a trade-off between the resolution and the circuit size.

As shown in Fig. 4.41, a type II PLL loses some of its earlier appeal in a fully digital architecture with a digitally controlled oscillator. Instead of a pole at dc, a low-pass filter with a transfer function of $L(z)$ is added. It uses a specific example of the loop filter that consists of four cascaded stages of single-pole IIR filters. Any one of the four filtering stages could be bypassed such that the resulting PLL order could be between the first and fifth.

Implementation of a single-pole IIR filter is shown in Fig. 4.51. The frequency characteristic and, consequently, the pole location are controlled by the attenuation

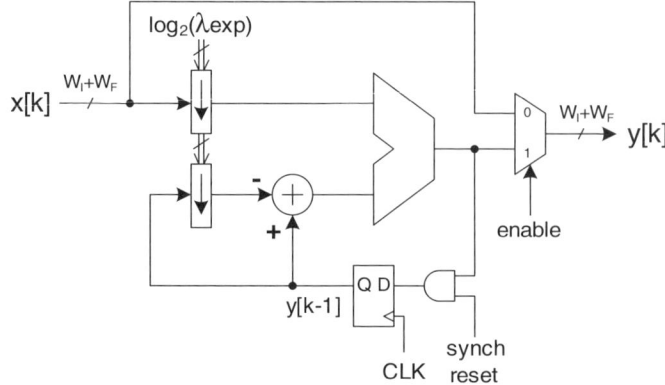

**Figure 4.51** Single-stage IIR filter.

factor $\lambda$, which is realized as a right-bit-shift operator. The time-domain equation is expressed as

$$y[k] = (1-\lambda)y[k-1] + \lambda x[k] \quad (4.89)$$

The $z$-domain transfer function is expressed as

$$H_{\text{iir1}}(z) = \frac{\lambda z}{z - (1-\lambda)} \quad (4.90)$$

and its $s$-domain representation is

$$H_{\text{iir1}}(s) = \frac{1 + s/f_R}{1 + s/\lambda f_R} \quad (4.91)$$

The linear frequency ($s = j\omega = j2\pi f$) representation is

$$H_{\text{iir1}}(f) = \frac{1 + j(2\pi f/f_R)}{1 + j(2\pi f/\lambda f_R)} \quad (4.92)$$

with the 3-dB cutoff frequency:

$$f_{\text{BW, iir1}} = \frac{\lambda}{2\pi} f_R \quad (4.93)$$

Using analog techniques, the maximum practically achievable order of a PLL is third [33,74,78], mainly for stability reasons, especially when process and temperature variations are taken into account. However, these restrictions do not exist with digital VLSI implementations and it is possible to create higher-order structures that would provide efficient noise reduction and accurate frequency response [79]. The all-digital loop renders a digital design of the possible loop filter, thus providing many benefits in terms of testability, flexibility, and portability to various processes.

Figure 4.52 shows an $s$-domain model of a type II higher-order ADPLL frequency synthesizer. Equation 4.79 is now modified to include the IIR filter factor of Eq. 4.91:

$$H_{\text{ol}}(s) = \left(\alpha + \frac{\varrho f_R}{s}\right) \frac{f_R}{s} \frac{1 + s/f_R}{1 + s/\lambda f_R} = \frac{\varrho f_R^2}{s} \frac{1 + s/(\varrho f_R/\alpha)}{s} \frac{1 + s/f_R}{1 + s/\lambda f_R} \quad (4.94)$$

It shows two poles at the origin $\omega_{p1} = \omega_{p2} = 0$, one pole at $\omega_{p3} = j\lambda f_R$, and two zeros at $\omega_{z1} = j(\varrho f_R/\alpha)$ and $\omega_{z2} = jf_R$.

The ADPLL implemented in GSM applications uses four independently controlled IIR stages. A four-pole IIR filter attenuates the reference and TDC quantization noise at the 80-dB/decade slope, primarily to meet the GSM spectral mask requirements at 400 kHz offset. Equation 4.94 is modified to reflect the four

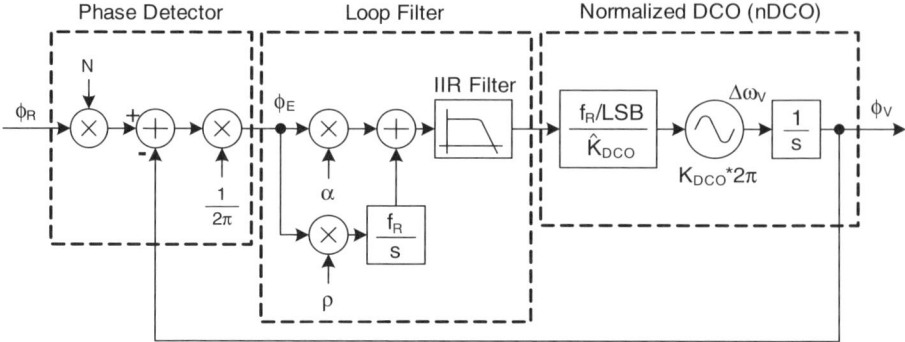

**Figure 4.52** Linearized equivalent s-domain model of a type II higher-order ADPLL.

cascaded single-stage IIR filters, each with an attenuation factor $\lambda_i$, where $i = 0 \cdots 3$:

$$H_{ol}(s) = \frac{\varrho f_R^2}{s} \frac{1 + s/(\varrho f_R/\alpha)}{s} \prod_{i=0}^{3} \frac{1 + s/f_R}{1 + s/\lambda_i f_R} \qquad (4.95)$$

The closed-loop transfer function for the reference is low-pass with the gain multiplier $N$:

$$H_{cl}(s) = N \frac{H_{ol}(s)}{1 + H_{ol}(s)} \qquad (4.96)$$

The closed-loop transfer function for the TDC is low-pass:

$$H_{cl,T}(s) = \frac{H_{ol}(s)}{1 + H_{ol}(s)} \qquad (4.97)$$

The closed-loop transfer function for the DCO is high-pass:

$$H_{cl,V}(s) = \frac{1}{1 + H_{ol}(s)} \qquad (4.98)$$

### 4.13.1 PLL Stability Analysis

For a BLUETOOTH frequency synthesizer, where a simple first or second-order loop filter could be employed, loop stability has not been an issue since it is easy to maintain it with reasonably intuitive values of loop gain factors. A type I ADPLL is unconditionally stable when $\alpha < 1$. A type II ADPLL is stable and exhibits acceptable peaking when $\zeta \geq 0.707$. A more sophisticated loop stability analysis is needed for GSM applications that require a higher-order loop filter, such as the one shown in Fig. 4.50. Conventional s-domain control loop theory tools are used below to carry it out.

The ADPLL open-loop transfer function follows Eq. 4.95. Figure 4.53 shows the magnitude and phase of the open-loop transfer function for the default ADPLL loop settings: $\alpha = 2^{-7}$, $\varrho = 2^{-15}$, $\lambda_i = 2^{-3}$ for $i = 0 \cdots 2$, and $\lambda_3 = 2^{-4}$. The stability analysis results in the following numbers:

1. Open-loop 0-dB point = 35.1 kHz
2. Open-loop $-180°$ point = 186 kHz
3. Phase margin = 47.8°
4. Gain margin = 18.6 dB
5. Closed-loop gain at 400 kHz = $-33.0$ dB

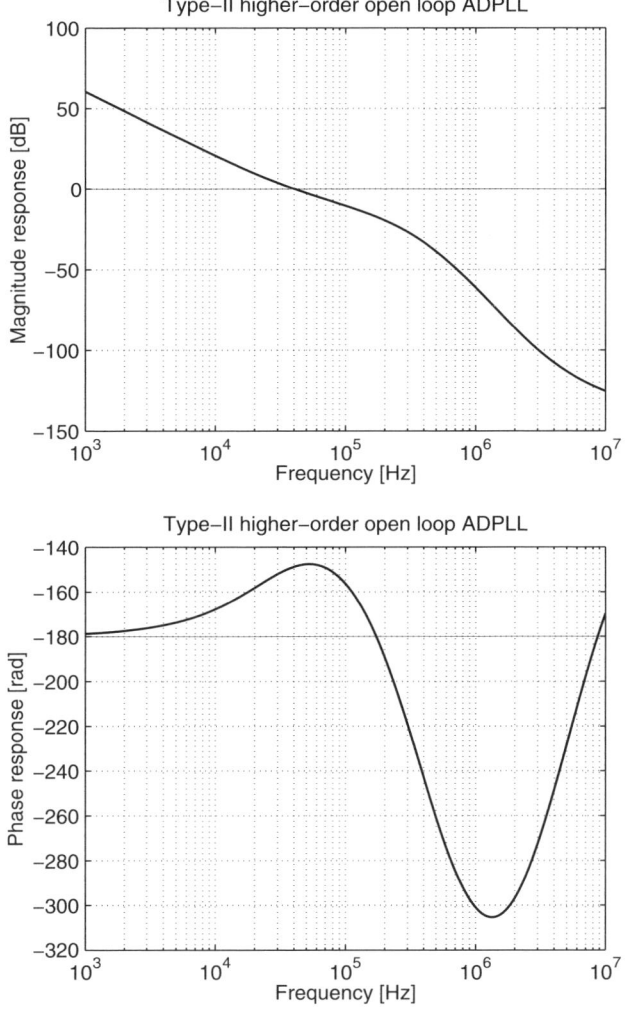

**Figure 4.53** Open-loop transfer function with default loop settings.

The closed-loop transfer function before the frequency-division ratio $N$ multiplication,

$$H_{cl}(s) = \frac{H_{ol}(s)}{1 + H_{ol}(s)} \tag{4.99}$$

is shown in Fig. 4.54. It corresponds to the TDC feed. For the FREF transfer function, $N = 1800\,\text{MHz}/26\,\text{MHz} = 69.23 = 36.8\,\text{dB}$. Figure 4.55 shows the magnitude and phase of the DCO-phase closed-loop transfer function.

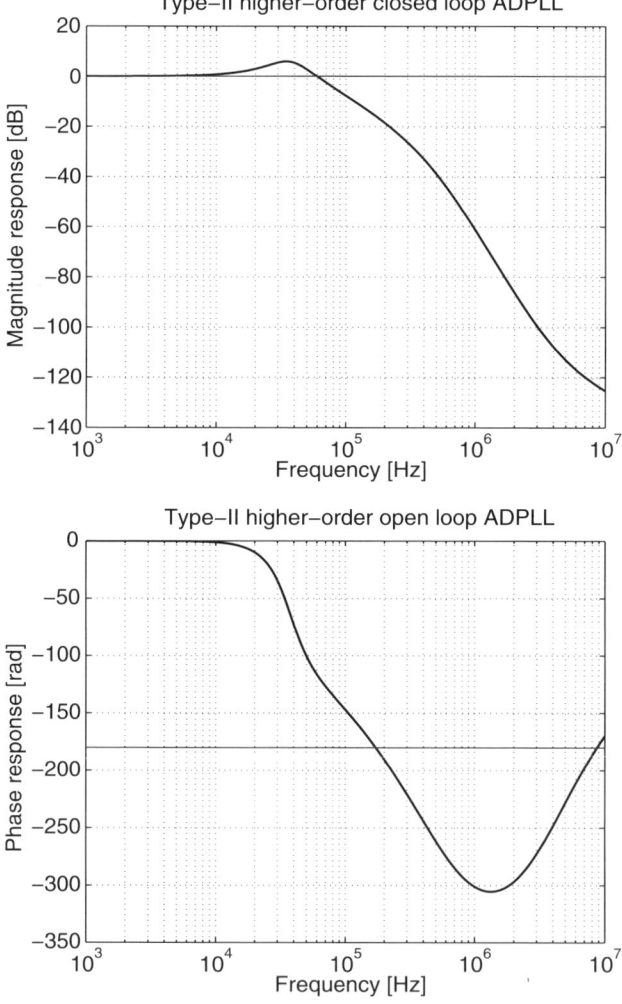

**Figure 4.54** Reference closed-loop transfer function with default loop settings.

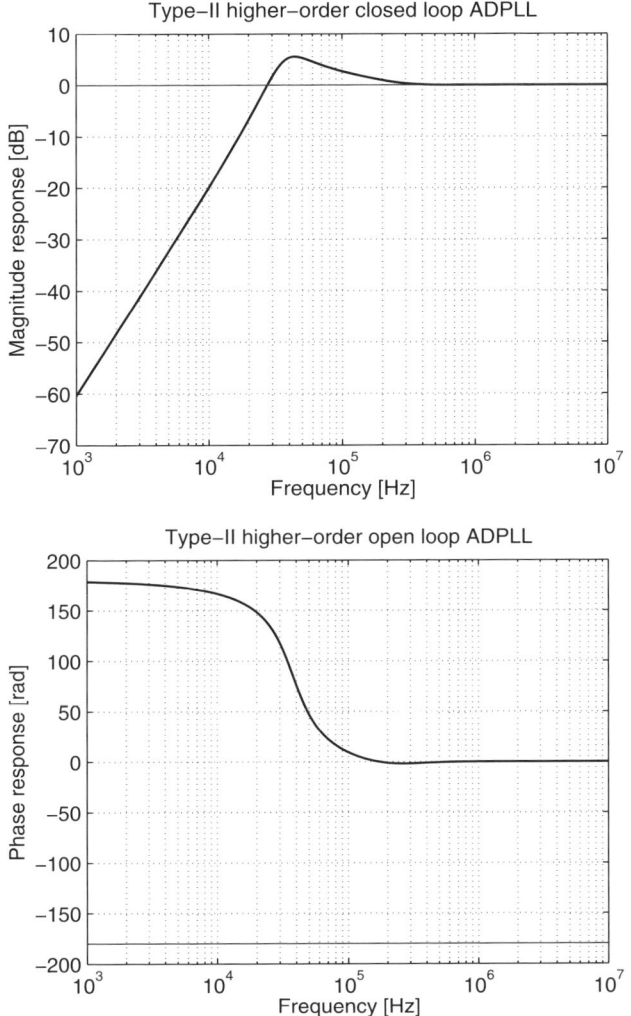

**Figure 4.55** DCO closed-loop transfer function with default loop settings.

## 4.14 NONLINEAR DIFFERENTIAL TERM OF AN ADPLL

A differential term $(1 - z^{-1})$ could be added to the phase detector output $\phi_E[k]$ to monitor its instantaneous variations. Due to its high-pass and hence noise-enhancement characteristics, the differential term has to be filtered in a nonlinear manner. This could be accomplished by a thresholder circuit that senses the absolute phase-error difference between the current and previous samples and activates a DCO correction for large phase-error steps, as shown in Fig. 4.56 for a type I PLL. The

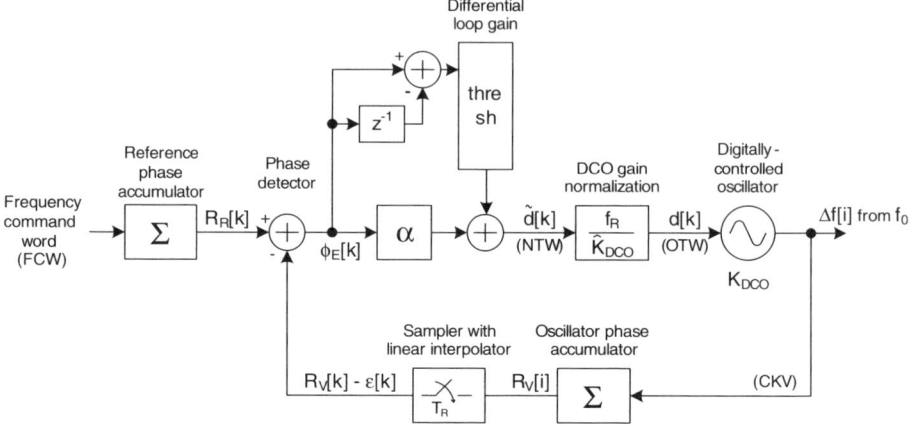

**Figure 4.56** Phase-domain ADPLL synthesizer with a nonlinear differential term.

differential term is very useful in handling situations in which an *occasional* rapid frequency perturbation occurs during the regular tracking operation when the PLL loop is settled and normally slower in response. The threshold should be set high enough to avoid being triggered by the distribution of thermal and flicker noise expected due to those loop components with continuous-domain characteristics.

These sudden changes in the oscillating frequency might be due to, for example, sudden supply voltage drop when the integrated digital baseband starts a new activity or the power amplifier starts ramping up. Relying on the proportional term to handle the sudden perturbation would normally require a long time, due to the narrow loop bandwidth. To filter out any transitory phase-error perturbations, which might not necessarily indicate a consistent change in the oscillating frequency, it is necessary to qualify the new phase error for a number of clock cycles. A gear-shifting procedure, described in Section 4.16, is normally required whenever the proportional loop gain $\alpha$ is modified.

### 4.14.1 Quality Monitoring of an RF Clock

The differential term of the phase error can be used to ascertain the quality of the generated RF clock CKV and to indicate the loss of PLL locking. Figure 4.57 shows a clock quality monitoring (CQM) *circuit*. Its operation is based on the observation that the phase noise at the ADPLL RF output has a *strong* correlation to the variance of the digital FREF-clocked phase-error signal $\phi_E$. To keep the hardware complexity to a minimum, a simplified first norm (maximum of absolute difference of consecutive samples) of the phase error is calculated instead of the second norm (rms squared). The CQM circuit asserts the lock maintenance notification signal as long as the instantaneous changes of the phase-error $\phi_E$ samples are within limits.

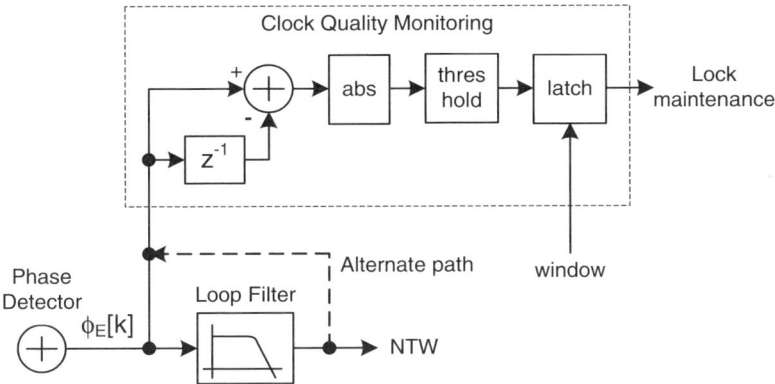

**Figure 4.57** Clock quality monitoring circuit.

## 4.15 DCO GAIN ESTIMATION USING A PLL

Incorrect estimation of $K_{DCO}$ gain affects the accuracy of the loop gain $\alpha$, which itself is not a major concern. The best indication of this is that in this implementation, $\alpha$ is register programmable as a negative exponent of the radix 2, which is definitely very coarse. It was found through system analysis that a finer granularity of the loop gain control was not needed. Why, then, the reader might ask, is this apparent stress on the DCO gain estimation? The main reason is the predictive loop operation for transmitter frequency modulation, introduced in Chapter 5. This method does require a *precise* knowledge of DCO gain. This estimate is used in the DCO gain-normalization multiplier $f_R/\widehat{K}_{DCO}$ of Figs. 3.1 and 4.30.

As mentioned in Section 3.2, the DCO gain estimate $\widehat{K}_{DCO}$ can be computed by harnessing the power of the existing phase-detection circuitry for the purpose of determining the oscillator frequency deviation $\Delta f_V$. The DCO frequency deviation $\Delta f_V$ can be calculated by observing the phase-error difference $\Delta \phi_E$ (expressed as a fraction of the DCO clock period) in the observation interval of the phase detector update, which is normally equal to the frequency reference clock period $T_R$. Equation 4.33 is rewritten here:

$$\Delta f_V = \frac{\Delta \phi_E}{T_R} = \Delta \phi_E f_R \qquad (4.100)$$

Equation 4.100 can be plugged into Eq. 3.3 to provide an estimated DCO gain:

$$\widehat{K}_{DCO}(f_V, \text{OTW}) = \frac{\Delta \phi_E}{\Delta \text{OTW}} f_R \qquad (4.101)$$

For a given DCO input OTW, Eq. 4.101 *theoretically* allows us to calculate the local value of the oscillator gain $K_{DCO}$ by observing the phase detector output $\Delta \phi_E$ as a response to the $\Delta \text{OTW}$ input perturbation in the preceding reference clock cycle.

Naturally, the reference frequency $f_R$ is the system parameter that is, for all practical purposes, known exactly.

Unfortunately, as mentioned in Section 4.6.2, the method of frequency estimation described above is a poor choice, due to the excessive TDC quantization for realistic values of $\Delta t_{res}$. Instead, the difference between the steady-state phase error values is more appropriate. Equation 4.36 captures the relationship, which is repeated here:

$$\Delta f_V = \phi_E \alpha f_R \qquad (4.102)$$

Equation 4.102 can be plugged into Eq. 3.3 to provide an estimated DCO gain:

$$\widehat{K}_{DCO}(f_V, \text{OTW}) = \frac{\Delta \phi_E \alpha}{\Delta \text{OTW}} f_R \qquad (4.103)$$

A significant advantage in operation can be obtained by noting that in a type I PLL the phase error $\phi_E$ is proportional to the relative oscillating frequency. Consequently, not only could the power of the phase detection circuitry be harnessed but also the averaging and adaptive capabilities of the PLL loop itself. Equation 4.101 can now be used with normal loop updates (unlike the general case) for an arbitrary number of FREF clock cycles. At the end of the measurement, the final $\Delta \phi_E$ and $\Delta \text{OTW}$ values are used. The loop itself provides the averaging and frequency quantization reduction.

In the analysis above, the OTW perturbation was the cause for the frequency deviation calculation by observing the resulting phase-error change. This order could be reversed beneficially by setting the frequency deviation a priori and observing the resulting change in the tuning word.

## 4.16 GEAR SHIFTING OF PLL GAIN

During the normal course of PLL operation, two unique intervals with conflicting requirements are readily distinguished. In the first interval, the goal is to acquire the desired frequency as soon as practically possible. The loop bandwidth needs to be made wide even at the expense of enhanced phase noise and spurs, which are quite unimportant at that time. In the second interval, which covers the actual TX and RX operations, the goal is to maintain or track the desired frequency that got acquired during the first phase. The loop bandwidth there has to be kept lower to minimize the phase noise and spurs of the reference path. In addition, higher-order loop operation might be advantageous.

These conflicting requirements for the PLL characteristic necessitate a mechanism by which the loop bandwidth could be shifted seamlessly once the acquisition phase is completed. The *gear-shifting* mechanism [80] described here can be used to achieve this. This idea has two embodiments:

1. Autonomous gear shifting, which is fully contained in the oscillator tracking gain block (GT). It is utilized in our example test chip implementation for normalized gain switchover while using tracking-bank varactors.

2. Extended gear shifting with *zero-phase restart* (ZPR), which involves multiple blocks. It implements the progressive refinement of DCO resolution as the operational frequency approaches the desired frequency. That idea was introduced in Fig. 2.9 and described in conjunction with Fig. 3.9.

This mechanism deals with the introduction of a gear shift or switchover of the normalized PLL gain constant $\alpha$. During frequency or phase acquisition, a larger loop gain constant, $\alpha_1$, is used such that the resulting phase error is within limits (see Eq. 4.37). After the frequency and phase are acquired, the phase error developed, which is a rough indication of the frequency offset, is in a steady state. The following two operations are performed simultaneously (in a synchronous digital design it means within the same clock cycle) while making a transition into tracking:

1. Add the dc offset to the phase error or the DCO tuning word.
2. Reduce the loop constant from $\alpha_1$ to $\alpha_2$.

Since operation 1 results in substantial lowering of the maximum phase error, $\alpha$ could now safely be reduced. It should be noted that in a type I PLL, the frequency deviation is directly proportional to the phase error and the gain constant, $\alpha$.

This gear-shifting idea makes an implementational sense primarily in digital systems. It is generally difficult to perform PLL gear shifting in analog circuits because of the imperfect matching and voltage or charge losses during switching, thus resulting in phase *hits* or perturbations whenever a sudden perturbation (gear shift) is introduced. The solution described here is fully digital and provides a hitless operation while not requiring a large dynamic range from the phase detector. It is a perfect match to the phase-domain all-digital PLL synthesizer architecture.

Although it is generally difficult to execute an instantaneous gear shifting properly in the analog domain, some attempts in continuous-mode operation have been demonstrated. For example, a time-continuous adaptive gear shifting for clock recovery applications is described in [81]. The idea there is to gradually reduce the loop gain based on the filtered phase variations. As the loop settles, the phase detector output produces fewer and fewer variations at its output, causing less charge to be stored on a capacitor. This is used to gradually reduce bias in the charge pump, thus reducing the overall loop gain. Unfortunately, since the charge-pump current is controlled dynamically, this could become an additional source of phase noise at the VCO input. As another example, in reference [82] the variable loop bandwidth is switched by changing the charge-pump current together with PLL filter parameters.

### 4.16.1 Autonomous Gear-Shifting Mechanism

Figure 4.58 shows the idea behind the PLL gear-shifting scheme. The asynchronous flip-flop controlled by the *tracking mode control* sequencer signal is a shorthand notation for a latching mechanism of the $\Delta\phi$ phase error adjustment. In practice, it would be implemented as a state machine with a synchronous reset that stores a

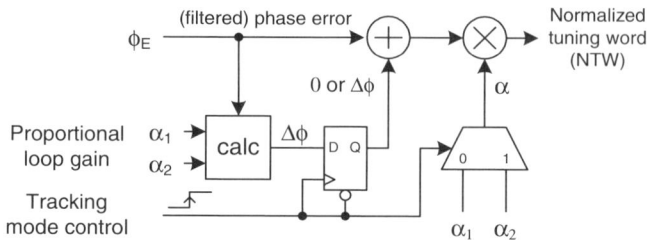

**Figure 4.58** Gear-shifting mechanism. (From [80], © 2005 IEEE.)

new $\Delta\phi$ value into a cleared register upon transition from acquisition (or fast tracking) to tracking bandwidth. During the fast-tracking mode, the PLL loop operates under the high loop bandwidth regime, which is controlled by the normalized loop gain $\alpha_1$. Just before the gear-shift switching instance, the phase error has a value of $\phi_1$ and the normalized tuning word has a value of $NTW_1 = \alpha_1 \phi_1$. At the gear-shift switching instance, the new phase-error value $\phi_2 = \phi_1 + \Delta\phi$ should be adjusted for the new lower tracking-mode loop gain value $\alpha_2$ such that there would be no frequency perturbation (i.e., $NTW_1 = NTW_2$) of the oscillator *before* and *after* the event:

$$\alpha_1 \phi_1 = \alpha_2(\phi_1 + \Delta\phi) \qquad (4.104)$$

The required phase-error adjustment $\Delta\phi$ of Eq. 4.105 is derived from Eq. 4.104:

$$\Delta\phi = \frac{\alpha_1 - \alpha_2}{\alpha_2}\phi_1 = \left(\frac{\alpha_1}{\alpha_2} - 1\right)\phi_1 = \frac{\alpha_1}{\alpha_2}\phi_1 - \phi_1 \qquad (4.105)$$

This $\Delta\phi$ value is then maintained as constant and added to the phase-error samples throughout the subsequent tracking-mode operation. It is very advantageous to restrict the ratio of normalized loop gains $\alpha_1/\alpha_2$ to power-of-2 values such that Eq. 4.105 simply reduces to the left-bit-shift operation of the phase error $\phi_1$ just before the gear-shift instance minus $\phi_1$ itself.

It should be noted that the "effective" center frequency $f_{V,\text{eff}}$ in the tracking mode is now much closer to the desired frequency than was the "raw" oscillator center frequency $f_V$ at the beginning of the acquisition-mode operation:

$$f_{V,\text{eff}} = f_V + \Delta\phi\, \alpha_2 K_{DCO} \qquad (4.106)$$

Practical realization of the gear-shifting mechanism is revealed in Fig. 4.59. For ease of implementation, the loop gain factors $\alpha_1$ and $\alpha_2$ are restricted to negative-power-of-2 integers such that the loop gain multipliers could be reduced merely to right-bit-shift operators. Since the final loop bandwidth needs to be narrower than during fast tracking, the first $\log_2(\alpha_1)$ bit shift is reused as the first stage of a

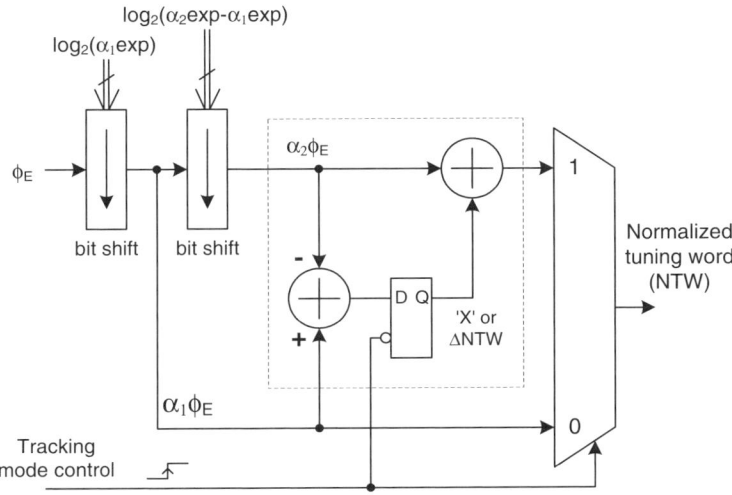

**Figure 4.59** Practical realization of a gear-shifting mechanism. (From [80], © 2005 IEEE.)

two-stage cascade in the final tracking mode. The second stage shifts by $\log_2(\alpha_2 - \alpha_1)$ bits. Consequently, the net effect is a shift by $\log_2(\alpha_2)$ bits in the tracking mode.

Equation 4.104, describing the normalized tuning word (NTW) *during* the gear-shifting instance, is now modified as

$$\alpha_1 \phi_1 = \alpha_2 \phi_1 + \Delta\text{NTW} \tag{4.107}$$

where $\phi_1$ is the phase-error sample value and $\Delta\text{NTW} = \phi_1\alpha_1 - \phi_1\alpha_2$ is an adjustment to the normalized tuning word so as to guarantee no frequency perturbation of the oscillator. The continuously calculated $\Delta\text{NTW}$ value is sampled during the gear-shift cycle when the tracking-mode control signal transitions from low to high and is maintained by the latch throughout the tracking mode.

### 4.16.1.1 Multiple Gear-Shifting Events

For the optimal acquisition performance, the loop bandwidth could be gear-shifted several times. Figure 4.60 shows a phase-error trajectory when a PLL performs settling an initial frequency error while undergoing two gear-shift events. The phase-error trajectory is represented by a thick curve being an average local value and two dotted lines denoting maximum and minimum bounds of the noise envelope. At the gear-shifting event, the last phase-error sample value becomes a starting point for the new trajectory. Obviously, the switching perturbation from the average trajectory could be as large as the noise distribution just before the event. Under the operation regime in which the TDC noise is dominant, the amount of phase-error noise is directly proportional to the loop bandwidth or loop gain factor. Hence each downshift of the loop gain in Fig. 4.60 reduces the phase-error variability. Continual reduction in

**146**   ALL-DIGITAL PHASE-LOCKED LOOP

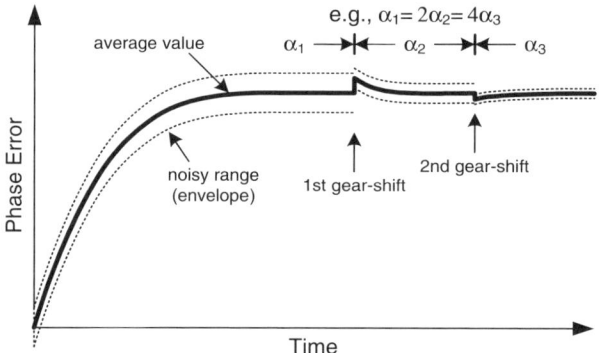

**Figure 4.60** Phase-error trajectory during frequency acquisition while undergoing two gear-shift events. (From [80], © 2005 IEEE.)

the loop bandwidth increases the DCO phase-noise contribution, so at a certain point the phase error signal starts to become more noisy.

Figure 4.59 could naturally be extended to the sequential reduction of three or more loop gain parameters $\alpha$. With each reduction of $\alpha$ as the loop gets closer to the desired frequency, the frequency dynamic range would be reduced, but the frequency resolution would become finer. One such implementation that supports two gear-shifting events is shown in Fig. 4.61. The first fast-tracking mode is invoked by

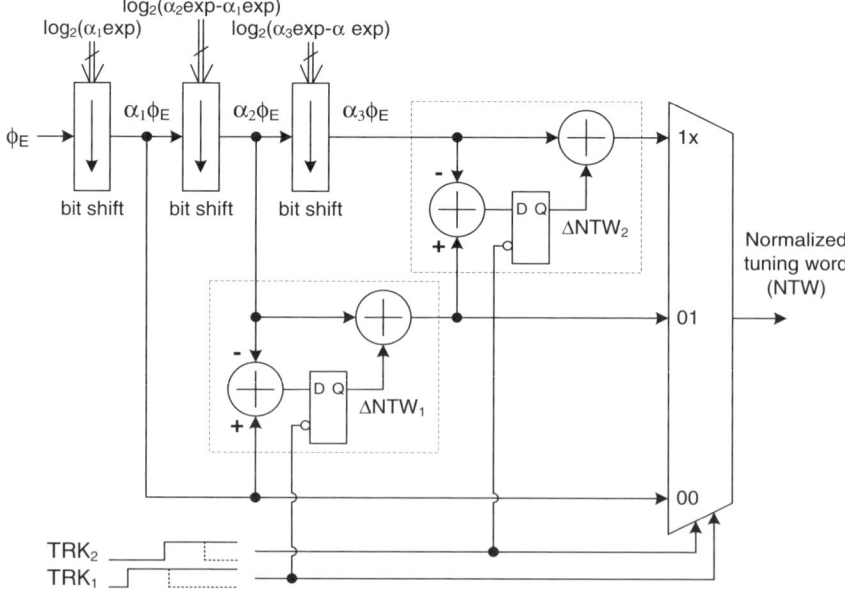

**Figure 4.61** Sequential reduction of PLL bandwidth through double gear shifting. (From [80], © 2005 IEEE.)

the TRK$_1$ control signal and is identical in operation to the scheme of Fig. 4.59. The second gear shifting is triggered by the rising TRK$_2$ control signal, while TRK$_1$ remains high. At that time, the new normalized tuning word adjustment $\Delta$NTW$_2$ is based on NTW during the switching instance:

$$\Delta\text{NTW}_2 = (\Delta\text{NTW}_1 + \alpha_2\phi_2) - \alpha_3\phi_2 \tag{4.108}$$

where $\phi_2$ is the phase-error sample value during the second gear shift of gain $\alpha_3$.

Figure 4.62 demonstrates an example of how the double gear-shifting operation of Fig. 4.61 could be extended arbitrarily to optimize the transition from acquisition to tracking. The quantization error during gear-shifting instances is minimized by constraining the gain reduction to 1 bit or one-half of the gain value. The tracking control signals establish time instances of 1-bit gain reductions and should last long enough to settle quantization errors of the preceding gains. The number of stages is determined by the exponent difference of the desired loop gains between the acquisition and tracking modes.

Extension to the gear-shifting concept is presented in Fig. 4.63. Instantaneous reduction of the PLL bandwidth could also be accomplished by changing the IIR filter pole location using the $\lambda$ factor, as described by Eq. 4.93. An advantage is taken here from the fact that switching $\lambda$ does not affect the baseline of the output, so no extra adjustment as required by the $\alpha$ gear-shift case is needed. It is suggested that the gear-shifting operation be a combination of $\alpha$ and $\lambda$ scaling.

Figure 4.64 shows a compact realization of the gear-shifting mechanism using feedback. It is inspired by Fig. 4.63, in which the hitless amplitude scaling by factor $\lambda$ (now $\alpha$) occurs infrequently at discrete gear-shifting events.

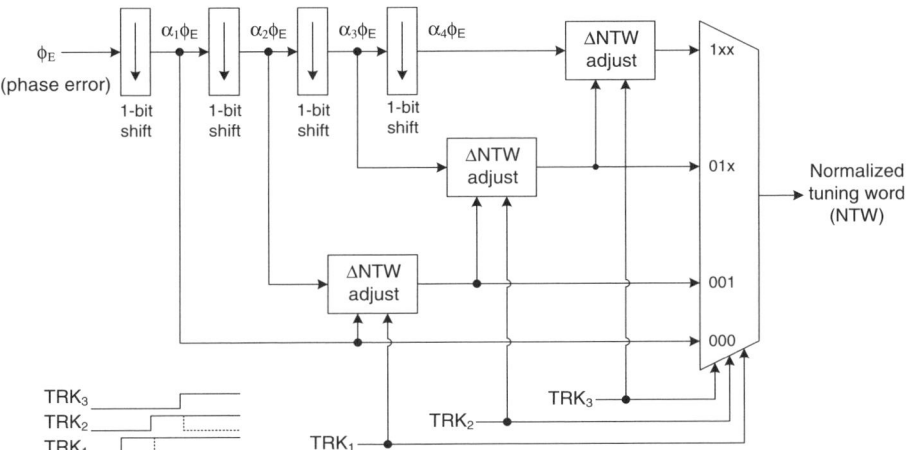

**Figure 4.62** Sequential reduction of PLL bandwidth through graduated 1-bit gear shifting.

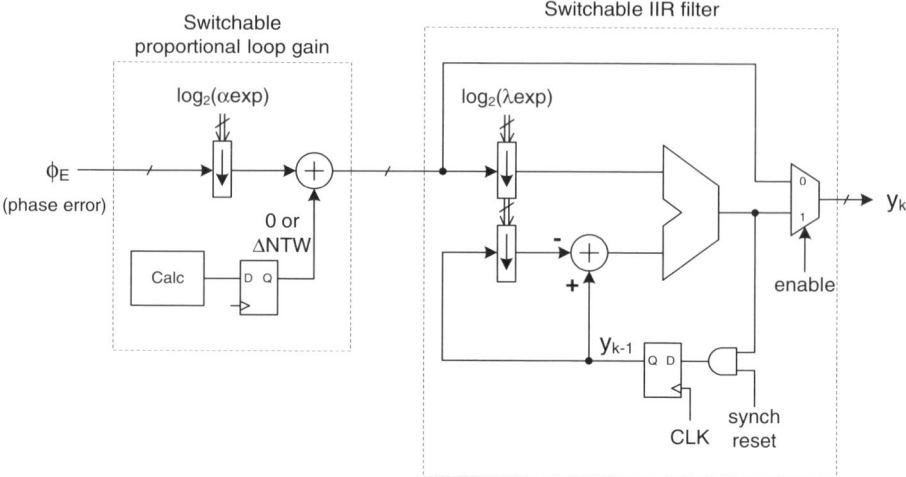

**Figure 4.63** Gear-shifting realization as a switchable proportional loop gain $\alpha$ or/and switchable IIR filter pole factor $\lambda$.

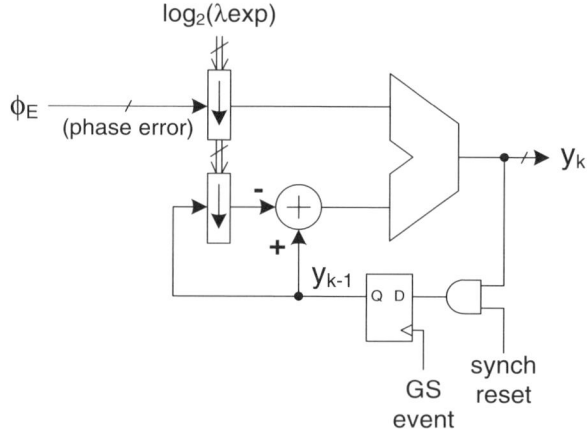

**Figure 4.64** Gear-shifting feedback realization using a switchable IIR filter.

## 4.16.2 Extended Gear-Shifting Scheme with Zero-Phase Restart

The gear-shift scheme described in Section 4.16.1 is self-contained in the DCO gain block. A further improvement could be made to the DCO operational range; however, this requires some modification to either or both sources of the phase-error signal: reference phase $R_R$ and variable phase $R_V$. Since $R_R$ operates at a much lower frequency than $R_V$, it is much easier in practice to perform any phase adjustment to the reference phase accumulator.

Since the gear-switching operation is hitless, there is no perturbation to the unadjusted phase error $\phi[k]$ (from the phase detector). If the phase-error value is at the maximum of the dynamic range at the end of the acquisition, it will remain there during tracking. The frequency range would be proportionately lower for the new $\alpha$, but the new "effective" center frequency is now closer to the desired frequency. The main idea behind the improved scheme is to make the effective center frequency after the gear-shifting switch to be exactly zero by an appropriate adjustment of the dc phase-error correction $\Delta\phi$ and either the reference phase $R_R$ or the DCO phase $R_V$.

The scheme described above could be implemented simply by performing the following steps during the gear-shifting operation:

1. Make $R_R$ equal to $R_V$ (or vice versa, but this would not be as advantageous) in order to bring $\phi_2$ to zero just after the gear-shift instance (still within the same clock cycle). This could be achieved practically by loading the variable accumulator value $R_V$ into the reference register $R_R$ during the gear-shift clock cycle and performing the regular adjustment by the frequency control word (FCW). The phase-error value $\phi_2$ would not be zero but be equal to FCW as normally expected follow-up to zero at the next clock cycle.

2. Modify $\Delta\phi$ of the original method to be expressed as

$$\Delta\phi = \frac{\alpha_1}{\alpha_2}\phi_1 \qquad (4.109)$$

It is very advantageous to restrict the acquisition-to-tracking normalized loop gain ratio $\alpha_1/\alpha_2$ to power-of-2 values such that the Eq. 4.109 simply reduces to the left-bit-shift operation of the phase error $\phi_1$. It should be noted that this improved method could be adapted to work for a higher-order PLL. In this case, the loop filter integrator would additionally have to be reset.

#### 4.16.2.1 Zero-Phase Restart Mechanism

The improved gear shifting with *zero-phase restart* (ZPR) is the basis of the DCO mode switchover operation of Fig. 2.9. During the active mode of operation, the new DCO tuning word is latched by the register with every clock cycle. Upon the DCO operational bank mode switchover, the last stored value of the tuning word is maintained by the register. Consequently, during regular operation, only one interface path can be active at a given time, whereas the previously executed modes maintain their final DCO control states. A ZPR mechanism is used to zero-out the phase detector output to avoid any discontinuities in the oscillator tuning word during mode switchover. A short explanation of the ZPR principle is as follows: At the mode switchover, the tuning word of the last mode corresponds to a certain value of the phase error. This tuning word is now frozen, so the phase-error value that maintains it is no longer needed. However, the new mode is always referenced to the new center frequency established by the last mode. Consequently, it operates on the *excess* phase error rather than absolute. Therefore, the old value of the phase error that corresponds to the frozen tuning word of the last mode would constantly have to be subtracted from the new phase error. A better solution is to use the proposed method of zero-phase restarting. In this way, a hitless progression through the three DCO operational modes is accomplished.

Figure 4.5 is a general block diagram of the phase detection mechanism. The phase detection is based on three signals: a reference phase PHR (also known as $R_R[k]$) calculated by the reference phase accumulator PR, a fractional error correction PHF (also known as $\varepsilon[k]$) calculated by a fractional error correction circuit (combination of TDC and PF), and a sampled variable-phase correction PHV_SMP (also known as $R_V[k]$) calculated by an integer-only variable-phase accumulator PV. The phase error PHE is calculated arithmetically (Eq. 4.16, repeated here for convenience) as

$$\phi_E[k] = R_R[k] - R_V[k] + \varepsilon[k] \tag{4.110}$$

with proper bit alignment to line up integer and fractional portions.

The modified reference phase accumulator circuit of Fig. 4.65 is used in conjunction with the PVT and acquisition bit controllers of Figs. 3.11 and 3.12, respectively, to restart the phase error PHE at the correct value during a mode change. The sequencer mode control signal (CTL_PLL_P or CTL_PLL_A) is monitored constantly for continuity by the oscillator PVT and acquisition control circuits in Figs. 3.11 and 3.12, respectively. During normal operation, the reference phase accumulator performs the following operation:

$$R_R[k] = R_R[k-1] + N \tag{4.111}$$

where $N \equiv FCW$, $R_R[k-1]$ is the previous clock-cycle value, and $R_R[k]$ is the current value. At the mode termination instance, the ZPR one-shot indication

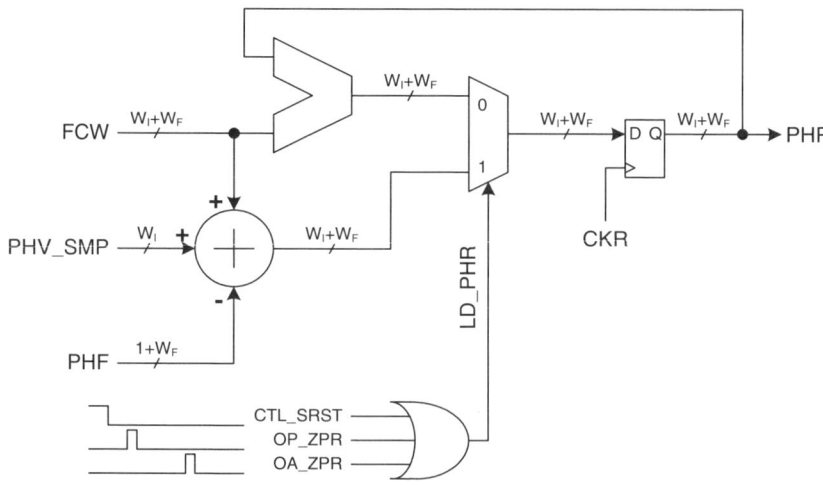

**Figure 4.65** Reference phase accumulator (PR) with zero-phase restart.

(OP_ZPR or OA_ZPR) is asserted in those blocks. This causes a single cycle deviation from the normal accumulation operation of the PR circuit, during which the register gets loaded with the following value:

$$R_R[k] = N + R_V[k-1] - \varepsilon[k-1] \quad (4.112)$$

Plugging Eq. 4.112 into Eq. 4.110 produces

$$\phi_E[k] = [N + R_V[k-1] - \varepsilon[k-1]] - R_V[k] + \varepsilon[k] \quad (4.113)$$
$$= N - ([R_V[k] - R_V[k-1]] - [\varepsilon[k] - \varepsilon[k-1]]) \quad (4.114)$$

Applying expectation operator on both sides yields

$$\mathcal{E}\{\phi_E[k]\} = N - \mathcal{E}\{[R_V[k] - R_V[k-1]] - [\varepsilon[k] - \varepsilon[k-1]]\} \quad (4.115)$$
$$= 0 \quad (4.116)$$

This is because the expected value of the variable-phase change (i.e., $R_V[k] - R_V[k-1]$), adjusted for the change in the fractional error correction, is equal to the division ratio $N$.

As a result of the zero-phase restart, referring again to Fig. 4.5 or Fig. 4.30 for a more comprehensive view, the phase error is forced to start again from a very small value close to zero. Any deviation from zero would be due to noise or nonlinearity of the TDC.

The ZPR mechanism is additionally utilized as a substitute for a synchronous reset (active on CTL_SRST) of the variable-phase accumulator PV. Referring back to Fig. 4.30, the PV digital incrementer operates at the ultrahigh clock rate of 2.4 GHz, and implementing a dedicated asynchronous or synchronous reset is expensive in terms of power dissipation and performance. Instead, advantage is taken from the fact that since the reference and variable phases operate on modulo arithmetic, their power-up absolute values do not matter—it is only their difference, phase error, that is propagated further. Consequently, performing the zero-phase restart at the power-up essentially accomplishes the task of synchronous reset.

**4.16.2.2 Simulation Runs** The zero-phase restart operation is demonstrated in Fig. 4.66. This plot consists of four subplots revealing the key ADPLL signals. The top three plots are subdivided further into three sequential operational modes: PVT, acquisition, and tracking. Time units on the $x$-axis are counts of FREF clock cycles.

1. "PHE": phase error $\phi_E$ in each of the operational modes.
2. "TUNE": oscillator tuning words (OTW). In the PVT and acquisition modes it is an integer-valued signal. In the tracking mode it is a fixed-point signal with 5 fractional bits.

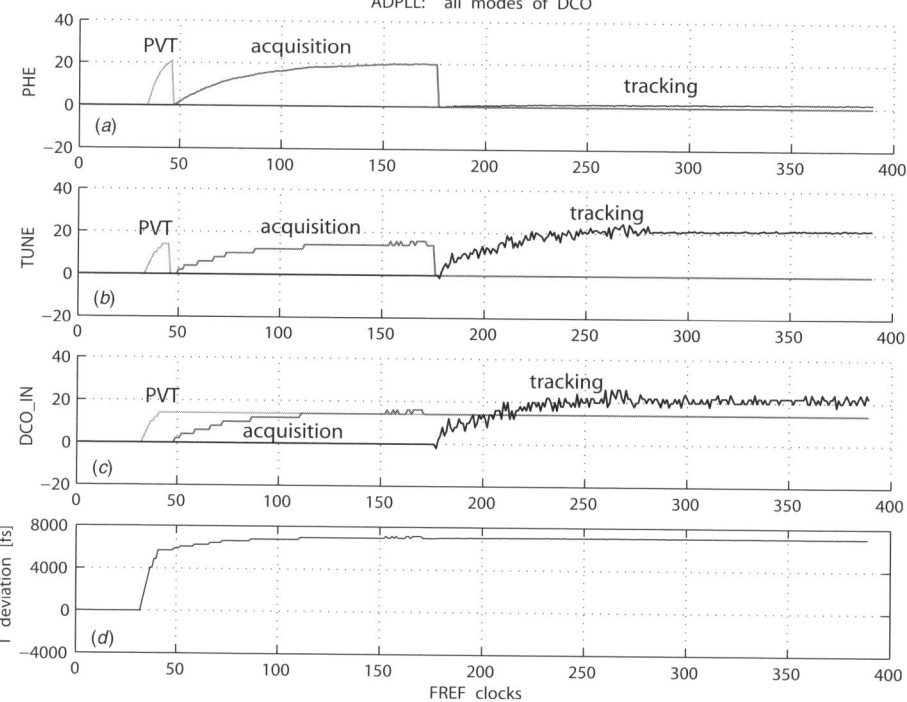

**Figure 4.66** Simulation plots demonstrating zero-phase restart.

3. "DCO_IN": DCO inputs. In the tracking mode, the signal displayed is an FREF-sampled version of the merged integer and fractional parts. Hence, it appears to have more noise than the OTW signal above. (The actual high-rate signal will be shown in a separate plot.)
4. "T deviation": DCO period deviation in femtoseconds that is proportional to the DCO frequency deviation (1 fs = 5.77 kHz from Table 2.1).

For the BLUETOOTH example, initially, the loop operates at channel 0 (2402 MHz). Around time 40, a new value of the frequency command word (FCW) that corresponds to channel 40 (2442 MHz) is entered. The loop approaches the new frequency first in the PVT mode, with a loop gain of $\alpha_P = 1/2^3$ and develops a tuning word of 14. Using entries from Table 2.2, this translates to a frequency deviation of about 32.5 MHz. At around time 48, the loop enters the acquisition mode with $\alpha_A = 1/2^5$ by performing the first zero-phase restart. The phase error of 20 developed (confirming Eq. 4.36: $20 \times 1/2^3 \times 13$ MHz = 32.5 MHz) drops instantly to zero while the PVT varactor bank gets frozen ("DCO–IN") and the acquisition bank varactors start evolving to acquire the remaining distance of 40 MHz − 32.5 MHz = 7.5 MHz. The second ZPR happens at time ∼170 (during the switchover from acquisition to tracking) and the phase error of 20 developed drops instantly to zero. At this time, the loop enters the fast-tracking mode with $\alpha_{TA} = 1/2^5$.

At time $\sim$280, after steady state is reached, the PLL transitions into the normal tracking mode with a reduced loop gain of $\alpha_T = 1/2^8$. The reduced loop bandwidth exhibits itself as a smoother and more stable DCO frequency deviation curve. During the gear-shifting switch, the instantaneous values of the phase error, tuning word, and DCO period are maintained, proving the correctness of implementing the hitless operation.

Figure 4.67 further demonstrates the ADPLL gear-shifting operation and is similar in form to Fig. 3.18 except for not performing transmitting modulation. The top plot shows the fixed-point tracking tuning word, and the bottom plot shows a corresponding instantaneous merged DCO input. Worth noting is the hitless and continuous switchover of the loop bandwidth at time $\sim$280.

Figure 4.68 demonstrates the settling time advantage of the double gear-shift operation of Fig. 4.61 vs. the single gear-shift operation of Fig. 4.59. The plots illustrate tuning word vs. clock cycles. Transition from acquisition to tracking happens at time $\sim$500. In both plots, the first gear shifting appears at time $\sim$700. In the double gear-shifting case, the bandwidth change is half of that in the left-hand plot. The second gear shifting (right-hand plot only) appears at time $\sim$900.

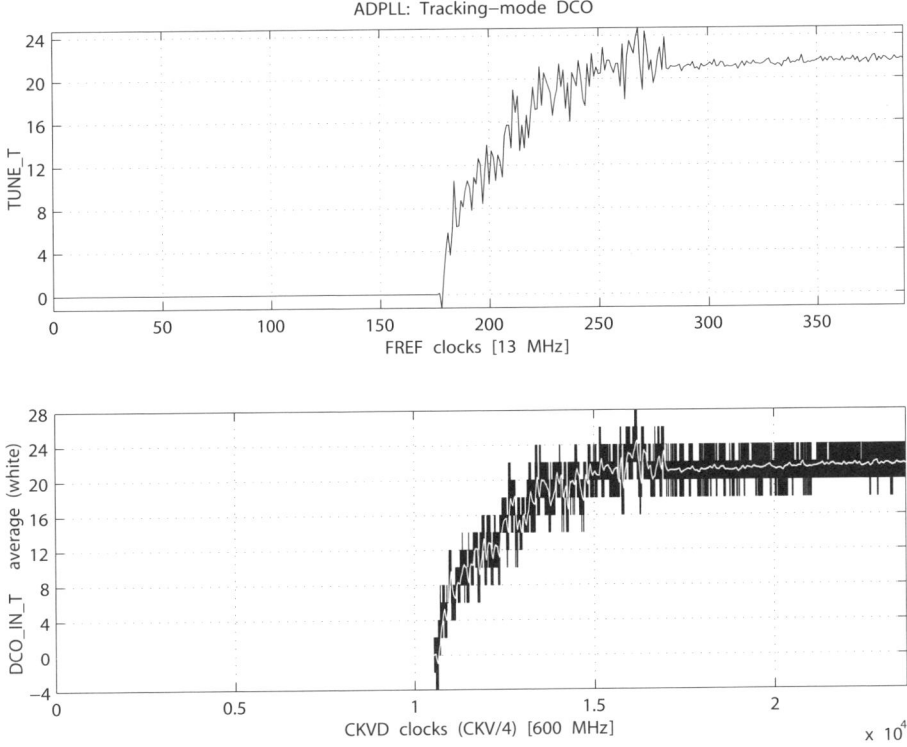

**Figure 4.67** Simulation plots demonstrating the correctness of gear shifting.

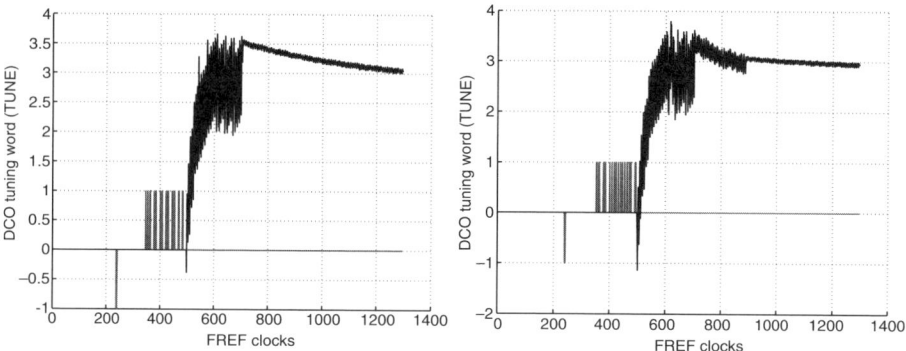

**Figure 4.68** Simulation plots demonstrating single (left) and double (right) gear shifting.

Performance improvement can be seen in the double gear shifting by comparing the slopes of the tuning words at time ~1000. The slope of the tuning word curve is relatively flat in the right-hand plot, whereas the slope in the left-hand plot is still pronounced at past 1200 clock cycles.

## 4.17 EDGE SKIPPING DITHERING SCHEME (OPTIONAL)

Figure 4.69 shows the whole-clock edge skipping method. The basic idea behind it is twofold: to decouple the tuning word calculation operation from applying it to the

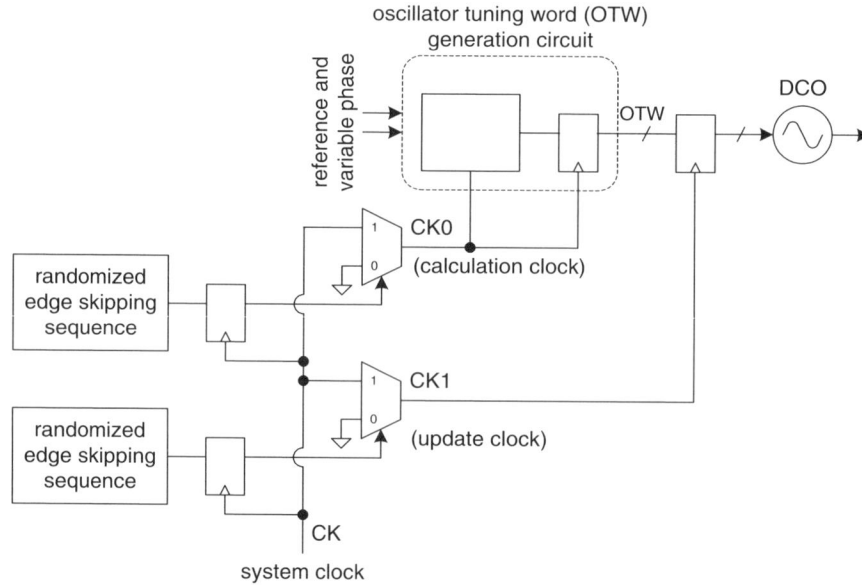

**Figure 4.69** Time dithering through whole-clock edge skipping.

oscillator, and to perform a full-cycle clock edge-skipping procedure to avoid dealing with a high-frequency oversampling clock.

Preferably, both "random" stream generators are coupled such that the update clock edges pass through only when the calculation clock edges are blocked. This will ensure that the DCO frequency is updated far from the digital logic activities.

If the silicon chip die contains a microprocessor and a DSP on the same substrate, which is often the case with modern RF transceivers, it is advantageous to clock it *synchronously* to the time-dithered update clock CKU. Two significant benefits could thus be obtained: First, randomly modulating the clock period prevents substrate noise with a strong periodical correlation to couple from the digital baseband to the RF section. Second, if the processor clock exhibits enough delay from the synthesizer update clock, the phase detection and tuning word adjustment operations occur during the "quiet" periods of the DSP.

## 4.18 SUMMARY

In this chapter we presented an all-digital phase-locked-loop (ADPLL)–based frequency synthesizer that builds on the normalized DCO described in Chapter 3. We revealed a novel phase correction mechanism that closes the loop around nDCO such that the oscillator's phase and frequency drift is corrected by means of the frequency reference. We also described the frequency reference retiming performed in such a way as to stochastically avoid metastability. A block diagram of the ADPLL described was presented in Fig. 4.30. It contains only two nonideal parameters which are not known exactly and where distortion and noise could be introduced: the DCO gain ($K_{DCO}$) and the time-to-digital converter resolution ($\Delta t_{res}$). The rest of the system is *exact* and is *completely* immune from any time- or amplitude-domain uncertainties and perturbations. For these reasons, the architecture presented appears to be highly competitive or even to exceed the performance of conventional RF synthesizers while consuming less power and area as well as enabling integration with the digital baseband.

## CHAPTER 5

# APPLICATION: ADPLL-BASED TRANSMITTER

The synthesizer described in Chapter 4 has a convenient method of digitally controlling the oscillating frequency through the *frequency command word* (FCW) that feeds the reference phase accumulator. The FCW fixed-point word could easily be augmented with dynamically changing modulating data to accomplish a frequency or phase (generally called *angle* in the communication theory) modulation of the synthesizer RF output. This chapter adds a necessary mechanism to accomplish this objective.

A two-point *direct* modulation scheme was proposed in reference [19] which performs phase compensation of the PLL by digitally integrating the modulating data bits transmitted and using the integrator output to shift the phase of the reference clock signal, while the Gaussian filtered data modulates the VCO frequency *directly*. However, this approach is quite analog in nature and requires precise component matching of not only the VCO but also the phase shifter. Another feedforward compensation method, which also requires precise knowledge of the VCO transfer function and other analog circuits, was proposed in reference [20]. It uses DSP to calculate the inverse of the VCO transfer function, which is obtained through lab measurements, followed by a high-precision DAC to pretune the VCO control voltage to the desired excursion. Even though the large frequency shifts could be performed quickly within the VCO accuracy, the problem of narrow PLL bandwidth remains for residual frequency offsets. Another major difficulty with this architecture is the VCO gain variation due to process and environmental changes.

---

*All-Digital Frequency Synthesizer in Deep-Submicron CMOS*, by Robert Bogdan Staszewski and Poras T. Balsara
Copyright © 2006 John Wiley & Sons, Inc.

In contrast, in this chapter we describe a solution that is digital in nature, consumes little hardware overhead, and requires only one component matching (i.e., DCO), which is done *just in time* in a digital manner with very fine resolution. Novel methods of both the direct frequency modulation and the calibration of the RF oscillator transfer function that avoid the problems above are described. These methods could be employed without regard to any specific digitally intensive synthesizer architecture.

We also introduce two other building blocks: a pulse-shaping filter (Section 5.3) and a power amplifier (Section 5.4) with possible digital amplitude modulation (Section 5.5). Although these two blocks are not the main focus of this book, they contain useful novel ideas. Furthermore, adding these blocks to the digital RF synthesizer, which already has direct frequency-modulation capability, would now allow us to complete the entire transmitter part of an RF transceiver for wireless digital communications. This would demonstrate use of the RF synthesizer described in this book.

## 5.1 DIRECT FREQUENCY MODULATION OF A DCO

The oscillating frequency could be controlled dynamically by adding the appropriately scaled modulating data $y[k] = \text{FCW}_{\text{data}}[k]$ directly to the quasistatic frequency command word $\text{FCW}_{\text{channel}}$ at the reference accumulator input that is normally used for channel selection:

$$\text{FCW}[k] = \text{FCW}_{\text{channel}}[k] + \text{FCW}_{\text{data}}[k] \qquad (5.1)$$

where $k$ is a discrete-time index set by FREF.

This idea is depicted in Fig. 1.7, which shows the front-end accumulator stage of a direct digital frequency synthesizer (DDFS), which is identical to the reference phase accumulator of the ADPLL architecture. Introducing the modulating data redefines the FCW, originally defined in Section 4.1, as the expected *instantaneous* frequency-division ratio of the desired synthesizer output and the reference frequency:

$$\text{FCW}[k] = \frac{\mathcal{E}(f_V[k])}{f_R} \qquad (5.2)$$

Generally, the direct frequency or phase transmit modulation of a PLL of an RF synthesizer is a challenging task. To attenuate the reference spurs and phase noise, the PLL bandwidth is usually kept low. This effectively prevents use of closed-loop modulation if the modulating data rate is not much smaller than the loop bandwidth [29,30]. However, the direct closed-loop modulation of the DCO frequency still would be considered a more cost-effective solution than the alternative of an image reject quadrature modulator.

## 5.1.1 Discrete-Time Frequency Modulation

The transmitter presented is capable of utilizing any reference clock for its operation as long as it can generate a sufficient number (usually $\geq 6$ for BLUETOOTH) of samples per symbol period in order to satisfy the Nyquist criterion and sufficiently attenuate the discrete-time signal spectral replicas. The only significant filter circuit in this all-discrete-time design is the RF oscillator, which provides 6 dB/octave attenuation due to its frequency/phase conversion. Since any wireless terminal will typically have some available reference clock, such as a crystal oscillator clock used as a frequency reference in the range 8 to 40 MHz for host cellular systems, an extra cost in implementing another oscillator could thus be avoided.

Spectral replicas of the discrete-time modulating signal appear at the DCO input every sampling rate frequency $f_R$, as shown in Fig. 5.1. They are attenuated through multiplication of the $\text{sinc}^2$ function due to the zero-order hold of the DCO input. The frequency spectrum $S_f(\omega)$ replicas are further attenuated by 6 dB/octave through the $1/s$ operation of the oscillator to finally appear at the RF output phase spectrum $S_\phi(\omega)$. If the sampling rate is high enough, the replicas will be sufficiently attenuated, making the RF signal undistinguishable from that created by conventional transmitters with continuous-time filtering at baseband.

## 5.1.2 Hybrid of Predictive/Closed PLL Operation

The PLL operation could be enhanced dramatically by taking advantage of the predictive capabilities of the all-digital PLL loop. The idea is as follows: The DCO oscillator does not necessarily have to follow the modulating FCW command with the normal PLL response. In this fully digital implementation, where the DCO control and resulting phase-error measurement are in numerical format, it is easy to predict the current $K_{\text{DCO}}$ gain of the oscillator simply by observing the past phase-error responses to the DCO or nDCO corrections. With a good estimate of the $K_{\text{DCO}}$ gain, the normal DCO control could be augmented with an "open-loop" instantaneous frequency jump estimate of the new FCW command. The resulting phase error should be very small and subject to the normal closed PLL correction transients.

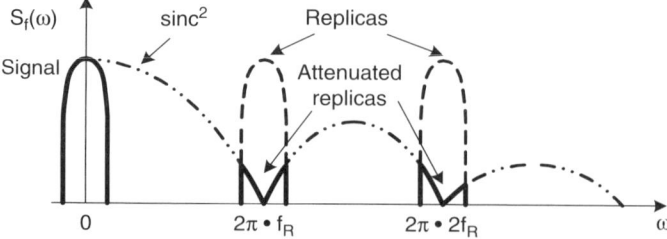

**Figure 5.1** Spectral replicas of the modulating signal and their filtering through zero-order hold. Additional 6 dB/octave filtering from $S_f(\omega)$ to $S_\phi(\omega)$ is provided inherently by the DCO. (From [83], © 2005 IEEE.)

Since the time response of this type I PLL loop is very fast (less than a few microseconds), the prediction feature is less important for channel hopping, where the time allowed is much greater. It is, however, essential to realize the direct frequency synthesizer modulation in the *Gaussian frequency shift keying* (GFSK) modulation scheme of BLUETOOTH or GSM[1] as well as the chip phase modulation of the 802.11b or Wideband-CDMA.

### 5.1.2.1 Direct Oscillator Modulation with PLL Compensation

The main idea is to modulate the DCO frequency directly in a feedforward manner such that it effectively removes the loop dynamics from the modulating transmitter path. However, the rest of the loop, including all error sources, operates under the normal closed-loop regime. Figure 5.2 shows the idea schematically.

The modulating data $y[k]$ at the upper feed immediately affects the oscillating frequency with an open-loop *frequency* impulse response function:

$$h_{\text{dir}}^f[0] = \frac{1}{\alpha} \alpha \frac{f_R}{\widehat{K}_{\text{DCO}}} K_{\text{DCO}} = f_R \frac{K_{\text{DCO}}}{\widehat{K}_{\text{DCO}}} = f_R r \quad (5.3)$$

expressed in Hz/LSB units. As defined in Section 3.3, the dimensionless ratio $r = K_{\text{DCO}}/\widehat{K}_{\text{DCO}}$ is a measure of the DCO gain estimation accuracy, and it is perfect if $r = 1$.

Unfortunately, the PLL will try to correct this perceived frequency perturbation integrated over the update period $1/f_R$. Its open-loop *phase* impulse response function (from upper $y[k]$ feed to $R_V[k]$ integrator output) on the next cycle is

$$h_{\text{dir}}[1] = f_R \frac{K_{\text{DCO}}}{\widehat{K}_{\text{DCO}}} \frac{1}{f_R} = \frac{K_{\text{DCO}}}{\widehat{K}_{\text{DCO}}} = r \quad (5.4)$$

expressed in cycles/LSB. If the DCO gain estimate $\widehat{K}_{\text{DCO}}$ is accurate, $h_{\text{dir}}[1] = 1$.

The transfer characteristic of a closed-loop PLL with only the upper feed is high-pass. The low-frequency components of the $y[k]$ data will be integrated in the variable accumulator, thus affecting the oscillator frequency baseline. It is necessary, therefore, to add a phase-compensating circuit $R_Y[k]$, as shown at the lower $y[k]$ feed; that would completely subtract the phase contribution of the upper $y[k]$ direct modulation feed into the PLL, provided that the DCO gain could be estimated correctly. The phase-compensating open-loop impulse response function (from lower $y[k]$ feed to the $R_Y[k]$ accumulator output) on the next cycle is

$$h_{\text{comp}}[1] = 1 \quad (5.5)$$

expressed in cycles/LSB. Consequently, if $\widehat{K}_{\text{DCO}} = K_{\text{DCO}}$ in Eq. 5.4 (i.e., $r = 1$), the loop will be compensated perfectly ($h_{\text{dir}}[1] = h_{\text{comp}}[1]$) and the feedforward

---

[1]GSM actually uses *Gaussian minimum shift keying* (GMSK), which is a special case of GFSK.

# 160 APPLICATION: ADPLL-BASED TRANSMITTER

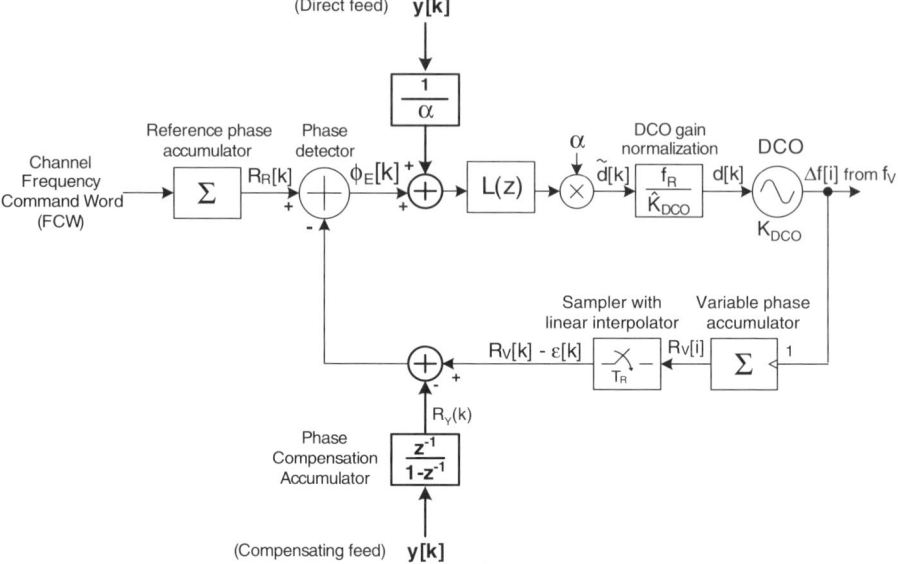

**Figure 5.2** Principle of direct oscillator modulation with a straightforward PLL compensation scheme.

modulation will be exact. In case of nonexact matching, the residual error $r - 1$ would have to undergo a normal high-pass loop response with the following $s$-domain transfer function, which has a single pole at $(\alpha/2\pi)f_R$ (Hz):

$$H(s) = \frac{r - 1}{1 + \alpha(f_R/s)} \quad (5.6)$$

The immediate and direct DCO frequency control $h_{\text{dir}}^f[k]$ (Eq. 5.3), made possible by accurate prediction of the DCO transfer function, is combined with the phase compensation $h_{\text{comp}}[k]$ (Eq. 5.5) of the PLL response $h_{\text{dir}}[k]$ (Eq. 5.4). The two factors constitute the hybrid of predictive/close PLL modulation method.

The analysis above is rather qualitative in nature and seeks cycle-by-cycle cancellation of the direct and compensation terms. A more thorough analysis of the loop behavior in the face of the DCO gain estimation error (i.e., when $r \neq 1$) is presented below.

The $z$-domain transfer function from both feeds of the modulating data $y[k]$ to the frequency deviation $\Delta f$ at the PLL output is

$$H(z) = \left[1 + \alpha L(z)\frac{1}{z - 1}\right]\frac{rf_R}{1 + r\alpha L(z)[1/(z - 1)]} \quad (5.7)$$

## 5.1 DIRECT FREQUENCY MODULATION OF A DCO

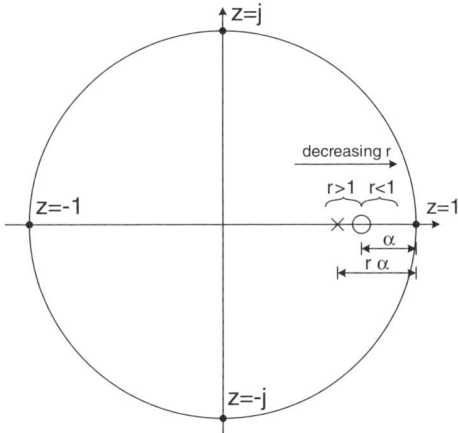

**Figure 5.3** Complex plane location of the $H(z)$ zero and pole movement with different values of the DCO gain estimate accuracy $r$. The loop filter $L(z)$ transfer function here is unity all-pass.

This could be simplified to

$$H(z) = rf_R \frac{z - [1 - \alpha L(z)]}{z - [1 - r\alpha L(z)]} \quad (5.8)$$

The PLL has a pole at $1 - r\alpha L(z)$ and a zero at $1 - \alpha L(z)$, as illustrated in Fig. 5.3. The dc gain is always unity, which could readily be seen by inspection:

$$H(z)|_{z=1} = f_R$$

Assuming that $L(z) = 1$, to simplify the analysis, the high-frequency gain is

$$H(z)|_{z=-1} = rf_R \frac{-2 + \alpha}{-2 + \alpha r}$$

which could be further simplified when a realistic approximation of $\alpha \ll 1$ is assumed.

$$H(z)|_{z=-1} \approx rf_R \quad (5.9)$$

Figure 5.4 shows the direct modulation transfer function $H(f)$ for various cases of the $K_{\text{DCO}}$ estimate of $r$: correct, $r = 1$; $K_{\text{DCO}}$ underestimated, $r > 1$; and $K_{\text{DCO}}$ overestimated, $r < 1$. For an incorrect $K_{\text{DCO}}$ estimate, the transfer function is either somewhat high-pass or low-pass. This is governed by the single pole and single zero locations at $r\alpha f_R/2\pi$ and $\alpha f_R/2\pi$ of linear frequency in hertz, respectively. The zero location happens to be the same as the PLL phase transfer function

## 162 APPLICATION: ADPLL-BASED TRANSMITTER

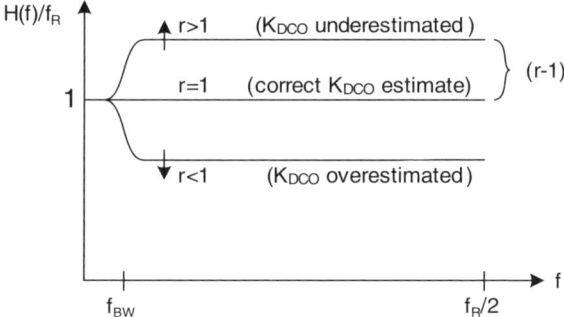

**Figure 5.4** Direct modulation transfer function $H(f)$, where $f = (z-1)f_R/2\pi$ is in hertz, with different values of the DCO gain estimate accuracy $r$.

loop bandwidth $f_{BW}$ of the reference noise or of the modulating data without the feedforward $y[k]$ entry. The feedforward $y[k]$ path could simply be viewed as placing a compensating zero in the pole's vicinity (see also Fig. 5.3). Only when the DCO gain estimate is correct do the pole and zero coincide, and the direct frequency modulation of the PLL exhibits truly wideband all-pass transfer characteristics up to the half of the sampling frequency $f_R$. With the reference frequency of 13 MHz commonly used in GSM cell phones, the bandwidth of this modulation method appears to be sufficient for any low- to mid-rate communication system.

### 5.1.2.2 Improved All-Digital Architecture

A technically more elegant variant of Fig. 5.2, depicted in Fig. 5.5, would merge the phase compensation accumulator $R_Y[k]$ with the reference phase accumulator $R_R[k]$. Now, the frequency command word is the sum of the channel and data signals, which is more intuitively appealing.

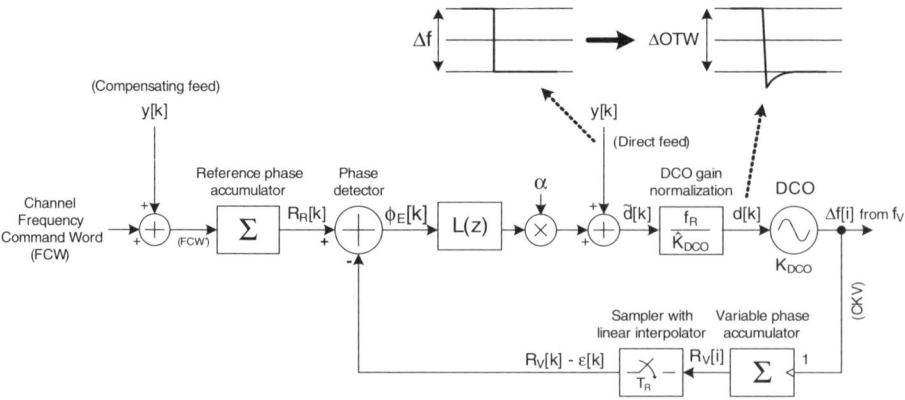

**Figure 5.5** Direct oscillator modulation architecture.

## 5.1 DIRECT FREQUENCY MODULATION OF A DCO

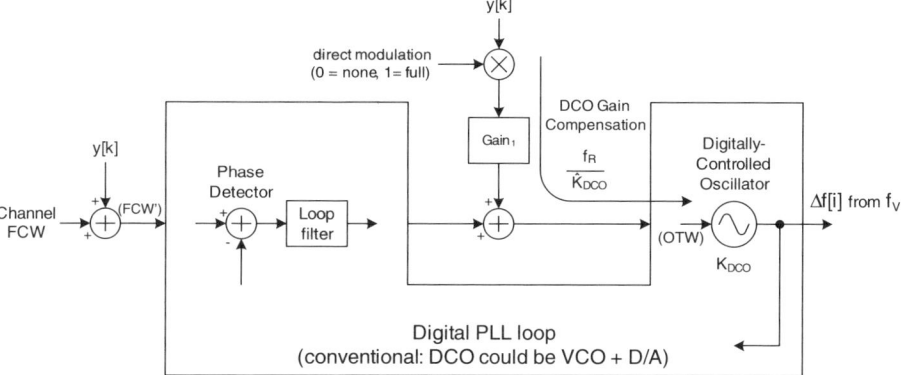

**Figure 5.6** Oscillator modulation scheme with a PLL compensation within a general digital PLL architecture.

### 5.1.2.3 Generalized Architecture
This direct oscillator modulation with the PLL compensating scheme works best in a digital implementation since almost perfect compensation could be achieved. It is also possible to make this idea work in a conventional analog-intensive PLL structure, as demonstrated in reference [19], although the benefits of this method are limited. The idea presented works equally well with higher-order digitally intensive PLLs.

Figure 5.6 shows how the direct oscillator modulation with PLL compensation scheme of Fig. 5.5 could fit into a general digital PLL structure. A conventional digital PLL might consist of a VCO, frequency prescaler and divider, phase detector, loop filter, and a digital-to-analog converter, which makes it possible to control the oscillating frequency through a digital word. The modulating data $y[k]$ are added dynamically to the channel frequency information to affect the frequency or phase of the oscillator output $f_{RF} = f_V$. This could be accomplished, for example, by controlling the frequency-division ratio of the fractional-$N$ PLL loop. The direct modulation structure is inserted somewhere between the loop filter and the oscillator. The gain of the direct modulating path from $y[k]$ to the oscillator input should be $f_R/\widehat{K}_{DCO}$ if $y[k]$ is expressed as the unitless fractional division ratio.

### 5.1.3 Effect of FREF/CKR Clock Misalignment

In the analysis above it is assumed that the DCO tuning word application instances coincide with the variable-phase sampling operation. In other words, $\Delta f[k]$ frequency deviation will be integrated over the $T_R = 1/f_R$ reference period. At the end of the sampling period, the integrated excess timing deviation of $\Delta T \text{DEV}[k] = \Delta f[k]T_R$ will be read out and a new $\Delta f[k+1]$ frequency deviation will be applied.

In practice, however, there is natural misalignment between the two operations. The variable-phase sampling is performed on the FREF edge, whereas the tuning word application occurs on the FREF retimed (CKR) edge. As shown in

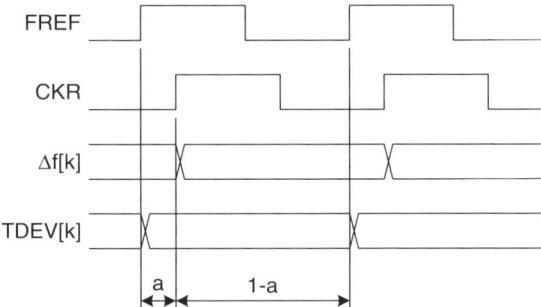

**Figure 5.7** Misalignment of the DCO tuning word application and the variable-phase sampling instances.

Fig. 4.24, CKR always transitions later than FREF. Consequently, as shown in Fig. 5.7, the integrated excess phase will contain some contribution from the preceding interval. This could be modeled as a weighted sum of the current and previous data samples:

$$\Delta T\text{DEV}[k] = (1-a)\Delta f[k] + a\Delta f[k-1] \tag{5.10}$$

where $a$ is the relative amount of overlap.

For the compensation to work correctly, a similarly weighted contribution to the reference phase accumulator must be used. Hence,

$$y'[k] = (1-a)y[k] + ay[k-1] \tag{5.11}$$

where $y'[k]$ is the adjusted $k$th sample of the compensation-feed data.

## 5.2 JUST-IN-TIME DCO GAIN CALCULATION

The DCO gain estimation $\widehat{K}_{\text{DCO}}$, as defined in Section 3.2, could be calculated conveniently and *just in time* at the beginning of every packet in a wireless transmitter [58]. If this gain is estimated correctly, the frequency synthesizer is able to shift frequency instantaneously. As alluded to in Section 4.15, the DCO gain could, alternatively, be estimated as the ratio of the *forced* oscillating frequency deviation $\Delta f_V$ to the steady-state change *observed* in the oscillator tuning word $\Delta$OTW:

$$\widehat{K}_{\text{DCO}}(f_V) = \frac{\Delta f_V}{\Delta \text{OTW}} \tag{5.12}$$

$\widehat{K}_{\text{DCO}}(f_V)$ is actually used in the denominator of the DCO gain normalization multiplier:

$$\frac{f_R}{\widehat{K}_{\text{DCO}}(f_V)} = \frac{f_R}{\Delta f_V}\Delta\text{OTW} \tag{5.13}$$

This is quite beneficial since the unknown OTW is in the numerator, and the inverse of the forced $\Delta f_V$ is known and could conveniently be precalculated. In this way, the use of dividers is avoided.

Referring to Fig. 6.12, at the end of the fast-tracking and the beginning of the regular-tracking PLL operation, there is a sudden frequency jump marking the beginning of the proper transmitting modulation mode. This $\Delta f_{\max}$ frequency jump corresponds to the maximum negative frequency deviation for a data bit value of 0 (which corresponds to the "−1.0" symbol) and equals

$$\Delta f_{\max} = \frac{m}{2} R \qquad (5.14)$$

where $m$ is the GFSK modulation index and $R$ is the data rate. (For BLUETOOTH, $m = 0.32$ and $R = 1$ Mb/s, resulting in $\Delta f_{\max} = 160$ kHz; for GSM, $m = 0.5$ and $R = 270.833$ kb/s, resulting in $\Delta f_{\max} = 67.708$ kHz.) Since the frequency jump is known precisely as commanded by the modulating data part of the frequency command word (FCW_DT), one needs to observe the tuning control word in the steady state in order to determine the DCO gain:

$$\widehat{K}_{\mathrm{DCO}}(f) = \frac{\Delta f_{\max}}{\Delta \mathrm{OTW}_{\max}} \qquad (5.15)$$

If the $K_{\mathrm{DCO}}$ gain is estimated correctly to start with, the precise frequency shift will be accomplished in one step, as shown in Fig. 5.8. However, if the $K_{\mathrm{DCO}}$ is not estimated accurately, the first frequency jump step will be off target by

$$\frac{K_{\mathrm{DCO}}}{\widehat{K}_{\mathrm{DCO}}} - 1 = r - 1$$

and one would require a number of clock cycles to correct the estimation error through the normal PLL dynamics, preferably in the fast-tracking mode, with the

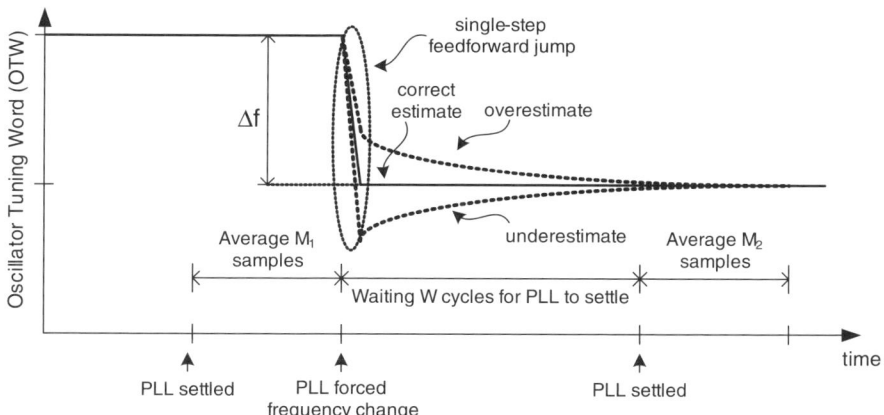

**Figure 5.8** DCO gain estimate by measuring the tuning word change in response to a fixed frequency jump. (From [58], © 2003 IEEE.)

**166**  APPLICATION: ADPLL-BASED TRANSMITTER

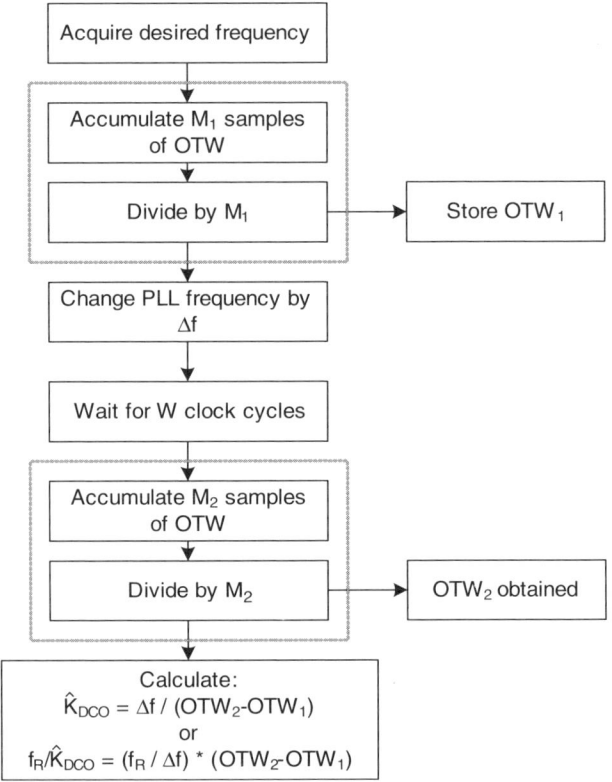

**Figure 5.9** DCO gain estimate flowchart. (From [58], © 2003 IEEE.)

transfer function described by Eq. 5.6. The $K_{DCO}$ gain could then simply be calculated as the ratio of $\Delta f_{max}$ to the *oscillator tuning word* difference. To lower the measurement variance, it might be required to average out the tuning inputs before and after the transition, as Fig. 5.8 reveals.

Figure 5.9 shows a DCO gain calculation flowchart. After the desired frequency is acquired, $M_1$ samples of oscillator tuning word OTW are averaged and the result is stored as $OTW_1$. Thereafter, a suitable frequency change $\Delta f$ is imposed and the system waits $W$ clock cycles for the PLL to settle. $M_2$ samples of OTW are then averaged and the result is stored as $OTW_2$. Finally, the DCO gain estimate $\widehat{K}_{DCO}$ or the normalizing gain $f_R/\widehat{K}_{DCO}$ is computed.

Figure 5.10 shows a hardware realization of the DCO normalizing gain $f_R/\widehat{K}_{DCO}$ estimation. All the memory elements (registers) are synchronously reset at the beginning of the operation by asserting the SRST control signal. At the appropriate times, $M_1$ and $M_2$ samples of OTW are summed and stored by accumulators 1 and 2, respectively. It is very convenient to limit $M_1$ and $M_2$ to power-of-2 integers, since the division operation now simplifies to a trivial right-bit-shift. The difference of averaged oscillator tuning words $\Delta OTW$ is multiplied by a constant $f_R/\Delta f$ to arrive at the DCO normalizing gain estimate.

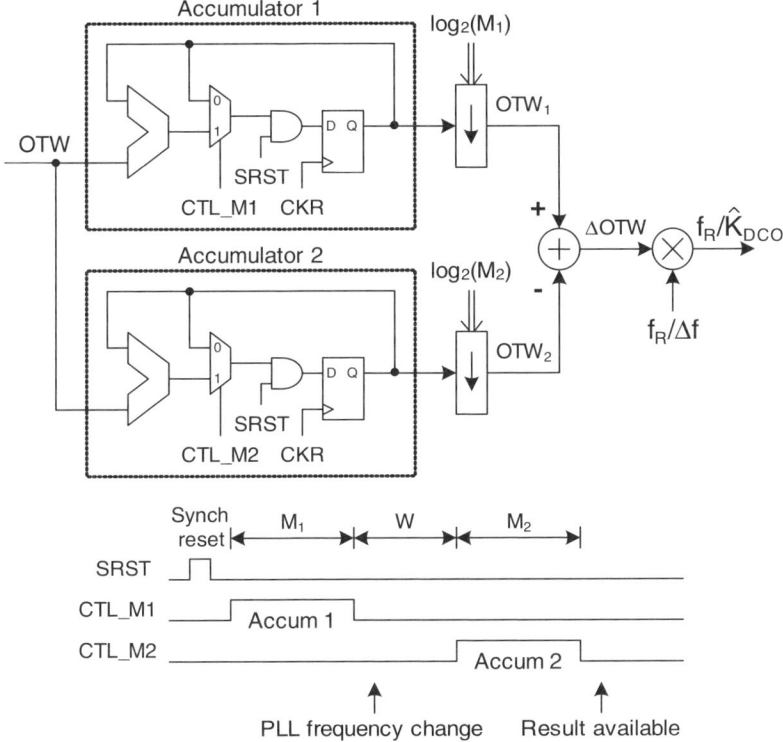

**Figure 5.10** Hardware realization of DCO normalizing gain estimation.

It should be noted that a frequency jump equal to the full modulation range is beneficial for two reasons: First, a little hardware overhead is required on top of the existing transmitter modulator circuitry to execute the frequency jump chosen. Second, measuring the local gain value around the expected operational range is bound to provide the most accurate estimate. To improve the estimate further, a larger frequency step of $2\Delta f_{max}$ covering the entire data modulation range could be performed. It should be noted that the algorithm could also be implemented in a combination of hardware and software using the existing DSP or microcontroller.

## 5.3 GFSK PULSE SHAPING OF TRANSMITTER DATA

As mentioned in Section 1.2.3, it is necessary to perform a pulse-shaping operation on the symbols transmitted to limit the bandwidth occupied by the modulated RF spectrum before they are emitted into the wireless channel. Figure 1.18 revealed the transmitter architecture chosen for the BLUETOOTH transmitter implemented to perform GFSK modulation. This section deals with the first two blocks of Fig. 1.18: the bit coder and the transmitting pulse-shaping filter $h(t)$.

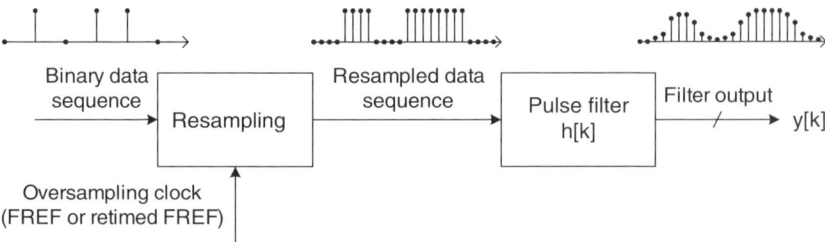

**Figure 5.11** Operation principle of a digital transmitter filter.

The idea of a pulse-shaping operation performed entirely in the digital domain is illustrated in Fig. 5.11 and is described mathematically in Appendix B. The binary input data are being oversampled by a commonly used 13 MHz frequency reference (FREF) clock generated by a crystal oscillator, which is an integer multiple of the 1 Mbps data (or symbol) rate. The oversampling clock is also called a *baseband clock*. In the GFSK system there is a one-to-one correspondence between data bits {0, 1} and symbols {−1, +1}, with $0 \to -1$ and $1 \to +1$. For this reason, the coding conversion is being performed implicitly.

Impulse response $h[k]$ of the Gaussian filter in the discrete-time domain is expressed as

$$h[k] = \frac{\sqrt{2\pi}}{\sqrt{\ln(2)}} \frac{BT_s}{T_s} \exp\left[-\left(\frac{\sqrt{2\pi}}{\sqrt{\ln(2)}} BT_s \frac{k}{\text{OSR}}\right)^2\right] \quad (5.16)$$

where $B$ is the 3-dB bandwidth, $T_s$ is the symbol period, and $\text{OSR} = f_R/(1/T_s)$ is the symbol oversampling ratio by the reference clock. For BLUETOOTH, $BT_s = 0.5$ and $T_s = 1\ \mu s$. For GSM, $BT_s = 0.3$ and $T_s = 3.692\ \mu s$.

Figure 5.12 plots the output of a transmitter filter with a symbol oversampling ratio of 8 for the data sequence "101110" with the initial state corresponding to bit "0".

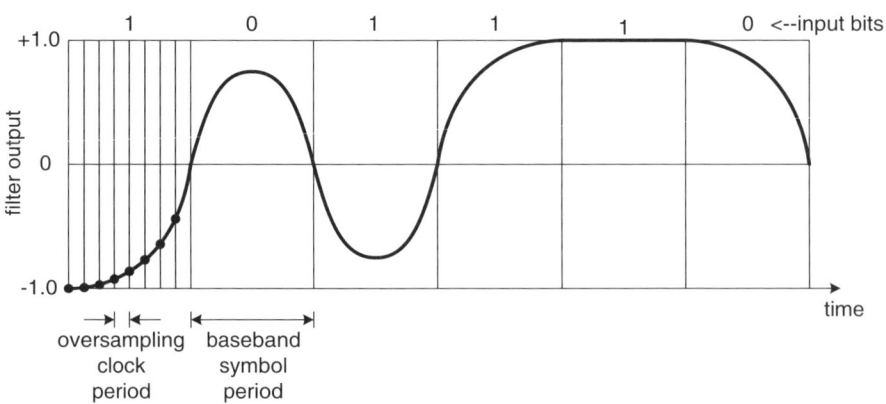

**Figure 5.12** Output waveform of a transmitter filter.

## 5.3 GFSK PULSE SHAPING OF TRANSMITTER DATA

The $+1$ and $-1$ filter output levels correspond to the peak frequency deviation of 160 kHz at the transmitter RF output according to the formula

$$\Delta f_{\text{pk}} = \frac{m}{2}\frac{1}{T_s} \qquad (5.17)$$

where $m$ is the modulation index and $T_s$ is the symbol period. For BLUETOOTH, $m = 0.32$ (nominal), $T_s = 1$ μs, and $\Delta f_{\text{pk}} = 160$ kHz; for GSM, $m = 0.5$, $T_s = 3.6923$ μs, and $\Delta f_{pk} = 67.708$ kHz.

It is quite inefficient to perform the pulse-shaping operation literally as shown in Fig. 5.11, as if it were simple FIR filtering. Namely, $h[k]$ has a number of properties that could greatly reduce the implementational complexity. First, the input data are not a fixed-point number but a 1-bit signal. Second, the input data are highly correlated since all the bits within symbol duration are the same, due to oversampling. Third, the coefficients are fixed. The observations above allow us to operate the actual $h[k]$ filter as a 3-bit state look-up table that selects the *cumulative coefficients* $C[k]$ (or its derivations):

$$C[k] = \sum_{l=0}^{k-1} h[l] \qquad (5.18)$$

with $C[0] = 0$. $C[k]$ is precalculated and stored in a look-up table with $k = 0 \cdots N - 1$ as the index. The power-of-2 integer $N = 64$ is the number of trajectory samples stored.

The state is based on the current and the preceding two symbols (more symbols are required for GSM, due to higher ISI)—it changes every symbol clock and has eight possible combinations. The data samples are precalculated for each state and are stored in a look-up table. They are read out on every FREF clock within the symbol duration. It is shown below that there is a high amount of data redundancy, such as symmetry, constant values for two states, and so on, so the storage requirements could be quite relaxed. In fact, only one state is stored in ROM-like fashion, and the other states are calculated automatically on power-up.

Figure 5.13 shows an implementation block diagram of the digital transmitter filter. It consists of a search (SRCH), phase calculation (PH), state tracking (STATE), and actual filter coefficient storage (FLT) subblocks. The search circuit starts with the "search start" control line asserted and the data bits input from the digital baseband at the low level. It then waits for the starting bit sequence "010" on the data input line that indicates the beginning of a transmitted bit stream. When the first bit arrives, this circuit determines where the pulse lies with the *retimed frequency reference* (CKR) clock granularity. The initial midpulse location is used for sampling of all subsequent bits. The bit stream originates in the digital baseband, which uses the same FREF clock, although with an arbitrary phase shift, as the transmitter. Therefore, once determined, the middle of the pulse holds for the entire packet duration.

# 170  APPLICATION: ADPLL-BASED TRANSMITTER

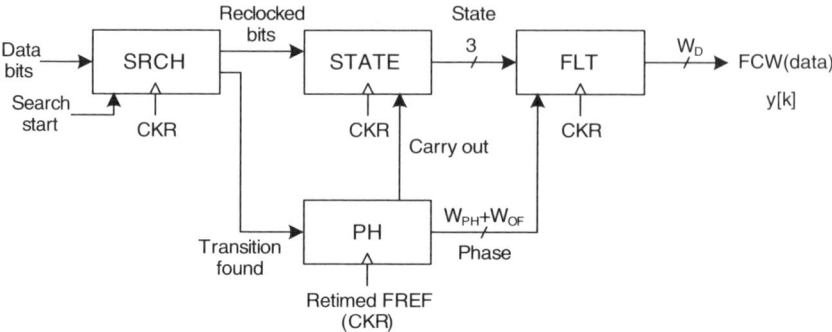

**Figure 5.13** Top-level structure of a transmitter filter.

Detection of the "010" starting bit sequence triggers the phase calculation circuit (with the data transition found synchronization signal being asserted), which keeps track of the relative location of each CKR clock with respect to symbol boundaries. The state tracker (STATE) is a simple shift register that keeps track of the preceding 2 bits, which combined with the current bit determine the state of the filter storage subblock. The filter coefficient storage subblock takes the state information and the phase and produces an output FCW(data) by means of a look-up table. The output conforms with the frequency command word (FCW) notation, where the LSB bit of the integer part corresponds to the reference frequency. The filtered data output is fed as the $y[k]$ input signal to the frequency modulator of Fig. 5.5.

Figure 5.14 illustrates curves for various state transitions based on the current symbol, DT[0], and the previous two symbol values, DT[−1] and DT[−2]. It expands Fig. 5.12 into a detailed development. Each state selects one of two

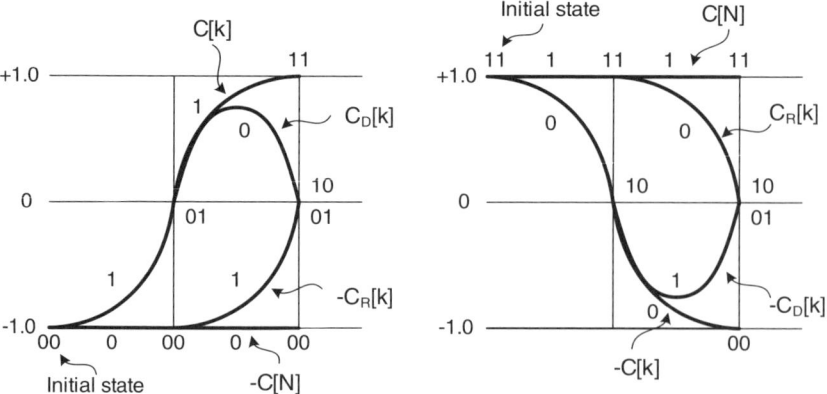

**Figure 5.14** Curves for various state transitions based on previous symbol values. (From [83], © 2005 IEEE.)

## 5.3 GFSK PULSE SHAPING OF TRANSMITTER DATA

**Table 5.1 Transmitter Filter Output Curves**

| DT[−2] | DT[−1] | DT[0] | Curve | Coefficient Set Operation |
|---|---|---|---|---|
| 0 | 0 | 0 | −C[N] | Negated maximum value |
| 0 | 0 | 1 | −C_R[k] | Negated reversed-order coefficients |
| 0 | 1 | 0 | C_D[k] | Di-bit coefficients |
| 0 | 1 | 1 | C[k] | Basic set of coefficients |
| 1 | 0 | 0 | −C[k] | Negated basic set of coefficients |
| 1 | 0 | 1 | −C_D[k] | Negated di-bit coefficients |
| 1 | 1 | 0 | C_R[k] | Reversed-order coefficients |
| 1 | 1 | 1 | C[N] | Maximum value |

candidate curves: input bit 0 and input bit 1. On the left plot, the initial state is "00" (state format: "DT[−2] DT[−1]"). On the right plot, the initial state is "11". In both cases the output curves are the same, but flipped over a horizontal axis. These horizontal and vertical symmetries reveal redundancy, which could be exploited to ease curve storage requirements.

The current output curve can easily be determined by the state field from the state circuit, as shown in Table 5.1. The fundamental state, from which curves of all other states are derived, is "011", where the right bit is the most recent. It is associated with curve C, which is defined as the cumulative coefficient $C[k]$ in Appendix B.

Curve $C[k]$ is a function of index $k$, where $0 \leq k \leq N-1$ and $N$ is the oversampling ratio or integer number of CKR clock cycles per symbol. It is stored in a filter memory with 8 bits per sample. From the $C[k]$ points, data points for all the other curves could be derived. The curve $C[N]$ is a shorthand notation for the constant maximum value of the accumulated coefficient, which corresponds to maximum frequency deviation. Its value is roughly equal or just slightly higher than $C[N-1]$ since the slope and the incremental contributions are very small at this point.

The reversed-order coefficients $C_R[k]$ are computed as

$$C_R[k] = C[N-k] \qquad (5.19)$$

where $C[N]$ is the maximum value of the accumulated coefficient as defined above. The di-bit (corresponding to the 010 data) coefficients $C_D[k]$ feature the highest amount of ISI and are computed as

$$C_D[k] = C[k] + C[N-k] - C[N-1] \qquad (5.20)$$

The remaining curves are the numeric negations of these curves. Only $C[k]$ coefficients are stored as "constants"; the rest are calculated on power-up through an arithmetic combinational logic.

Figure 5.15 shows the evolution of all possible state trajectories starting from a state of 0. It is a waveform generalization of the plot shown in Fig. 5.12. During each symbol segment, a curve is selected based on the current symbol bit (0 or 1)

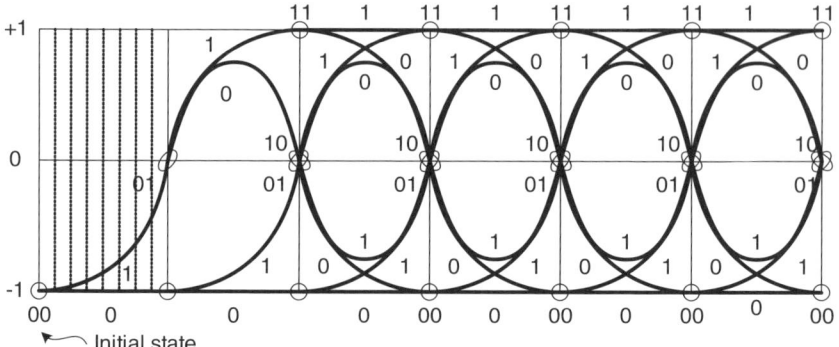

**Figure 5.15** Trellis of a GFSK transmitter filter.

and the 2-bit state, which depends on the preceding two symbols. Since there are four states and symbol alphabet is of size 2, there are a total of eight distinct curves. The waveform with the most intersymbol interference would be a "101010..." pattern, and the waveform with the output stuck at zero would be "000000...".

Application of the foregoing filter for GMSK (used in GSM) would require considerably more effort, due mainly to the substantially higher ISI that spreads over four symbols (see Appendix B). Two more memory bits would need to be added.

### 5.3.1 Interpolative Filter Operation

The pulse-shaping filter shown in Fig. 5.13 requires an integer ratio of the reference frequency to the symbol data rate. Most wireless systems are built on this assumption and specify that the crystal oscillator is a multiple of the data rate. However, relaxing this constraint by allowing the use of a frequency reference from any crystal would make the architecture very portable. The interpolative transmitter filter architecture presented in this section is capable of utilizing any reference clock for its operation as long as it can generate a sufficient number (usually, $\geq 6$ for BLUETOOTH) of samples per symbol period in order to satisfy the Nyquist criterion and sufficiently attenuate the discrete-time signal spectral replicas. The only significant filter circuit in this all-discrete-time design is the RF oscillator, which provides 6 dB/octave attenuation, due to its frequency/phase conversion. Since any wireless terminal will typically have some available reference clock, such as a crystal oscillator clock used as a frequency reference in the range 8 to 40 MHz for host cellular systems, an extra cost in implementing another oscillator could thus be avoided.

A baseband clock for pulse-shape filtering is derived through a simple fractional-$N$ division of the reference frequency. This saves area and power since it is no longer required to create a low-jitter clock for baseband symbol generation and modulating data. It is especially advantageous when the available reference frequency is not an integer multiple of the symbol rate. The transmitter described is realized without explicit analog filtering.

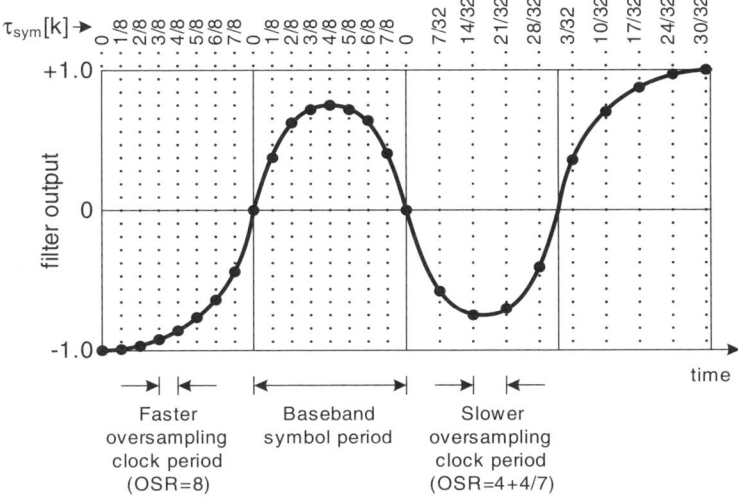

**Figure 5.16** Output waveform of a transmitter filter covering four symbols. The OSR value changes in the middle. (From [83], © 2005 IEEE.)

Figure 5.16 illustrates the output of a transmitter filter. The frequency offset from the RF carrier during transmission is proportional to the filter output, such that the "+1.0" and "−1.0" levels correspond to the peak frequency deviation of ±160 kHz in BLUETOOTH and ±67.71 kHz in GSM. The oversampling ratio changes from OSR = 8 to OSR = $4\frac{4}{7}$ = 32/7 in the middle of the four symbols. As long as the Nyquist criterion is satisfied and the signal spectral replicas are attenuated sufficiently (Fig. 5.1), the demodulated and filtered RF signal at the receiver will be indistinguishable for different oversampling values. In fact, it will be indistinguishable from the continuous-time modulating waveform, which, due to its nondiscrete nature, does not exhibit spectral replicas. Since the Nyquist theorem does not require synchronism to the source signal, the samples do not need to coincide with the symbol boundaries (e.g., between symbols 3 and 4 in Fig. 5.16), so a noninteger OSR should work equally well. This principle is a foundation of the interpolative pulse-shape filtering method, in which no real baseband symbol clock is ever generated.

Let $t[k]$ be a timestamp (in seconds) of the $k$th sample of the modulating data $y[k]$. The $T_s$-normalized unitless timestamps $\tau[k]$ are proportional to the inverse of the OSR, which is a programmable register in this design:

$$\tau[k] = \frac{t[k]}{T_s} = \sum_{l=1}^{k} \frac{1}{\text{OSR}} = \frac{k}{\text{OSR}} \quad (5.21)$$

In other words, the generally fractional oversampling ratio controls the traversal speed along the Fig. 5.16 continuous-time trajectory. Since $\tau[k]$ can grow without

## 174 APPLICATION: ADPLL-BASED TRANSMITTER

bound, a new variable is needed:

$$\tau_{\text{sym}}[k] = \tau[k] - \lfloor \tau[k] \rfloor \tag{5.22}$$

where $\lfloor \bullet \rfloor$ is the floor operator. $\tau_{\text{sym}}[k]$ tracks the $T_s$-normalized offset from the previous symbol boundary, and it indicates the relative sample position within the symbol. The fixed-point $\tau_{\text{sym}}[k]$ is shown in Fig. 5.16 and is used as an argument to $C[k]$ and its derivations. $\tau_{\text{sym}s}[k]$ is obtained as an output of an overflowing accumulator in which the carry-out output indicates crossing of the symbol boundary. Of course, the accumulator must not have any residual error, so its start has to be synchronized with the symbol boundary.

Implementation of the interpolative pulse-shaping filter is illustrated in Fig. 5.17. Detection of the first symbol transition in the data input triggers the symbol offset calculation circuit (1/OSR accumulator), which keeps track of the relative location of each sample or FREF clock with respect to symbol boundaries (i.e., the $\tau_{\text{sym}}[k]$ variable of Eq. 5.22). The carry-out of the accumulator generates the symbol clock. The three-symbol state memory is a simple shift register that determines

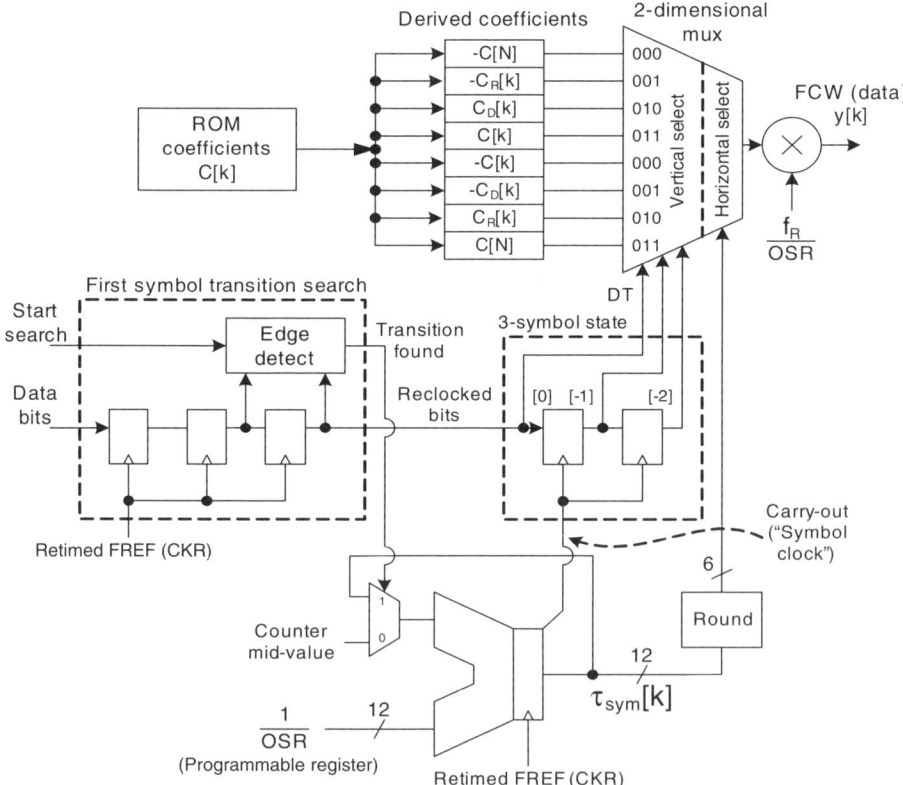

**Figure 5.17** Interpolative transmitter pulse-shaping filter. (From [83], © 2005 IEEE.)

the vertical selection of the two-dimensional mux of look-up-table precalculated coefficients. The data samples are precalculated for each state and are stored in a look-up table. They are read out on every retimed FREF clock within the symbol duration. Because of a high amount of data redundancy, such as symmetry, constant values for two states, and so on, the storage requirements are quite relaxed. In fact, only one state (i.e., the 64-step-response $C[k]$ coefficients that cover one symbol) is stored in ROM-like fashion, and the other states are calculated automatically on power-up. The interpolation is performed with a zero-order hold to the closest coefficient through the rounding operation of the $\tau_{sym}[k]$ accumulator output. The filter output is fed as the $y[k]$ input signal to the frequency synthesizer in Fig. 5.29.

## 5.4 POWER AMPLIFIER

This section deals with the last stage on the integrated transmitter path: the power amplifier (PA). The purpose of a PA in a BLUETOOTH system is to deliver several milliwatts of RF power to the antenna in an efficient manner. In a GSM system, this block would serve as a *pre-power amplifier* (pre-PA or PPA), delivering several milliwatts of RF power to an external PA with an output on the order of 1 W.

Power amplifiers have traditionally been categorized into classes A, B, C, D, E, and F [4]. Classes A, B, and C are considered classical in the sense that both the input and output waveforms are sinusoidal. Voiding this assumption with class E and F operation leads to higher performance and efficiency.

The class E stage is a nonlinear amplifier that potentially achieves 100% efficiency while delivering full power. It has been shown to be best suited in a low-voltage environment [85]. An ideal schematic is shown in Fig. 5.18. Transistor $M_1$ is used here as an on/off switch. RFC (radio-frequency choke) is a large external inductor (usually about 100 nH) that acts as a bidirectional current source at radio frequencies and connects the switch to the supply voltage $V_{DD}$. $C_1$ is a capacitance connected in parallel to the switch and includes the parasitic capacitance of $M_1$. The $C_2$–$L_1$ filter circuit is tuned to the first harmonic of the input frequency and passes a sinusoidal current only to the load $R_L$.

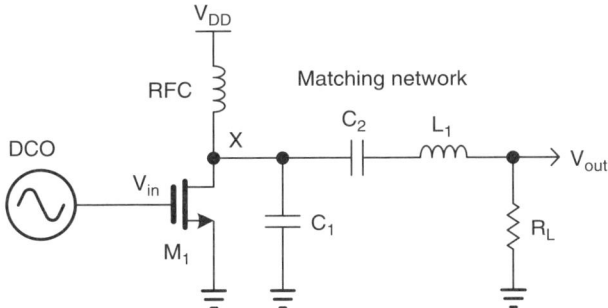

**Figure 5.18** Class E power amplifier.

The values of $C_1$, $C_2$, $L_1$, and $R_L$ are chosen such that $V_X$ satisfies three conditions [4]:

1. As the switch turns off, $V_X$ remains low long enough for the drain current to drop to zero.
2. $V_X$ reaches zero just before the switch turns on.
3. $dV_X/dt$ is near zero when the switch turns on.

In this case, as almost universally in GHz-range applications, the load resistance $R_L$ is 50 Ω. Inductor $L_1$ is realized as a bond wire of 3 nH value. $C_2$ is an external 1.5 pF capacitor. $C_1$ is an internal metal-to-metal 1.4 pF capacitor. The $M_1$ transistor is a 32-finger NMOS of size $W/L = 2.5/0.15$.

During the time when the switch is closed, the voltage across it is zero. During the time when the switch is open, the current through it is zero. Since the voltage and current of the switch do not overlap, the power dissipation of the switch is ideally zero. When the switch turns off, the current through RFC splits between the two branches containing $C_1$ and $R_L$. The capacitance $C_1$ starts charging and slowly produces a voltage across the switch. Satisfying condition 1 is quite easy and is guaranteed by $C_1$. Without $C_1$, $V_X$ could rise as $V_{in}$ is dropped, introducing substantial power loss in $M_1$. When the switch turns on, any charge stored on $C_1$ will be discharged to the ground, resulting in a power loss. To avoid this, the circuit must be designed to satisfy conditions 2 and 3 such that the voltage across $M_1$ reaches zero at the turn-on time and stays there for some time.

After the switch turns off, the load network operates as a damped second-order system with overdamped, underdamped, or critically damped response. If the quality factor $Q$ of the network makes it critically damped, the drain voltage of $M_1$ will follow the $V_X$ curve in Fig. 5.19. This will satisfy conditions 2 and 3. If the network response is underdamped, there would be a dying oscillatory response of $V_X$, and condition 3 could not be met. If the network response is overdamped, $V_X$ might not reach zero by the time $M_1$ turns on. It should be noted that due to the inverting nature of the amplifier, the input and output waveforms are shifted by 180°.

In the ideal situation mentioned above, the efficiency of a class E amplifier is 100%. However, in practice, the switch has a finite on-resistance, and the transition

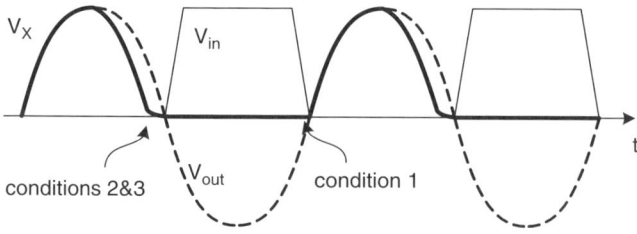

**Figure 5.19** Waveforms of a class E PA.

times from the off-state to the on-state and vice versa are not negligible. Both of these factors result in power dissipation in the switch and reduce efficiency [85].

The class E power amplifier is appropriate for the architecture desired, for the following reasons:

- Its low-voltage operation is ideally suited for deep-submicron CMOS. The end-stage transistor operates as a switch. Unlike in class A, B, and C stages where the transistor acts as a current source and $V_{ds}$ must be controlled precisely at all times to be high enough to avoid entering the triode region, $V_{ds}$ here can be arbitrarily low and no control is necessary. The only requirement is that $V_{gs}$ must be able to go higher than the threshold voltage for the transistor to turn on.

- The transistor switch works best with digital input waveforms, preferably with sharp rise and fall times. Contrast it with classical PAs, which require sinusoidal inputs. This is where the deep-submicron strengths lie. The DCO output is already in a digital format. The duty cycle of the input waveform can conveniently control the output amplitude and power.

- The class E stage is preferred over class F, which is similar to class E but has an additional filtering network to create a high-impedance load at the transistor drain for the second or third harmonics, thus sharpening the edges. The filtering network requires an extra $LC$ tank, which is quite area-expensive in a deep-submicron process. In addition, class F amplifiers have in practice consistently shown worse performance than class E amplifiers [85].

- The power efficiency is theoretically 100%, but in practice 80 to 90% have been reported consistently [85]. The efficiency does not degrade substantially with the output power.

Since the targeted output power for BLUETOOTH or GSM (pre-PA) applications is only several milliwatts, the efficiency is not as important as meeting the basic design specifications, which in itself is quite a challenge at a 1.5-V supply. In this case it is still advantageous to operate the power amplifier with a digital switch, even though the class E conditions might not be fully met.

## 5.5 DIGITAL AMPLITUDE MODULATION

As mentioned in Section 5.4, the output power of a class E RF power amplifier could be controlled by changing the duty cycle or pulse width of its RF digital input. The pulse width controls how long the switch is turned on during the RF cycle and, consequently, how much energy gets transferred to the load $R_L$. This idea, shown in Fig. 5.20, is used for the RF amplitude and power control transmitted. In the BLUETOOTH test chip implemented, only static RF power control is required, and this is accomplished here through the RF waveform amplitude control. When combined with direct all-digital phase modulation, the method allows for implementation of a polar transmitter addressing, for example, BLUETOOTH extended data rate (EDR), EDGE, or Wideband CDMA (WCDMA) wireless comunication standards.

**178**   APPLICATION: ADPLL-BASED TRANSMITTER

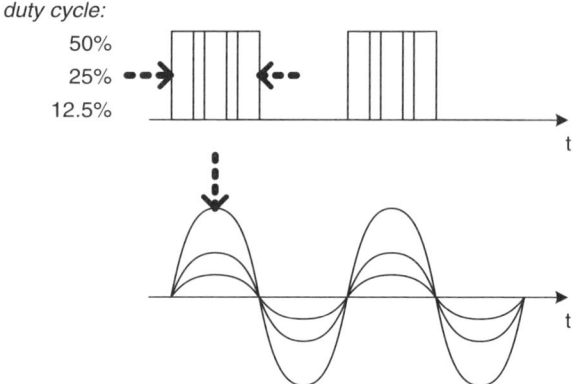

**Figure 5.20**   Output power control through the duty cycle of a class E PA input.

At the time the techniques described here were introduced, there were no reports on using this type of pulsewidth modulation in RF applications. A well-known reference [6] states that this idea is "fairly useless at the gigahertz carrier frequencies of cellular telephones.... difficult to use pulsewidth modulation once carrier frequencies exceed roughly 10 MHz." This work demonstrates that 2.4-GHz operation can be achieved successfuly, mainly due to the ultrafast speed of operation of digital logic gates in this modern deep-submicron process.

Figure 5.21 shows an example of an amplitude modulation implementation using a digital pulse slimmer method. The digital oscillator output clock CKV is met at the digital AND gate with its delayed replica. The delay path could be constructed of a string of inverters or buffers in which the delay could be controlled through a current-starving mechanism or variable capacitative load. In this implementation, delay control through a variable power supply voltage was chosen. The AND gate output is connected to a class E power amplifier input PA_IN. Depending on the relative time delay of the two paths, the timing and duty cycle of the AND gate output could be controlled. The duty cycle or pulsewidth variation directly affects the turn-on time of the PA digital switch, thus establishing the RF output amplitude. The amplitude vs. pulsewidth relationship is quite linear, except for the very narrow input pulse, which might not have enough energy to turn on the switch reliably. This nonlinear region of operation could descriptively be called a *dead zone*, a reference to a commonly used term in conventional phase detectors. The dead zone could be avoided entirely at a system level by choosing modulation techniques that guarantee a certain minimum level of the signal envelope. For example, GFSK and GMSK are *constant*-envelope modulation schemes. Offset-8PSK is a modulation technique used in GSM-EDGE that purposefully rotates the I-Q constellation with every symbol so as to avoid the origin. These methods have long been employed to improve efficiency of power amplifiers and to facilitate the use of a saturation mode of operation.

The timing diagram on the bottom right of Fig. 5.21 shows two regions of operation with different behavior at the leading and lagging output edges with

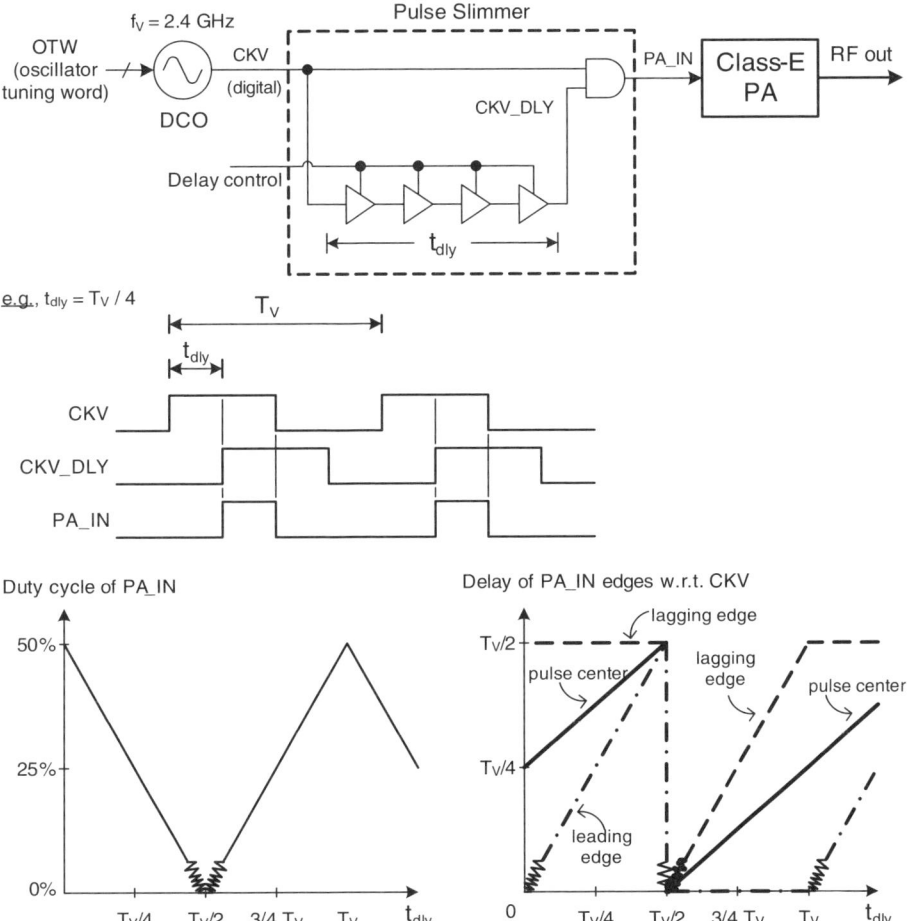

**Figure 5.21** Digital amplitude control through pulsewidth modulation.

respect to the $t_{dly}$ value of the delay path. In the first region, the leading edge of the output traverses but the lagging edge does not. A reversed operation takes place in the second region (dashed line). Since the pulse position is determined by where its center lies, neither provides orthogonality of the phase modulation in the oscillator and of the amplitude modulation in the oscillator pulse slimmer circuit. Consequently, phase adjustment is necessary with amplitude change. This is not a difficult task since the phase control is done in the digital domain through manipulation of the *oscillator tuning word* (OTW).

To save power and reduce jitter due to the long chain of buffers or inverters in the delay path, it might sometimes be beneficial to use an inverted CKV_DLY signal, which is equivalent to an extra half-cycle ($T_V/2$) periodic shift. This could be accomplished through either feeding the delay path from the inverted CKV clock output, or inverting the CKV_DLY signal itself. It is important to note that the maximum required amount of delay is never greater than half of the CKV clock cycle since the negated CKV (of the opposite phase) could always be used.

### 5.5.1 Discrete Pulse-Slimming Control

The implementation of Fig. 5.21 is not completely digital since it requires a *delay control* signal, which is an analog voltage. It can be done digitally as follows: The coarse delay control is performed by adding and subtracting a number of the inverters or buffers in the delay path. Finer delay control could be implemented by selecting taps of a delay line.

Figure 5.22 shows a delay path example of four buffer delay stages. The buffer delay could be the same in each stage. In this case it would result in a total of five possible delay values, from 0 to 4. A better solution might be to have a binary-weighted arrangement of the buffer delays. In this case it would result in a total of 16 possible delay values, from 0 to 15, not including the fixed multiplexer delays. The "effective" delay could be expressed as

$$t_{\text{dly}} = \sum_{j=0}^{N-1} d_j t_{d,0} \cdot 2^j = t_{d,0} \sum_{j=0}^{N-1} d_j \cdot 2^j \tag{5.23}$$

where $N (= 4$ here) is the number of binary-weighted stages, $d_j$ is the $j$th control word bit, and $t_{d,0}$ is the basic element delay of weight $2^0$. Each next stage contains twice the amount of the delay, which could conveniently be realized as doubling the number of inverters or buffers.

The delay control word must be synchronized to the CKV clock to avoid changing it while the signal is still propagating. A method similar to the DCO synchronous sampling and timing adjustment of Fig. 3.2 could be used.

The delay buffer arrangement of Fig. 5.22 is preferred over a transversal delay line configuration in which a large multiplexer selects various taps of a delay line comprised of a string of inverters or buffers, as shown in Fig. 5.23. This is due primarily to the difficulties of building a larger, fast multiplexer with equalized delays for various inputs.

An alternative method to increase the effective delay resolution below that of a single inverter/buffer would be to change the number of inverters dynamically at a

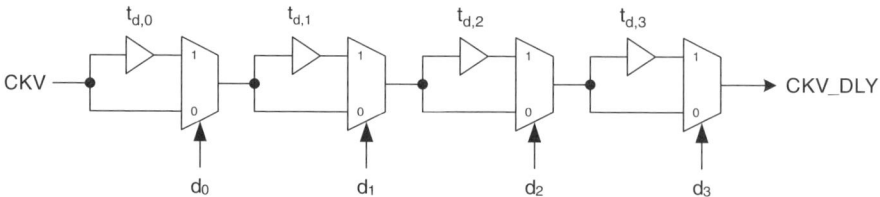

**Figure 5.22** Discrete delay control of a delay path.

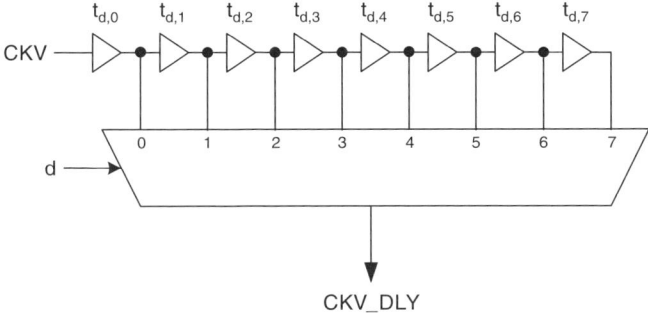

**Figure 5.23** Discrete delay control of the delay path with a larger multiplexer.

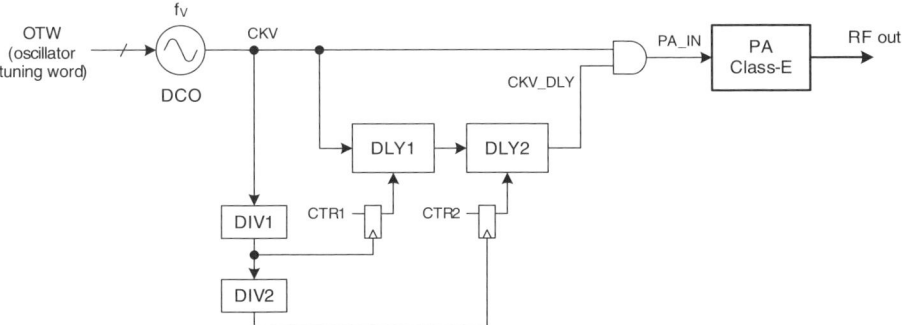

**Figure 5.24** Discrete delay of PWM with additional high-speed dithering.

rate much higher than the symbol rate. The time-averaged delay value of the number of inverters could thus be controlled with a fractional resolution, as illustrated in Fig. 1.21 for a fractional frequency-division ratio. Here again, a $\Sigma\Delta$ digital dithering stream is a good choice, due to its noise-shaping properties. It should be noted here that a binary-weighted delay control would not work very well with high-speed dithering. However, the delay path could be cascaded into a lower-rate binary-weighted structure and a higher-rate unit-weighted structure that would be subject to dithering. Such an implementation is shown in Fig. 5.24 with high-speed delay dithering DLY1 and low-speed delay selection DLY2. DIV1 and DIV2 are CKV clock edge dividers and might be implemented as power-of-2 numbers.

### 5.5.2 Regulation of Transmitting Power

The dynamic amplitude modulation method could be used in its simplest form to statically regulate the output power of the class E power amplifier. It is done in a very efficient manner by injecting enough energy into the PA with every oscillator cycle to achieve the desired output amplitude or power. This is the main application

## 182  APPLICATION: ADPLL-BASED TRANSMITTER

of pulsewidth modulation for the BLUETOOTH transmitter described, which does not require dynamic amplitude modulation. However, the idea of dynamic amplitude control is investigated here for other applications that would require it. Examples are the BLUETOOTH extended data rate (EDR), 802.11b, and EDGE.

### 5.5.3 Tuning Word Adjustment

The polar coordinate representation of the QAM modulation was first illustrated in Fig. 1.17. Amplitude modulation techniques, such as the pulse-slimming method described in Section 5.5, do not always result in control fully orthogonal to phase/frequency modulation. To compensate for the unintended but known phase perturbation, an appropriate correction to the modulated phase could be performed digitally.

A block diagram of such an implementation is shown in Fig. 5.25. Since phase is an integral of frequency,

$$\phi(t) = 2\pi \int_{-\infty}^{t} f(\tau)\, d\tau \tag{5.24}$$

the DCO phase modulation is accomplished through a timed frequency adjustment. In a discrete-time system, frequency control is performed only at update intervals, usually determined by the frequency reference clock edges of period $T_R$. Equation 5.24 is now rewritten for discrete-time operation:

$$\phi[k] = 2\pi \sum_{-\infty}^{k} f[k] T_R \tag{5.25}$$

where $k$ is a time index. To simplify the analysis, Eqs. 5.24 and 5.25 could be interpreted as pertaining to the excess phase and amplitude quantities.

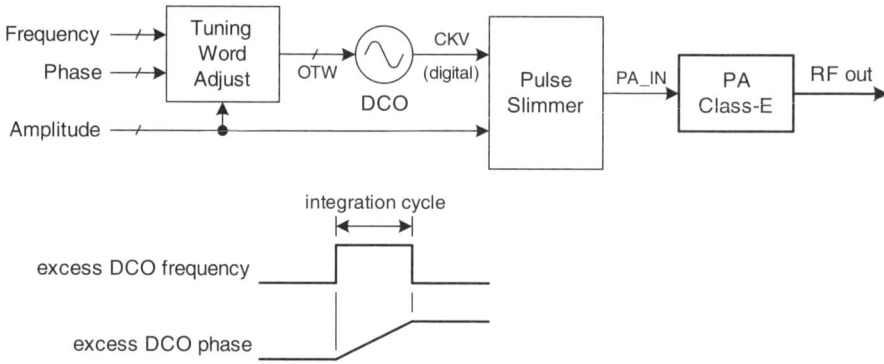

**Figure 5.25** QAM modulation through tuning word adjustment.

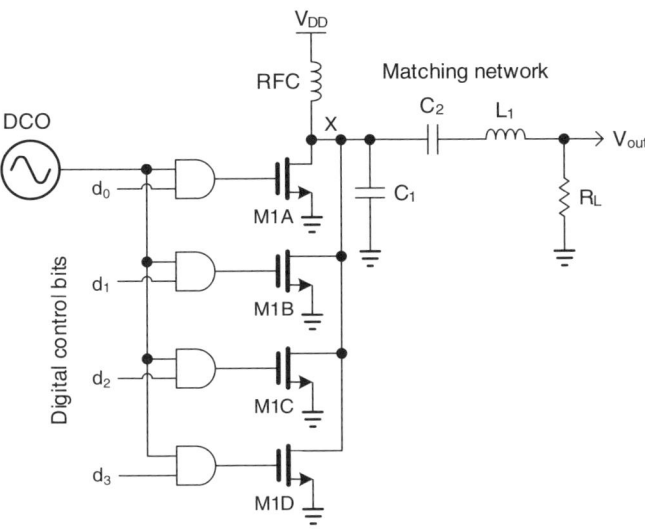

**Figure 5.26** Fully digital amplitude control through multiple-switch transistors.

The magnitude command modulates the PA output amplitude using one of the methods presented above. However, as shown in Fig. 5.21, the side effect of the pulse-slimming method is that the pulse center travels with the edge delay. Fortunately, this pulse center location is easy to predict, especially in the fully digital control environment. The proposed correction of the pulse center dislocation is to change the DCO frequency for a single clock cycle (Fig. 5.25) such that the resulting phase is equivalent to the pulse center shift predicted.

### 5.5.4 Fully Digital Amplitude Control

Figure 5.26 reveals a fully digital method of controlling the amplitude of an RF power amplifier by dynamically regulating the number of active transistor switches [67]. It includes unit-weighted multiple transistor switches that are used to provide coarse digital amplitude modulation. The number of active transistors is regulated through an AND gate realized as a complementary pass gate with a pull-down transistor. A fine amplitude modulation is accomplished through high-speed $\Sigma\Delta$ dithering of a separate bank of transistors.

## 5.6 GOING FORWARD: POLAR TRANSMITTER

A new paradigm facing analog and RF designers of deep-submicron CMOS circuits was formulated in Section 1.4.2 and is repeated below:

> In a deep-submicron CMOS process, time-domain resolution of a digital signal edge transition is superior to voltage resolution of analog signals.

## 184  APPLICATION: ADPLL-BASED TRANSMITTER

A successful design approach in this environment would exploit the paradigm by emphasizing the following:

- Fast switching characteristics or high $f_T$ values (50 ps and 80 GHz in this process, respectively) of MOS transistors: high-speed clocks and/or fine control of timing transitions
- High density of digital logic (150 kilogates/mm$^2$ in this process), making digital functions extremely inexpensive
- Small device geometries and precise device matching made possible by the fine lithography

while avoiding the following:

- Biasing currents that are commonly used in analog designs
- Reliance on voltage resolution
- Nonstandard devices that are not needed for memory and digital circuits

Figure 5.27 illustrates an application of the new paradigm to an RF wireless transmitter performing an *arbitrary* quadrature amplitude modulation (QAM). The low cost of digital logic allows for sophisticated digital signal processing. The tiny, well-matched devices allow for precise high-resolution conversions from digital domain to analog domain. The use of ultrahigh-speed clocks (i.e., high oversampling ratios) can eliminate the need for subsequent dedicated reconstruction filtering of spectral replicas and switching transients, so that only the natural filtering of an oscillator (1/s, due to the frequency-to-phase conversion) and a matching power amplifier and antenna filter network are relied upon. Since the converters utilize DCO clocks that are of high spectral purity, the sampling jitter is very small. The sampling jitter is not affected significantly by modulation, since as

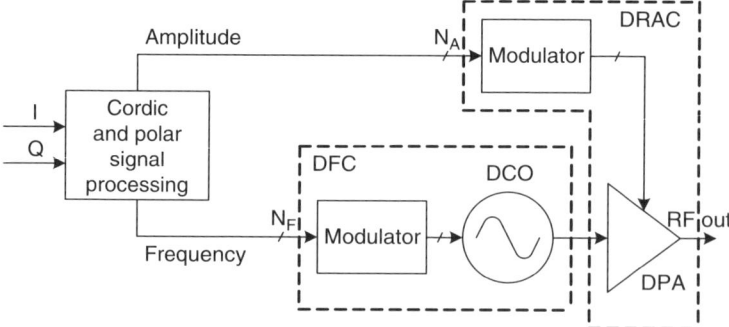

**Figure 5.27** Polar transmitter based on digitally controlled oscillator (DCO) and digitally controlled power amplifier (DPA) circuits. For simplicity, the all-digital PLL around the DCO is not shown.

## 5.6 GOING FORWARD: POLAR TRANSMITTER

shown in Section 6.5.1, the jitter due to modulation is not greater than the oscillator thermal jitter. The conversion functions covered are phase/frequency and amplitude modulations of an RF carrier realized using digitally controlled oscillator (DCO) and digitally controlled power amplifier (DPA) circuits, respectively. They are digitally intensive equivalents of the conventional voltage-controlled oscillator (VCO) and power amplifier driver circuits. Due to the fine feature size and high switching speed of modern CMOS technology, the respective *digital-to-frequency conversion* (DFC) and *digital-to-RF-amplitude conversion* (DRAC) transfer functions could be made very linear and of high dynamic range.

The architecture chosen is polar, as it implements the amplitude and phase modulations in separate paths. The I and Q samples of the Cartesian coordinate system generated in the digital baseband (DBB) processor are converted through a CORDIC algorithm into amplitude and phase samples of the polar coordinate system. The phase is then differentiated to obtain frequency deviation. The polar signals are subsequently conditioned through signal processing to increase the sampling rate sufficiently to reduce the quantization noise density and lessen the effects of modulating spectrum replicas. The frequency deviation output signal is fed into the DCO-based $N_F$-bit DFC, which produces a *phase-modulated* (PM) digital carrier:

$$y_{PM}(t) = \text{sgn}[\cos(\omega_0 t + \theta[k])] \quad (5.26)$$

where, $\text{sgn}(x) = 1$ for $x \geq 0$ and $\text{sgn}(x) = -1$ for $x < 0$, $\omega_0 = 2\pi f_0$ is the angular RF carrier frequency and $\theta[k]$ is the modulating baseband phase of the $k$th sample. The phase $\theta(t) = \int_{-\infty}^{t} f(t)\, dt$ is an integral of frequency deviation, where $t = kT_0$, with $T_0$ being the sampling period.

The *amplitude modulation* (AM) signal controls the envelope of the phase-modulated carrier by means of the DPA-based $N_A$-bit DRAC. Higher-order harmonics of the digital carrier are filtered out by a matching network, so that the sgn(·) operator is dropped. The composite DPA output contains the desired RF output spectrum:

$$y_{RF}(t) = a[k]\cos(\omega_0 t + \theta[k]) \quad (5.27)$$

where $a[k]$ is the modulating baseband amplitude of the $k$th sample.

Despite their commonalities, there are important differences between the two conversion functions. Due to the narrowband nature of the communication system, the DFC operating range is small but has fine resolution. The DRAC operating range, on the other hand, is almost full scale but not as precise. In addition, the phase-modulating path features an additional $1/s$ filtering caused by frequency-to-phase conversion of the oscillator. Of course, signal processing and delay between the AM and PM paths should be matched: otherwise, the recombined composite signal will be distorted. Fortunately, the matching invariability to the process, voltage, and temperature changes is guaranteed by the accurate clock-cycle characteristics of digital circuits. The group delay of DCO and DPA circuits is relatively small (tens of picoseconds, due to the high $f_T$ value of deep-submicron CMOS devices) compared with the tolerable range (tens of nanoseconds).

The DFC and DRAC are key functions of the all-digital transmitter, which does not use any current biasing or dedicated analog continuous-time filtering in the signal path. To improve matching, linearity, switching noise, and operational speed, the operating conversion cells (bit to frequency or RF carrier amplitude) are realized primarily as unit weighting. Due to the excellent device-matching characteristics in a deep-submicron CMOS process, it is relatively easy to guarantee 7-bit conversion resolution in one iteration cycle without resorting to elaborate layout schemes. The DFC and DRAC architectures are presented below. It should be noted that there appear to be no other reports in the literature of all-digital wireless RF transmitters. Spectral replicas of discrete-time modulating signals appear at the DCO and DPA inputs at integer multiples of the sampling rate frequency $f_R$, as shown in Fig. 5.1.

### 5.6.1 Generic Modulator

The two modulators in Fig. 5.27 could be considered a digital front end of a generic *digital-to-analog converter* (DAC), where *analog* here stands either for the frequency or the RF carrier amplitude. For the reasons stated above, the cell elements of physical converters are unit weighted. Consequently, the simplest realization of a modulator would be a binary-to-unit-weighted converter.

Unfortunately, the arrangement above would not be practical, due to the limited resolution of the conversion process. For example, the 12-kHz frequency step of the DFC is not adequate for GSM modulation, where the peak frequency deviation is 67.7 kHz. Similarly for the amplitude modulation, the 6-bit amplitude resolution is also too coarse. In this design, finer conversion resolution is achieved through high-speed dithering of the finest conversion cell elements, as shown in Fig. 5.28. The $N$-bit digital fixed-point input is split into $M$ integer (higher-order) bits and $N - M$ fractional (lower-order) bits. The integer word sets the number of activated

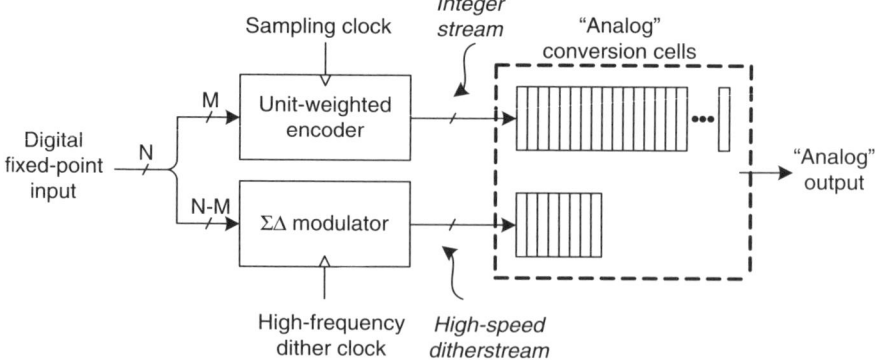

**Figure 5.28** Digital modulator as part of a generic DAC.

conversion elements. The fractional word is fed to a $\Sigma\Delta$ modulator that produces a high-speed dithering stream whose average value is equal to the fractional part of the fixed-point input word.

It should be noted that this DAC architecture is similar to that in reference [87] but with significant differences to allow higher-frequency operation at reduced power consumption. The lower-rate wide-bus-width integer stream is never merged in the digital domain with the higher-rate dithering stream. The final stream addition is done in the device cell domain: In the DCO, the varactor capacitances are added, and in the DPA, the transistor resistances and driving strengths are added. In this way, high-speed operation is constrained to a small portion of the circuit, thus saving the current consumption. A detailed description of the DCO modulator for BLUETOOTH was given in reference [56].

### 5.6.2 Polar TX Realization

Figure 5.29 shows the realization of the RF polar transmitter of Fig. 5.27 that is used in future designs. The BLUETOOTH transmitter presented does not use dynamic amplitude modulation, so an *amplitude control word* (ACW) is statically controlled to set the output power.

The transmitter is based on an all-digital PLL (ADPLL) frequency synthesizer with a digital direct frequency modulation capability. It uses digital design and circuit techniques from the ground up. At its heart lies a digitally controlled oscillator (DCO), which deliberately avoids analog tuning voltage controls. A digitally

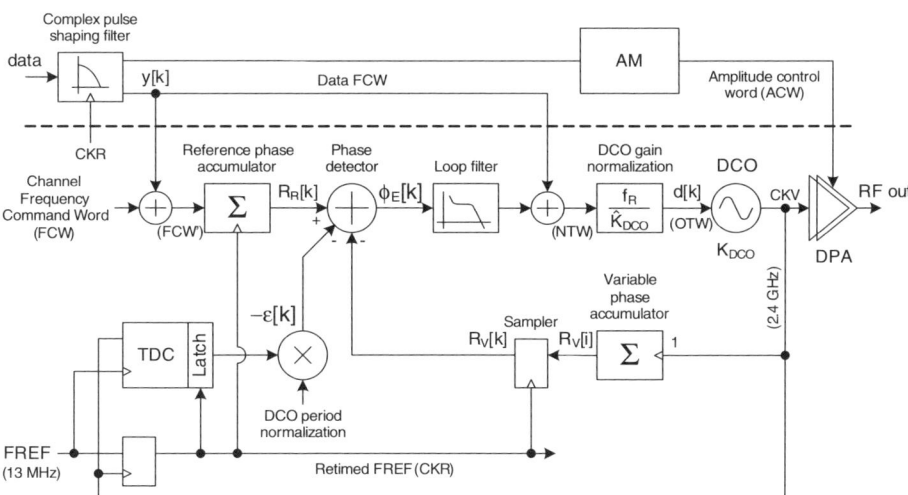

**Figure 5.29** Synchronous phase-domain all-digital PLL-based polar transmitter. Only the DCO tracking bank varactors are shown. The PVT and acquisition bank varactors have their own normalizing multipliers, which are inactive during normal operation. (From [67], © 2004 IEEE.)

controlled power amplifier (DPA) provides digital amplitude modulation. The DCO produces at its output a single-bit digital variable clock (CKV) in the RF band. In the feedforward path, the CKV clock toggles an array of NMOS transistor switches constituting a near-class E digitally controlled RF power amplifier (DPA) that is followed by a matching network and terminated with an antenna. In the feedback path, the CKV clock is used for phase detection and reference retiming. The channel and data frequency control words are in frequency command word (FCW) format, defined as the fractional frequency-division ratio $N$ with fine frequency resolution limited only by the FCW wordlength.

## 5.7 SUMMARY

In this chapter we suggested an application for the all-digital PLL frequency synthesizer described in Chapter 4 by adding a phase/frequency modulating capability. It was shown that mere adjustment of a frequency control word would subject it to the PLL bandwidth, which might attenuate higher-frequency components of the modulating transmitter data. A predictive modulation mechanism with a closed PLL operation that would solve this problem was described. A solution to the related problem of estimating the DCO gain was discussed.

It should be noted at this point that the feedforward phase/frequency adjustment capability might find useful applications in other areas that are not related to the transmitter modulation. For example, a DSP processor might request the internal clock generation unit to change its clock rate immediately while entering or leaving a power-saving mode. In another example, it might be necessary in a hard-disk drive read channel to shift the clock phase of a time-base generator "instantly" by $180°$. Instead of multiplexing a delay line tap, one could shift the right amount of frequency within a clock cycle.

Finally, a highly efficient class E power amplifier and a power regulation circuitry are added to the DCO output to demonstrate the full transmitter for a wireless communication. We also described a novel amplitude modulation method that could conveniently be implemented in digital fashion. This would culminate in an all-digital polar transmitter.

# CHAPTER 6

# BEHAVIORAL MODELING AND SIMULATION

With the first demonstration of a fully digital frequency synthesizer and transmitter for wireless applications, a need has arisen to model and simulate RF components using the same simulation engine as that used for the digital back end, which at present is likely to contain over a million gates. In this way, complex interactions and performance of an entire *system-on-chip* (SoC) *integrated circuit* (IC) could be validated and verified prior to tape-out. Here are some examples of these complex interactions:

1. Effect of the TDC resolution and nonlinearity on the close-in PLL phase noise performance and generated spurs
2. Effect of the DCO phase noise on PLL phase noise performance and spurs generated, especially when the PLL contains a higher-order digital loop filter and operates in fractional-$N$ mode
3. Effect of the DCO frequency resolution on the close-in phase noise of a PLL
4. Effect of the $\Sigma\Delta$ DCO dithering on the far-out phase noise
5. Effect of the DCO varactor mismatches on the modulated spectrum
6. Effect of the DPA resolution and nonlinearity on the RF output spectrum

Although SPICE-based simulation tools are extremely useful for small RF circuits containing several components (such as an RF oscillator), their slow simulation performance prevents us from investigating larger circuits (such as an RF

---

*All-Digital Frequency Synthesizer in Deep-Submicron CMOS*, by Robert Bogdan Staszewski and Poras T. Balsara
Copyright © 2006 John Wiley & Sons, Inc.

oscillator with a PLL and a transmitter or receiver). In fact, using the techniques presented, we were able to determine that the entire transmitter meets the RF specifications prior to tape-out. This level of validation seems now to be a requirement given the $1 million price tag for a reticle set in the most advanced deep-submicron CMOS process.

There have been a variety of communication channel modeling methods. At the pure system level there are C and MATLAB models, which are highly abstract and have very weak links to the actual hardware. On the opposite side of the spectrum, the system could be modeled at a very low level entirely in SPICE for analog-intensive systems or in a SPICE and Verilog (or VHDL) combination, with a varying degree of link between the two disparate simulation engines (e.g., a SPICE and Verilog co-simulation backplane). Establishing a link to a non-event-driven engine such as SPICE results in a hefty price for a simulation performance, thus making it impossible to determine the very basic figure of merit of a communication channel: the bit error rate.

In this chapter we describe a system modeling and simulation environment that is based on a standard single-core simulator (e.g., VHDL). Emphasis is put on modeling the oscillator and the time-to-digital converter. Other digital blocks in the loop are straightforward and can be modeled using standard HDL modeling techniques. The system presented is well suited for digitally intensive applications with a fair amount of analog circuit content. Extensive links to a file system for pre- and postprocessing facilitate a rich simulation and analysis environment. The main advantage of a single simulation engine is that it allows seamless integration of all hardware abstraction levels in a uniform environment. The single most important feature of a standard VHDL language, which makes it far superior to Verilog for mixed-signal designs, is its support of real or floating-point type signals. Widespread simulation and synthesis support of a standard VHDL language makes it possible for a complex communication system to achieve the goal to "build what we simulate, and simulate what we build." Simulator performance, stability, multi-vendor support, mature standard, and widespread use are all advantages of this environment.

## 6.1  SIMULATION METHODOLOGY

VHDL is based on an event-driven simulation engine. The simulator proceeds to the timestamp of the next event if all the activities associated with the current timestamp are exhausted. It is a very efficient method since the simulation activity (hence, computation time) is spent only on an as-needed basis. This is in sharp contrast with some other system-level simulators, which are based on an oversampled clock, such as Simulink or SPW (the Signal Processing Worksystem, by CoWare). In that environment, the simulation engine has to transverse all the equally spaced timestamps that oversample the signals sufficiently.

Operating in an oversampled domain is less problematic with baseband signals and systems or in an environment with a single clock. Even two clock domains are not an issue if their frequency ratio is a small integer. In that case, the higher-frequency clock is the common denominator. The operation becomes quite unwieldy, however, if the clocks are not easily related such that their common denominator clock is at a very high frequency.

Another environment exposing the inefficiency of an oversampled domain simulator is a narrowband RF system. Oversampling the BLUETOOTH RF waveform of a 2.4-GHz carrier generates an excessive amount of activity in light of the fact that the information is contained only in the 1-Mbps symbols. Considering eight samples per sinusoidal RF cycle, one 1-μs symbol would contain as many as 19,200 samples! Hence, a very efficient RF wave representation method in which only positive zero-crossing timestamps ($t_k$, $t_{k+1}$, $t_{k+2}$ in Fig. 1.2) are used was chosen.

## 6.2 DIGITAL BLOCKS

The digital blocks are modeled at various abstraction levels, as shown in Table 6.1. As an example, level 1 representation of an GFSK transmit filter might describe the behavior with a simple direct-form finite-impulse response (FIR) filter equation using real numbers for input, output, coefficients, and all the intermediate signals. Level 2 representation of the actual implementation with accumulated coefficients might show the top-level structure of major building blocks, which are then modeled behaviorally on the bit level using integers. Second-order effects, such as LSB truncation and rounding as well as MSB clipping, are included. Level 3 representation has the same I/O behavior as level 2, but its RTL representation drives the synthesis of gate connectivity. Each higher level of modeling is expected to improve the simulation time by an order of magnitude.

The unified approach ensures the interoperability of various abstraction levels within a single simulation environment. This allows simulating a system with a mixture of synthesized blocks and those that are still at the mathematical description levels.

**Table 6.1  VHDL Modeling Abstraction Levels**

| | |
|---|---|
| Level 1 | Mathematical equations and high-level behavioral description. Parameterized for easy analytic "what-if" questions. Optimized for simulation speed and flexibility. Fast enough to replace MATLAB for a bit error rate analysis. Includes important hardware-related nonidealities and second-order effects; "_l1.vhd" file name suffix. |
| Level 2 | Mathematical equations in the integer domain, implying the underlying architectural structure. 100% pin compatible with level 3. Can be used for top-level connectivity verification; "_l2.vhd" file name suffix. |
| Level 3 | Synthesizable register transfer level (RTL). |
| Level 4 | Gate-level netlist produced by synthesis tools. It could also be the actual gate-level netlist extracted from an auto place and route tool (APR). Accompanied by the cell and wire timing information (Vital or SDF). |

**192**  BEHAVIORAL MODELING AND SIMULATION

## 6.3 SUPPORT OF DIGITAL STREAM PROCESSING

Digital communication channel evaluation usually requires processing a long stream of digital data. Some measurements, such as the bit error rate, require as many as tens of millions of data bits; therefore, fast, efficient algorithms to store, retrieve, and access the data are necessary. Throughout this system, a time-causal operation and a linear complexity order is forced on the processing algorithms. As a result, the temporary storage requirement (RAM or disk swap space) is constant, and the simulation time is only linearly proportional to the digital stream length.

## 6.4 RANDOM NUMBER GENERATOR

A pseudorandom number generator is needed to create a long stream of digital input stimulus, to model electronic thermal and flicker $1/f$ noise, and to add jitter and wander deviations to the oscillator clock. In most HDL tools, system-supplied pseudorandom number generators do not usually have good random properties. A typical implementation of the *rand*(•) ANSI C function call or *uniform*(•) procedure call of the IEEE math_real VHDL package uses the linear congruential method, which although very efficient and fast, suffers from sequential correlation on successive calls. There is a danger that using it might skew the evaluation results for the performance of a communication system by, for example, not exercising all possible paths through a Viterbi detector. A good uniformly distributed random number generator with virtually no sequential correlation was built in VHDL based on Park and Miller's algorithm with Bays–Durham shuffle as described in reference [88]. A random number generator with Gaussian (normal) distribution was build based on the Box–Muller method as described in reference [88].

## 6.5 TIME-DOMAIN MODELING OF DCO PHASE NOISE

Phase noise in a DCO oscillator can be modeled using jitter and wander constructs. The flat electronic thermal noise of the $1/\omega^0$ region in Fig. 1.4 is modeled as a nonaccumulative jitter. The $1/\omega^2$ region of the upconverted thermal noise, on the other hand, is modeled as an accumulative wander.

### 6.5.1 Modeling Oscillator Jitter

Figure 6.1 illustrates a modeling principle of timing jitter. The $1T_0, 2T_0, 3T_0,$ and $4T_0$ timestamps are the ideal equidistant rising-edge events for an oscillator operating at frequency $f_0 = 1/T_0$. The ideal oscillator output might pass through a physical buffer that adds random fluctuations to its delay. The actual timestamps of the physical buffer output could be described mathematically as an additive random error at each occurrence of an ideal timestamp. These timing errors do not influence one another. If the random error is due to thermal electronic noise, edge timing

## 6.5 TIME-DOMAIN MODELING OF DCO PHASE NOISE

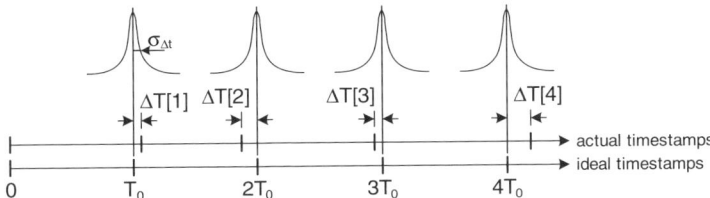

**Figure 6.1** Development of a nonaccumulative timing deviation.

deviations are said to be *independent and identically distributed* (IID) and are usually modeled as an *additive white Gaussian noise* (AWGN). In Fig. 6.1 we show the error probability curve from each ideal timestamp. The timing deviation $\text{TDEV}_j$ in the jitter case is the difference between the actual timestamps $t_j[i]$:

$$t_j[i] = iT_0 + \Delta t[i] \tag{6.1}$$

and the ideal timestamps at $iT_0$:

$$\text{TDEV}_j[i] = \Delta t[i] \tag{6.2}$$

The *period deviation*, $p_1[i] = t_j[i] - t_j[i-1]$, a term commonly used in jitter measurements, is statistically twice the timestamp deviation. This is due to the fact that an instantaneous period is perturbed from both sides and makes the neighboring errors not entirely independent. The period deviations could be modeled by passing the timestamp deviations through a two-tap FIR filter with $\{1, -1\}$ coefficients.

The relationship between the time and frequency domains of the jitter could be obtained as follows: The noise floor $\mathcal{L}$ is a single-sided spectral density. It needs to be multiplied by 2 in order to arrive at the $S_\phi$ (rad$^2$/Hz) double-sided spectral density in Eq. 1.5. Since the flat spectrum of the jitter in the discrete-time model extends to the Nyquist frequency, $S_\phi$ is multiplied by $f_0/2$ to arrive at the total power (rad$^2$). The rms jitter (rad) is its square root value. The radian quantity is converted to the timestamp deviation in seconds by multiplying it by the normalizing factor $T_0/2\pi$:

$$\sigma_{\Delta t} = \frac{T_0}{2\pi}\sqrt{\mathcal{L} f_0} \tag{6.3}$$

For BLUETOOTH, substituting $f_0 = 2.4\,\text{GHz}$, $T_0 = 417\,\text{ps}$, and $\mathcal{L} = 10^{-150\text{dB}/10} = 1 \times 10^{-15}\,\text{rad}^2/\text{Hz}$, it amounts to $\sigma_{\Delta t} = 103\,\text{fs}$. This value of Gaussian jitter is used by default in all VHDL simulations. In the GSM case, the noise floor has to be at least below $-164\,\text{dBc/Hz}$. Substituting $f_0 = 0.9\,\text{GHz}$, $T_0 = 1111\,\text{ps}$, and $\mathcal{L} = 10^{-164\text{dB}/10} = 4 \times 10^{-17}\,\text{rad}^2/\text{Hz}$, the rms jitter is $\sigma_{\Delta t} = 33.4\,\text{fs}$. Equation 6.3 has the following inverse relationship:

$$\mathcal{L} = (2\pi)^2 (\sigma_{\Delta t})^2 f_0 \tag{6.4}$$

## 6.5.2 Modeling Oscillator Wander

Figure 6.2 illustrates the modeling principle of timing wander, which is also called *accumulative jitter*. This could be visualized by a physical oscillator of nominal period $T_0$ whose actual period varies slightly from one cycle to the next, due to, for example, thermal noise effects internal to the oscillator. In contrast to the jitter case, here each transition timestamp depends on all previous period deviations. This simply acknowledges the fact that ideal timestamps exist only in theory and that the only memory of a given transition is the preceding transition timestamp. This behavior is modeled as a random walk. The timing deviation $\text{TDEV}_w$ in the wander case is the difference between the actual timestamps $t_w[i]$:

$$t_w[i] = iT_0 + \sum_{l=1}^{i} \Delta T[l] \tag{6.5}$$

and the ideal timestamps at $iT_0$:

$$\text{TDEV}_w[i] = \sum_{l=1}^{i} \Delta T[l] \tag{6.6}$$

It should be noted that other references use different terms. For example, reference [89] uses the term *absolute jitter*. Unlike the nonaccumulative case, the period deviation here is equivalent to the timestamp deviation.

Equation 1.11 is used to relate the wander component mathematically in the time and frequency domains. For example, based on lab measurements of a larger number of IC chips, a DCO phase-noise level of $-105$ dBc/Hz at 500 kHz offset was conservatively assumed. Equation 1.11 is now transformed into

$$\sigma_{\Delta T} = T_0 \Delta\omega \sqrt{\frac{\mathcal{L}\{\Delta\omega\}}{2\pi\omega_0}} = \frac{\Delta f}{f_0} \sqrt{T_0} \sqrt{\mathcal{L}\{\Delta f\}} \tag{6.7}$$

This equation is confirmed in reference [90]. For BLUETOOTH, substituting $f_0 = 2.4$ GHz, $T_0 = 417$ ps, $\Delta f = 500$ kHz, $\mathcal{L}\{\Delta f\} = 10^{-105\text{dB}/10} = 3.16 \times 10^{-11} \text{rad}^2/\text{Hz}$, results

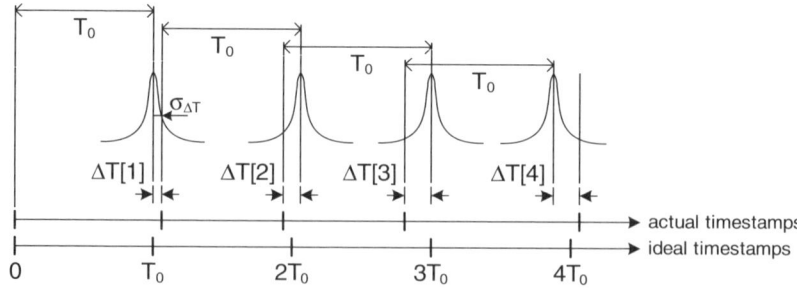

**Figure 6.2** Development of an accumulative timing deviation.

in $\sigma_{\Delta T} = 24$ fs. This value of Gaussian wander is used by default in all VHDL simulations. For GSM, substituting $f_0 = 0.9$ GHz, $T_0 = 1111$ ps, and $\mathcal{L} = 10^{-164\text{dB}/10} = 4 \times 10^{-17}$ rad$^2$/Hz at $\Delta f = 20$ MHz offset, the rms wander is $\sigma_{\Delta t} = 4.674$ fs. The inverse relationship of Eq. 6.7,

$$\mathcal{L}\{\Delta f\} = \frac{\sigma_{\Delta T}^2 f_0^3}{\Delta f^2} = \frac{(\sigma_{\Delta T}/T_0)^2 f_0}{\Delta f^2} \quad (6.8)$$

shows 20 dB/decade attenuation of the phase noise.

### 6.5.3 Modeling Oscillator Flicker (1/f) Noise

In previous work, $1/f$ noise has been modeled in the time domain by means of FIR and IIR filters [86]. Filter coefficients are derived depending on the type of noise that is modeled. The main disadvantage is that for high filter sampling rates, if the $1/f$ noise has to be described over several decades, the number of filter coefficients required becomes very large. For example, for a filter operating at 1 GHz to be able to describe the $1/f$ noise to 1 kHz would require 100,000 filter coefficients.

In this work, $1/f$ noise is constructed by passing white noise through several first-order low-pass filters. Each filter shapes different regions of the noise spectrum, to yield a composite output having the desired $1/f$ response of slope,

$$\text{slope} = \frac{A_{\text{dB}}}{r} \quad (6.9)$$

equal to 10 dB/decade or 3 dB/octave. The $1/f$ noise is then up-converted by the oscillator according to Eq. 1.10, which results in the final 30 dB/decade slope of the phase noise.

Figure 6.3 illustrates the modeling principle. The frequency over which the $1/f$ noise is modeled is divided into several regions. The frequency boundaries between regions serve as the corner frequency of the filters used in the noise-shaping process. The ratio of dc gains between the successive filters,

$$A = 10^{A_{\text{dB}}/20} \quad (6.10)$$

is selected such that the ratio of the corresponding neighboring corner frequencies,

$$r = \frac{f_{c,k+1}}{f_{c,k}} \quad (6.11)$$

satisfies Eq. 6.9. As seen from Fig. 6.3, the composite response of the filters yields a $1/f$ noise characteristic having the desired slope of 10 dB/decade.

Figure 6.4 illustrates a time-domain model for generating $1/f$ noise. White Gaussian unit-variance noise source $x[i]$ is input to each filter. The outputs of all

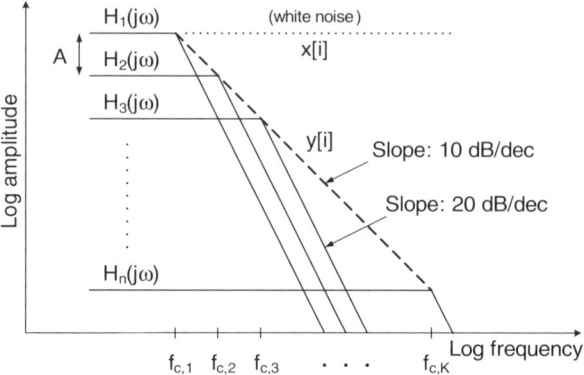

**Figure 6.3** Composite response of single-pole low-pass filters. (From [3], © 2005 IEEE.)

the filters are summed to obtain the shaped $1/f$ noise response over the frequency interval ranging from $f_{c,1}$ to $f_{c,K}$. Each filter, $h_k[i]$, is modeled as a first-order IIR filter described by the difference equation:

$$y_k[i] = (1 - a_k)y_k[i-1] + a_k A^{-(k-1)} x[i] \qquad (6.12)$$

where $k$ is the filter index, $k = 1, 2, \ldots, K$, and $A$ is the linear value of the attenuation factor given in Eq. 6.10 and $a_k$ is defined by

$$a_k = 2\pi \frac{f_{c,k}}{f_s} \qquad (6.13)$$

where $f_{c,k}$ is the corner frequency of the filter $k$ and $f_s$ is the common sampling rate.

If the shaping filters are implemented using a multirate approach and the corner frequencies are scaled with multirate frequencies, the filters will have the same feedback coefficients. The multirate approach both simplifies the design of these filters and reduces computational complexity. Consider the structure illustrated in Fig. 6.5.

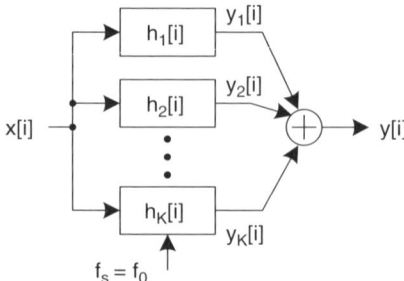

**Figure 6.4** Time-domain model for $1/f$ noise generation. (From [3], © 2005 IEEE.)

## 6.5 TIME-DOMAIN MODELING OF DCO PHASE NOISE

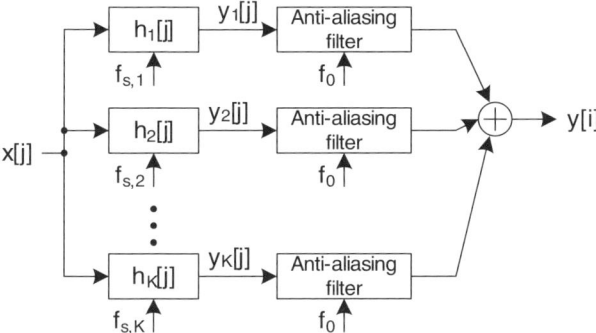

**Figure 6.5** Multirate $1/f$ noise configuration. (From [3], © 2005 IEEE.)

Each of the filters is clocked by a down-divided version of the oscillator frequency, $f_0$, such that

$$f_{s,k} = \frac{f_0}{M \prod_{l=0}^{K-k} r^l} \tag{6.14}$$

where $K$ is the total number of filter segments. The last filter section has the highest sampling frequency, $f_0/M$. Each subsequent section has a sampling frequency equal to that of the following section divided by $A$, such that

$$f_{s,k} = \frac{f_{s,k+1}}{r} \tag{6.15}$$

Aliases created due to the use of the lower sampling rates are removed by antialiasing filters located at the outputs of each filter. The antialiasing filters are clocked at the RF output frequency $f_0$ and include a decimation scaling gain of $\sqrt{f_{s,k}/f_0}$.

If the corner frequencies are related by the same sampling ratio, so that

$$f_{c,k} = \frac{f_{c,k+1}}{r} \tag{6.16}$$

all the filters will have the same feedback coefficients. For a first-order IIR filter, the magnitude of the feedback coefficient can be given by

$$a_k = 2\pi \frac{f_{c,k}}{f_{s,k}} \tag{6.17}$$

This formula clearly illustrates that if $f_s$ and $f_c$ scale by the same ratio, the filter coefficients remain constant.

**198** BEHAVIORAL MODELING AND SIMULATION

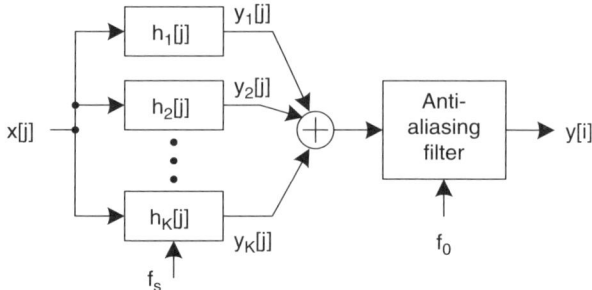

**Figure 6.6** Multirate $1/f$ noise configuration with a single clock domain for all noise-shaping filters. (From [3], © 2005 IEEE.)

A preferred implementation that does not require as many clock domains as in the preceding scheme is illustrated in Fig. 6.6. In this configuration, all the $1/f$ noise-shaping filters are driven at the same low clocking frequency $f_s$, and the aliased images are removed by an antialiasing filter located at the cumulative output, which is clocked at the RF clock rate $f_0$. The number of antialiasing filter stages used in an implementation depends on the amount of filtering required in the aliased band.

Figure 6.7 illustrates the spectral response at the DCO output for an implementation in which the $1/f$ filters have cutoff frequencies at $f_{c,1} = 100\,\text{Hz}$, $f_{c,2} = 1\,\text{kHz}$, $f_{c,3} = 10\,\text{kHz}$, $f_{c,4} = 100\,\text{kHz}$, and $f_{c,5} = 1\,\text{MHz}$. This implementation runs at an RF clock rate of $f_0 = 2400\,\text{MHz}$, and the $1/f$ filters are clocked at $f_s = 24\,\text{MHz}$. An antialiasing filter located at the output of the accumulator is clocked at the

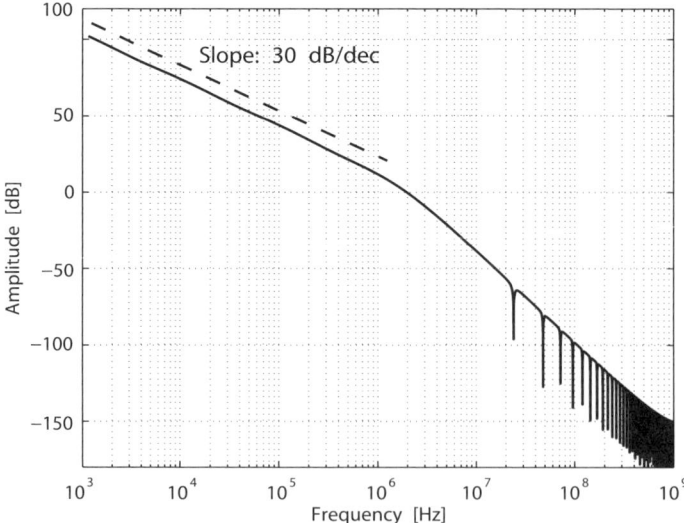

**Figure 6.7** Multirate $1/f$ noise response with sinc filtering. (From [3], © 2005 IEEE.)

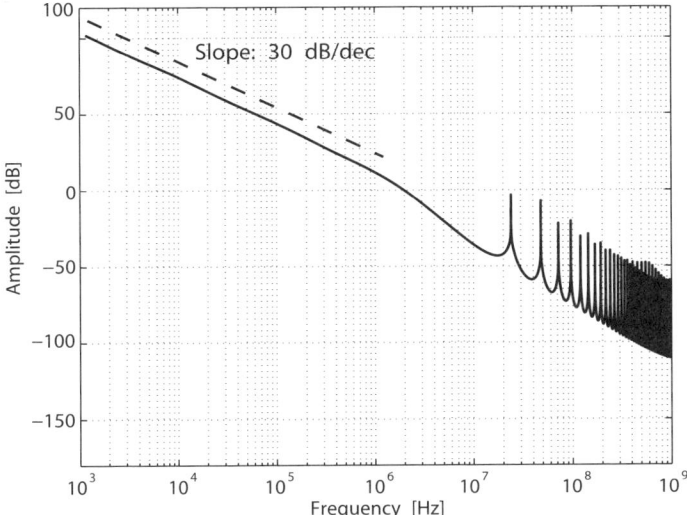

**Figure 6.8** Multirate $1/f$ noise response without sinc filtering. (From [3], © 2005 IEEE.)

RF clock rate and has a cutoff frequency of 2 MHz. The nulls at multiples of the $1/f$ filter sampling clock arise due to the sinc response that is created by the inherent sample-and-hold characteristic introduced by the use of multiple clock frequencies.

The sinc response not only provides an extra 20 dB/decade rolloff for frequencies above the $1/f$ filter clocking rate, but also nulls out the aliased frequency components that are present at multiples of this frequency. Figure 6.8 illustrates the filtered response without sinc filtering. As can be seen in this figure, in the absence of a sinc filter, aliased spurs appear at multiples of the $1/f$ clocking rate.

It should be emphasized that the construction method of the oscillator $1/f$ noise does not differ fundamentally from the wander method. In the former case, the appropriately filtered Gaussian noise $y[i]$ affects the oscillating frequency deviation $\Delta f[i]$. In the latter case, filtering is not used; white Gaussian noise $x[i]$ is used instead, as illustrated in Fig. 6.3. In both cases, the frequency-to-phase conversion of Eq. 1.10 gives an additional 20 dB/decade of slope.

Equation 6.7 relates the $1/\omega^2$ phase noise $\mathcal{L}\{\Delta\omega\}$ to the timing wander $\sigma_{\Delta T}$. In the case of shaped $1/f$ noise, phase noise at the corner frequencies is 3 dB less, due to the first-order filtering, than that given above. For the lowest corner frequency point on the $1/f$ noise curve in Fig. 6.3, which corresponds to $f_{c,1}$, we calculate the standard deviation using the formula

$$\sigma_{\Delta T, 1/f} = \frac{\Delta f_{c,1}}{f_0} \sqrt{T_0} \sqrt{2\mathcal{L}\{\Delta\omega_{c,1}\}} \quad (6.18)$$

The factor of $\sqrt{2}$ has been included to account for the 3-dB reduction in phase noise. However, due to the correlative nature of the noise-shaping filter outputs because of their common input, a further correction has to be made to $\mathcal{L}\{\Delta\omega_{c,1}\}$. For the case

$r = 10$, we need to subtract 5.5 dB. If each filter is fed from a separate and uncorrelated noise source, the additional correction would only be $-1.5$ dB.

### 6.5.4 Clock Edge Divider Effects

An RF oscillator is often followed by a clock edge divider, as shown in Fig. 6.9. This is motivated by the following:

1. Quadrature-based receiver architecture requires precise generation of four $90°$-spaced clocks (Fig. 1.13). The most straightforward method is to run the oscillator at double the frequency and use a quadrature divider.
2. Coupling of the strong RF power amplifier output back into the oscillator is decreased (see Fig. 1.11). In this case only the second harmonic of the PA output, which is much weaker, can affect the oscillator.
3. The most optimal quality factor, $Q$, of the RF inductor might happen to lie far above the operational RF band, so it is advantageous to operate the oscillator at high frequency and divide it down.

The analysis below examines how the addition of a clock edge divider affects the phase noise modeling of the oscillator. In case of flat noise, the jitter contributions are uncorrelated. The same value of rms jitter is to be used for an oscillator core operating at double the frequency. Since the jitter contributions are uncorrelated, it is acceptable to discard every other edge. The rms jitter value in time units does not change in the case of clock edge divisions. Another way of looking into this: Aliasing by 2 doubles the noise spectral density in $rad^2/Hz$, and since the period is now halved, the noise value in time units does not change. Hence, for the jitter case,

$$\sigma_{\Delta_t,0} = \sigma_{\Delta_t,1} \tag{6.19}$$

The equivalent rms wander $\sigma_{\Delta_T,0}$ at the oscillator core is lower by $\sqrt{N}$ than the wander $\sigma_{\Delta_T,1}$ at the output of the edge divider, where $N$ is the division ratio, according

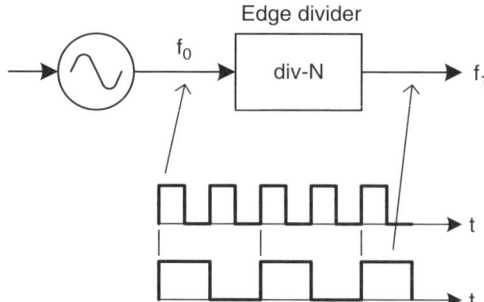

**Figure 6.9** DCO with an edge divider. In this timing example, $N = 2$. (From [3], © 2005 IEEE.)

to the law of power addition of independent and identically distributed random variables. Hence, for the wander case,

$$\sigma_{\Delta T,0} = \frac{1}{\sqrt{N}} \sigma_{\Delta T,1} \tag{6.20}$$

### 6.5.5 VHDL Model Realization of a DCO

For event-driven oscillator implementation in VHDL, Eq. 6.1 is rewritten here as

$$t_j[i] = \sum_{l=1}^{i} (T_0 + \Delta t[l] - \Delta t[l-1]) \tag{6.21}$$

with $\Delta t[0] = 0$, whereas Eq. 6.5 is rewritten here as

$$t_w[i] = \sum_{l=1}^{i} (T_0 + \Delta T[l]) \tag{6.22}$$

where $\Delta t[i]$ and $\Delta T[i]$ are Gaussian-distributed random variables with $\sigma_{\Delta t}$ (Eq. 6.3) and $\sigma_{\Delta T}$ (Eq. 6.7) standard deviations, respectively. The nonaccumulative and accumulative nature of the noise is readily seen from these two equations. The jitter and wander (including the $1/f$ noise) contributions, $\Delta t[i]$ and $\Delta T[i]$, are normally combined into one equation:

$$t_{jw}[i] = \sum_{l=1}^{i} (T_0 + \Delta t[l] - \Delta t[l-1] + \Delta T[l]) \tag{6.23}$$

A diagram of a VHDL model of a DCO is shown in Fig. 6.10. Referring to the implementational block diagram of the DCO gain paths in Fig. 3.9, DCO_IN_P, DCO_IN_A, DCO_IN_TI, and DCO_IN_TF are the digital std_logic_vector inputs deviating the DCO oscillating frequency by controlling the $LC$ tank capacitance of the PVT, acquisition, tracking-integer, and tracking-fractional varactor banks, respectively. Signed-number integer representations of these inputs are multiplied by their respective time-unit deviations from the natural period: DCO_QUANT_P, DCO_QUANT_A, and DCO_QUANT_T. They are VHDL generics rounded off to the closest femtosecond, whose calculated values were shown in Table 2.2 for the middle of the BLUETOOTH band. Their outputs are then summed up to create a composite period deviation signal of VHDL time-type. This signal is then subtracted from the natural or center oscillating period DCO_PER_0.[1] This time signal controls the instantaneous period of DCO oscillation. The actual code that implements the *period-controlled oscillator* (PCO) is given in Section C.2. For performance reasons, foreign C function calls are used

---
[1]Period deviation is opposite in sign to frequency deviation.

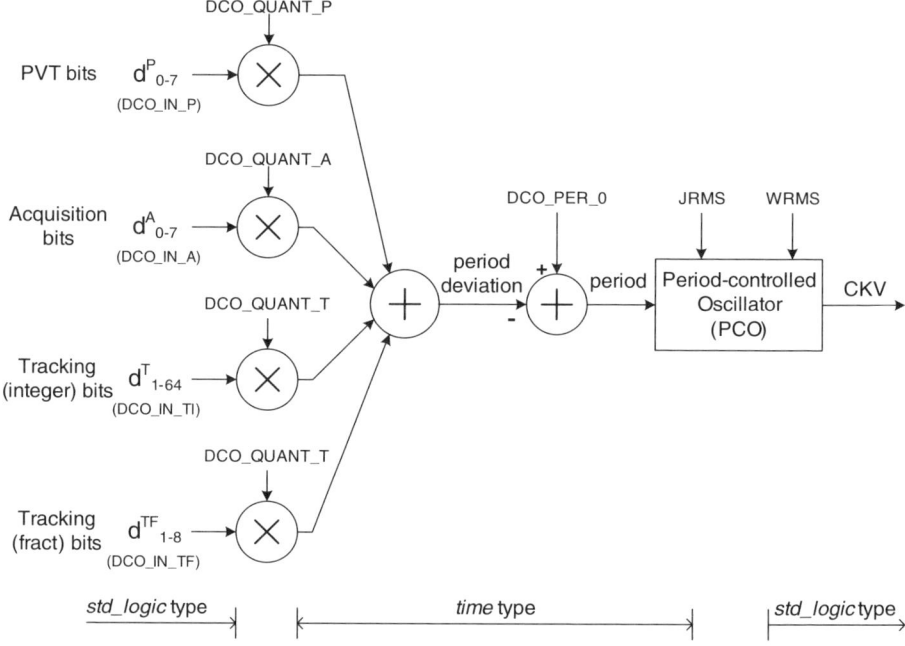

**Figure 6.10** DCO time-domain model in VHDL. (From [3], © 2005 IEEE.)

to calculate jitter and wander perturbations, which would be computationally expensive if implemented entirely in VHDL.

Level 2 VHDL code of a DCO oscillator is given in Section C.1. It uses digital tuning input ports of integer type. The conversion from std_logic_vector to integer is performed at the higher level in the synthesizable connectivity-only DCO wrapper.

### 6.5.6 Support of Physical $K_{DCO}$

The relationship that ties the frequency step $\Delta f^T$ of a DCO to the switchable capacitance $\Delta C^T$ is governed by Eq. 2.14 and is repeated here for convenience:

$$\Delta f^T(f) = -2\pi^2(L\Delta C^T)f^3 \qquad (6.24)$$

where $L$ is the *LC* tank inductance. Both $L$ and $\Delta C^T$ are constants for a stable PVT. They are the only unknowns and subject to the PVT variations on the right-hand side of the equation, so it makes sense to group them together as a product. The resonating frequency $f$ is controlled by the total capacitance $C$ of the *LC* tank. Of course, $f$ is known precisely (within a *frequency control word* FCW resolution) by virtue of the ADPLL loop operation. Since $K_{DCO} \equiv |\Delta f^T|$, Eq. 6.24 is rewritten as

$$K_{DCO}(f) = 2\pi^2(L\Delta C^T)f^3 \qquad (6.25)$$

which shows a strong (i.e., cubic) dependency on the *LC* tank resonating frequency. Taking the derivative of $K_{DCO}$ results in

$$\frac{dK_{DCO}(f)}{df} = 6\pi^2(L\Delta C^T)f^2 = 3\frac{K_{DCO}(f)}{f} \quad (6.26)$$

From this we derive the $K_{DCO}$ relative change with respect to the relative frequency deviation:

$$\frac{\Delta K_{DCO}(f)}{K_{DCO}(f)} = 3\frac{\Delta f}{f} \quad (6.27)$$

Let $K_{DCO}(f_0)$ be the DCO gain at a certain $f_0$. Changing the *LC* tank frequency by virtue of tuning the total capacitance *C* will result in a new $K_{DCO}(f)$ of

$$K_{DCO}(f) = K_{DCO}(f_0)\frac{f^3}{f_0^3} \quad (6.28)$$

Translating $K_{DCO}(f)$ (hertz) to the corresponding $\Delta T$ (seconds) using the linearity equation Eq. 2.20 simplifies to

$$\Delta T(f) = \frac{K_{DCO}(f_0)}{f_0^3}\frac{1}{T} \quad (6.29)$$

where *T* is being tracked as the period in the DCO model in Fig. 6.10.

## 6.6 MODELING METASTABILITY IN FLIP-FLOPS

Conventional synchronous digital design deals with metastability conveniently by specifying and meeting the setup and hold time of sequential components such as flip-flops and latches. A major inconvenience in ADPLL architecture is the need to deal with metastability as a *normal* occurrence expected in the course of device operation. Like any other system with mutually asynchronous clocks, this architecture requires a fair amount of attention to avoid *synchronization failures*, which are said to occur if the system attempts to use a signal while it is in a metastable state. A common method for dealing with synchronization failures is to increase *mean time between failures* (MTBF) sufficiently by cascading synchronizers such that occasional errors are no longer considered a problem. However, unlike conventional systems incorporating synchronizers in which the sufficient solution is to *resolve* metastability, this architecture also requires that the metastability be *stochastically avoided*.

The following list summarizes two ADPLL areas requiring special attention to metastability in the design, modeling, and simulation phases.

1. Sampling of the delayed replica of the DCO clock by the frequency reference in the time-to-digital converter (TDC); requires metastability resolution; described in Section 4.6.

2. Frequency reference retiming by the DCO clock in the clock retiming circuit; requires metastability avoidance; described in Section 4.7.3.

Attempts have been made to model the metastable behavior of flip-flops and latches in a hardware description language such as VHDL. In reference [91] an exponential model of metastable behavior is described and implemented for a fundamental element of a set/reset latch. Unfortunately, the model is quite complex, thus degrading the simulation performance. In this book we describe a much simpler but effective method to model metastability.

Metastability is modeled in critical flip-flops by continuous inspection of the timing relationship between the data input (D) and clock (CLK) pins and producing an *unknown* output X on the data output (Q) pin if the D-to-CLK skew falls within the forbidden metastable window. Referring to Fig. 4.22, the metastable window is defined as an $x$-axis region (D-to-CLK timing skew) such that the CLK-to-Q delay on the $y$-axis is longer by a certain amount than the nominal CLK-to-Q delay. For example, if the nominal CLK-to-Q delay is 100 ps when the D-to-CLK timing is far from critical, the metastability window would be 20 ps if one can tolerate CLK-to-Q delay increase by 90 ps. If one can tolerate a higher CLK-to-Q increase of 170 ps, the metastable window would drop by half to 10 ps. A question could be asked as to how far this window can extend. The limitation lies in the fact that for a tight D-to-CLK skew, the noise or other statistical uncertainty, such as jitter, could arbitrarily resolve the output such that the input data is missed. Therefore, for a conventional definition of setup time, not only must the output be free of any metastable condition, but the input data have to be captured correctly. For this reason, the setup and hold times are conservatively defined in standard-cell libraries for an output delay increase of 10 or 20% over nominal. The specific nature of TDC vector capturing does not require this restrictive constraint. Here, any output-level resolution is satisfactory for proper operation as long as it is not metastable at the time of capture, and consequently, the metastable window could be made arbitrarily small. In fact, in the design implemented, this timing window is made narrower than 1 fs.

The timing diagram of a flip-flop is redrawn (for multiple cycles) in Fig. 6.11 to expound further on this idea. The $x$-axis denotes the periodic D-to-CLK relationship with the repetition period equal to the CLK clock period of $T_0$. The $y$-axis denotes the CLK-to-Q delay, which nominally equals $t_{CLK-Q(\text{nom})}$ but increases exponentially as data transition closer to the capturing clock. Here, for simplicity, it is shown that the exact metastable point where the flip-flop delay becomes very large corresponds to the perfect alignment of clock and data. In practical circuits, this need not be the case, and as Fig. 4.22 of the tactical flip-flop reveals, the metastable point is skewed by 47 ps, with the data lagging the clock. This particular condition is subject to the specific circuit implementation. Just to demonstrate this point, the standard high-performance flip-flop in Fig. 4.21 shows that the metastable window lies on the opposite side, where the data lead the clock. This plot actually reveals two metastable windows, separated by about 65 ps, due to the asymmetry between the rise and fall transitions of a conventional master–slave fully static CMOS topology. This single aspect of the traditional flip-flop makes it unusable

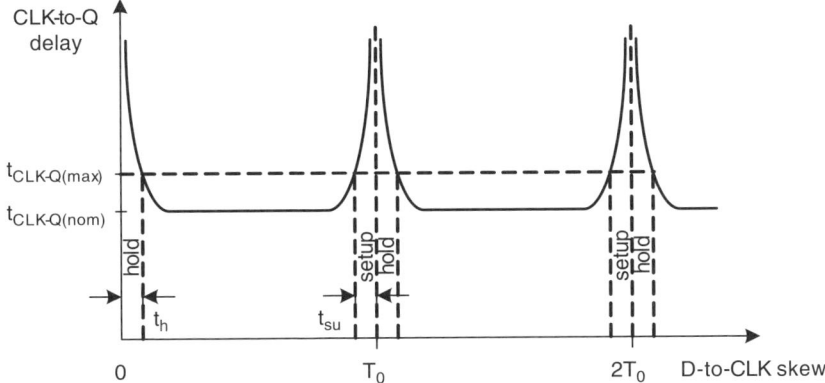

**Figure 6.11** Timing diagram of a flip-flop.

for TDC implementation, which requires resolution in the range 20 to 40 ps while maintaining good linearity. Consequently, a symmetric sense-amplifier-based flip-flop appears to be a better choice.

The setup and hold violation windows ($t_{su}$ and $t_h$, respectively) are centered around the clock edge and are defined for D-to-CLK conditions such that $t_{CLK-Q} > t_{CLK-Q(max)}$, where $t_{CLK-Q(max)}$ is the maximum allowed flip-flop output delay.

The timing periodicity shown in Fig. 6.11 is a general case where the clock multiplicity factor (i.e., the distance in number of edges between the launching and capturing clocks), could be greater than 1. Generally, any timing relationship is valid as long as the D-to-CLK skew does not fall into the forbidden regions.

In our system, an advantage is taken of the fact that VHDL already supports a nine-valued digital bit type `std_logic`, which is an IEEE standard, and one of its levels is X, defined as "forcing unknown." Referring to Fig. 4.14, the TDC flip-flops are modeled such that an X is generated on a Q output to indicate a metastable region of its D-to-CLK timing. The X could then be detected in the pseudo-thermometer-code edge detector and replaced with a randomly picked 0 or 1. It is the nature of this circuit that the metastable condition will be resolved within one full FREF clock cycle. However, due to noise, the resolution outcome is not known at the time of sampling. Therefore, a statistically probable binary result appears to be a good modeling choice for this phenomenon. For example, if the Q vector is "00111X0000...", there is a 50% chance of resolving it to either "0011100000..." or "0011110000", with the respective decoded TDC_RISE outcome of 5 or 6. The measurement error is thus contained to a single LSB.

Metastability modeling of the clock retiming circuit of Fig. 4.24 serves another purpose. These registers also generate an X on the Q output whenever the D-to-CLK timing relationship is in the forbidden region. However, no resolution attempt is carried out. This is in accordance with the intended operation of the clock retiming circuit, which should avoid the metastable candidate altogether. At the bottom of Fig. 4.24 are shown two potential metastable events and how the circuit keeps away from them. Of course, the MUX select signal that originates in TDC needs to be resolved in the same manner as with the pseudo-thermometer-code detector.

The VHDL code for the tactical flip-flop is presented in Section C.3. A TDC output decoder capable of handling metastable inputs appears in Section C.4.

## 6.7 SIMULATION RESULTS

### 6.7.1 Time-Domain Simulations

Figure 6.12 shows a composite trajectory plot of instantaneous frequency deviation while illustrating operation of various PLL modes. The $x$-axis is the time evolution in CKV clock units (about 417 ps/cycle). The $y$-axis is the frequency deviation from

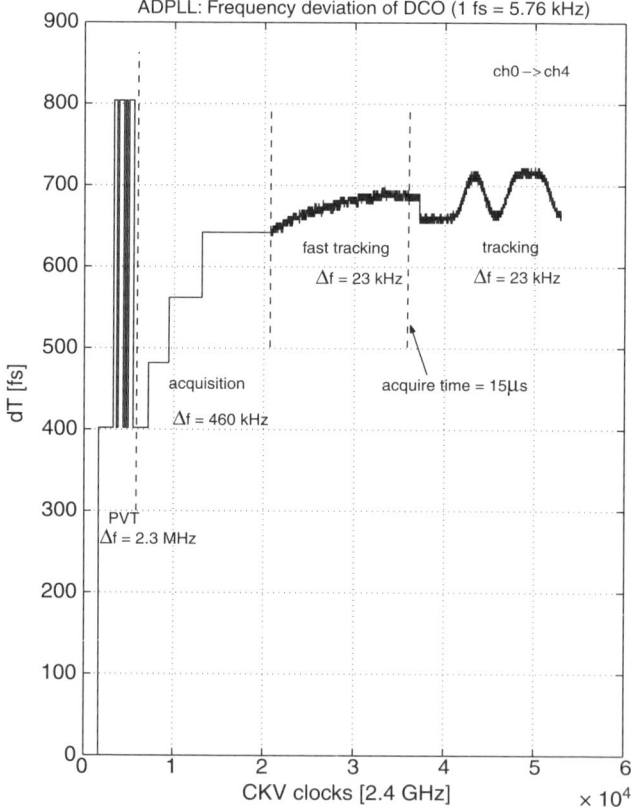

**Figure 6.12** Simulation plot of transmitter modulation at 2.4-GHz RF output; $y$-axis: $\Delta f$ in femtoseconds (1 fs = 5.75 kHz); $x$-axis: time in a 417-ps RF clock periods. (From [3], © 2005 IEEE.)

an initial value of 2402 MHz (channel 0) expressed in femtoseconds, where 1 fs corresponds to 5.77 kHz (see Table 2.1).

The initial starting point is the center frequency set to channel zero. At power-up, a "cold start" to channel 4 at 4 MHz away is initiated. The ADPLL operates first in the PVT mode by enabling the *PVT oscillator controller* (OP). This controller makes very coarse 2.3-MHz adjustments to the frequency, which are equivalent to 399-fs adjustments to the oscillator period. Next, the output of the PVT controller is put on hold and the *acquisition oscillator controller* (OA) is enabled. The acquisition controller quickly brings the frequency near the channel selected in 460-kHz steps. After acquisition of the channel is complete, the output of the OA controller is put on hold and the *integer tracking oscillator controller* OTI and *fractional tracking oscillator controller* OTF are enabled. The finest selection of the channel requested can be accomplished only by using tracking bank varactors with all the resolution enhancement techniques possible for this capacitor bank. The dynamic range of this mode has to cover the frequency resolution grid of the preceding acquisition mode. In the fast-tracking mode, the frequency steps are the finest (less than 1 kHz), but the loop bandwidth could be as wide as in the acquisition mode. A tracking mode featuring narrow loop bandwidth completes the channel acquisition and frequency locking.

The entire locking process takes 15 μs with a reference frequency of 13 MHz (about 36,000 CKV cycles or 196 FREF cycles). Upon reaching an acquisition steady state, GFSK data modulation takes place.

The plot also shows how the PLL deals naturally with the frequency quantization effects of an oscillator in the PVT mode by chattering (dithering) between the upper and lower frequency levels allowed for the frequency desired. This phenomenon was described in Section 2.6. As the simulations reveal, a closed-loop PLL system performs an inherent time-dithering operation if the DCO frequency granularity is finite. The mechanism is as follows: If the long-term average oscillating frequency lies between two neighboring steps, the PLL will operate at the lower frequency for some time until the accumulated phase error exceeds the resolution threshold. At that point the PLL will switch to the higher frequency until the accumulated negative phase error forces it to go back to the first lower frequency. This type of dithering of a single quantized level is a slow process, on the order of FREF clock cycles, and it is not observed in the acquisition mode. It is because the synthesizer does not stay there long enough for the loop to accumulate enough phase to trigger a phase detector output change that would make a corrective action.

The closed-loop low-speed dithering mechanism described above should be distinguished from open-loop $\Sigma\Delta$ high-speed dithering in the tracking mode.

### 6.7.2 Frequency-Deviation Simulations

Figure 6.13 shows a time-domain plot of an instantaneous frequency deviation at the transmitter RF output when the $1/f^2$ noise (wander) of the oscillator is turned off. However, the DCO still contains an electronic thermal noise floor (jitter) of $-150$ dBc. Since the VHDL simulator is foreign to the frequency concept, the

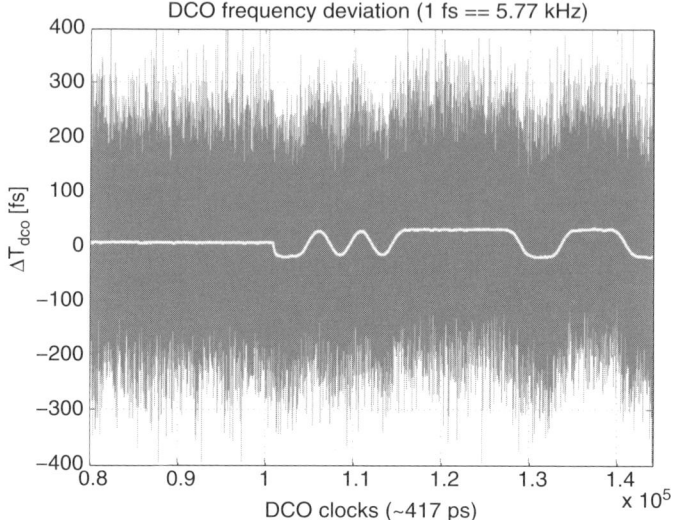

**Figure 6.13** Instantaneous period deviation of CKV for $\alpha = 1/2^8$ with a phase-noise floor of only $-150$ dBc; the thick white line is a leaky integration.

$y$-axis represents the related period deviation in femtoseconds. According to Table 2.1, 1 fs of the DCO period deviation is equivalent to 5.77 kHz of the DCO frequency deviation at the beginning of the BLUETOOTH band. A full symbol deviation of 160 kHz translates to about 28 fs. It might seem a surprise that signal content is completely sunk within noise whose peaks are an order-of-magnitude larger than the signal peaks. Fortunately, performing simple low-pass filtering through a leaky integration (first-order IIR filter with unity gain at dc) reveals a very clean modulating signal. The reason is simple: White electronic thermal noise has most of its energy contained in higher frequencies. Due to the sampling nature of a discrete-time oscillator, its noise extends from dc to one-half of the RF frequency of 2.4 GHz. Since 99.9% of the signal energy is contained in the 1-MHz band, the noise portion that cannot be distinguished from the signal by virtue of falling into the signal band is

$$\frac{1 \text{ MHz}}{2400 \text{ MHz}/2} = 0.00083 = 0.083\% \tag{6.30}$$

The out-of-band components are easily filtered out.

Figure 6.14 shows a similar time-domain plot but with the $1/f^2$ noise now turned on. The maximum frequency deviation peaks due to the noise are about the same, but the filtered component shows some distortion. In this case, the $1/f^2$ component contains a large energy component at lower frequencies that cannot be separated through linear filtering.

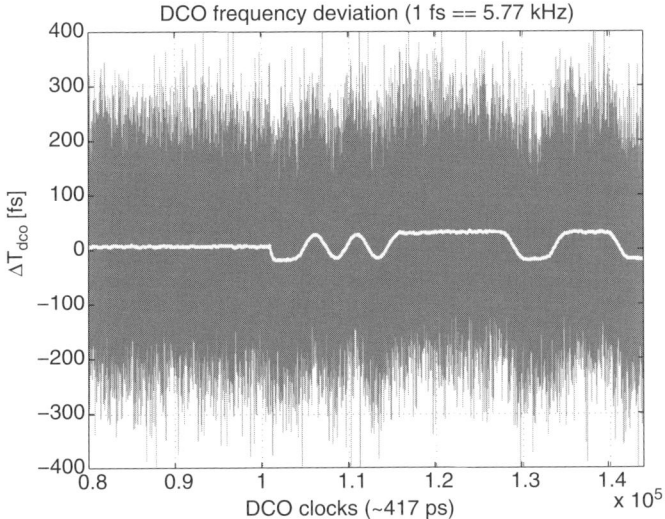

**Figure 6.14** Instantaneous period deviation of CKV for $\alpha = 1/2^8$ with a phase noise floor of $-150$ dBc and $1/f^2$ noise of $-105$ dBc at 500 kHz; the thick white line is a leaky integration.

The DCO jitter source is turned off in Fig. 6.15, leaving only the wander component, which is quite small in amplitude ($\Delta T$) terms but distorts the signal mainly through lower-frequency components.

### 6.7.3 Phase-Domain Simulations of Transmitters

Figure 6.16 (left) shows a power spectral density of the GFSK digital filter output $y(t)$ of Fig. 5.13. The filter coefficients were designed for the nominal BLUETOOTH bandwidth–symbol period product $BT = 0.32$. The filter output is in *frequency control word* (FCW) format (defined in Eq. 4.9), which is interpreted as a carrier frequency deviation. A nominal modulation index of $h = 0.5$ is used for FCW construction. Figure 6.16 (right) shows the power spectral density of a GFSK-modulated RF carrier such as the one described by Eq. 1.18. Its center lies at the carrier frequency.

### 6.7.4 Synthesizer Phase-Noise Simulations

Figure 6.17 shows the synthesizer phase-noise spectrum for a proportional loop gain setting of $\alpha = 2^{-8}$. The reference frequency is 13 MHz, so the PLL bandwidth is calculated as $(13 \text{ MHz}) (2^{-8}/2\pi) = 8086 \text{ Hz}$. The FREF phase noise (modeled in a similar manner as in the DCO) is set to $-130$ dBc/Hz, which corresponds roughly to an inexpensive, readily available crystal oscillator. The FREF phase noise could also be normalized to the DCO clock cycle, which is equivalent to

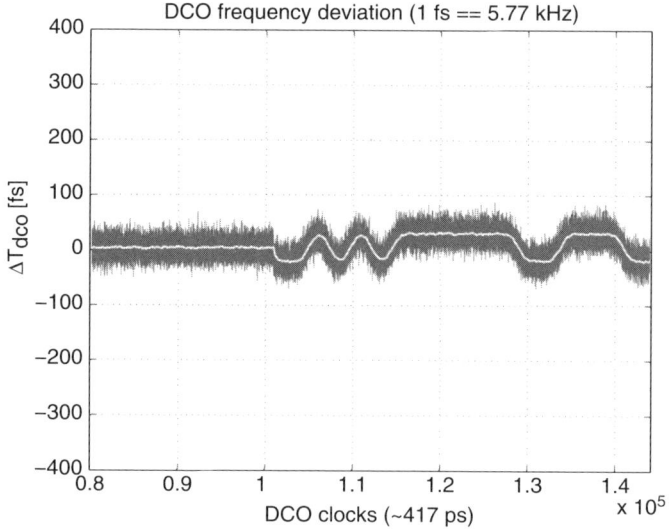

**Figure 6.15** Instantaneous period deviation of CKV for $\alpha = 1/2^8$ with $1/f^2$ noise of only $-105$ dBc at 500 kHz; the thick white line is a leaky integration.

multiplication by the frequency-division ratio:

$$-130\,\text{dBc/Hz} + 20\log(2402\,\text{MHz}/13\,\text{MHz}) = -130 + 45.33$$
$$= -84.7\,\text{dBc/Hz} \quad (6.31)$$

Figure 6.17 shows two distinct regions of the up-converted thermal noise of a $-20$ dB/decade slope, which was introduced in Fig. 1.4. The first region contains noise frequency components within the loop bandwidth that are corrected by the

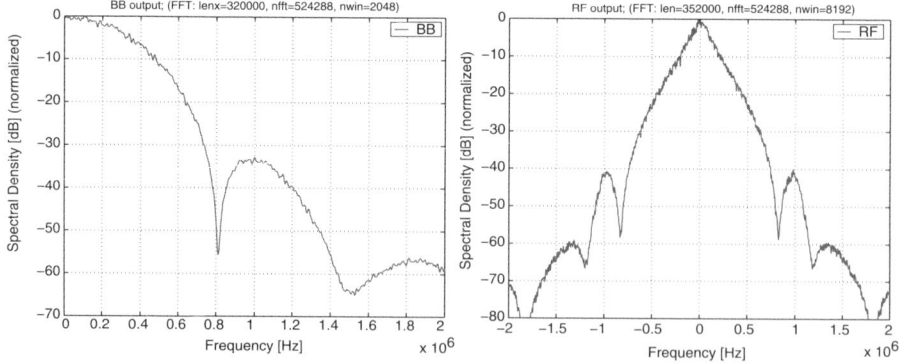

**Figure 6.16** GFSK-modulated spectra: (left) pulse-shaping filter output; (right) RF carrier output with center at the carrier frequency.

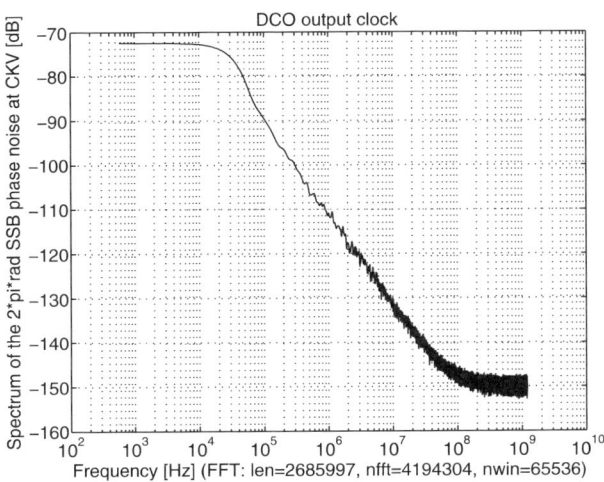

**Figure 6.17** Simulated spectrum of a CKV clock for $\alpha = 1/2^8$.

PLL. Since the loop is of type I with a slope of 20 dB/decade (see Eq. 4.50), the noise characteristic becomes flat. In the second region lie frequency components outside the loop that could not be corrected and which exhibit the original $-20$ dB/decade slope. The electronic noise floor of $-150$ dBc/Hz at high frequencies confirms Eq. 6.3. The phase-noise readout of $-105$ dBc/Hz at 500-kHz offset confirms Eq. 6.7.

Figure 6.18 shows simulated phase noise for $\alpha = 2^{-6}$. To be noted is the wider and lower flat region resulting from wider PLL bandwidth.

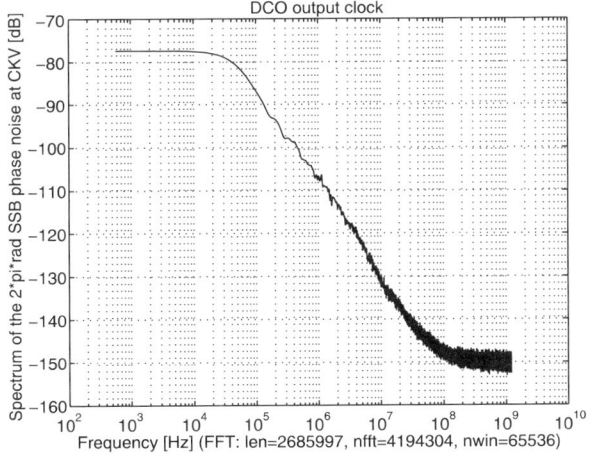

**Figure 6.18** Simulated spectrum of a CKV clock for $\alpha = 1/2^6$.

**212**  BEHAVIORAL MODELING AND SIMULATION

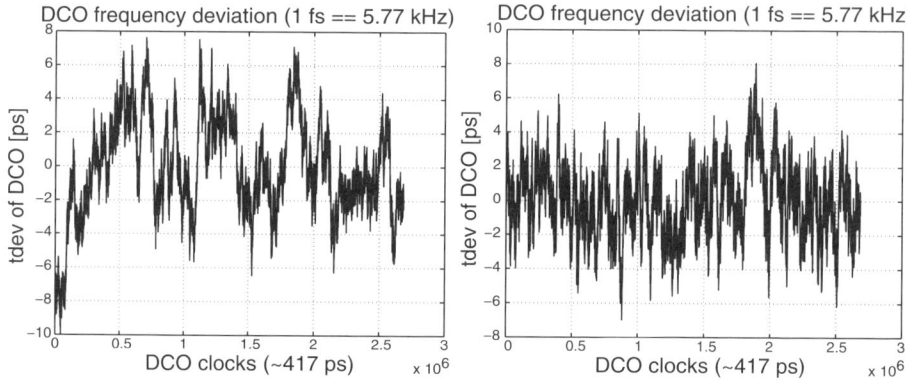

**Figure 6.19** Simulation plot of TDEV for $\alpha = 1/2^8$ (left) and $\alpha = 1/2^6$ (right).

Figure 6.19 shows the timing deviation TDEV (random walk from the ideal timing instances) evolution for $\alpha = 2^{-8}$ and $\alpha = 2^{-6}$, respectively. As demonstrated above in the frequency-domain plots, the wider PLL bandwidth removes more lower-frequency DCO phase noise components.

## 6.8 SUMMARY

In this chapter we described simulation and modeling methodology for an all-digital RF frequency synthesizer and transmitter. The complete model uses time-domain event-driven constructs. A novel method of modeling the up-converted $1/f$ noise is presented. The examples apply to a digitally controlled oscillator, even though the modeling techniques are readily extendable to a general class of oscillators. The simulator proposed is an event-driven VHDL engine. Sharing the simulation engine for an RF transmitter with a digital back end enables us to seamlessly employ fully digital frequency synthesizers using sophisticated DSP algorithms, realized in the most advanced deep-submicron digital CMOS processes. We have demonstrated the feasibility and attractiveness of employing event-driven simulation methodology for RF circuits within large system-on-chip designs.

# CHAPTER 7

# IMPLEMENTATION AND EXPERIMENTAL RESULTS

In this chapter, implementation of a frequency synthesizer is described together with a full top-level BLUETOOTH transmitter core. First, a top-level block diagram is presented and all major constituent blocks are listed. Subsequently, chip micrographs and an IC chip evaluation board are described. Next, the characterization data obtained from an ADPLL-based BLUETOOTH transmitter are presented. The key performance measures for an unmodulated synthesizer are its phase noise and spurious tone output. The synthesizer, without the frequency modulation capability, could also be used as a *local oscillator* (LO) to perform frequency translation in a receiver path.

## 7.1 DSP AND ITS RF INTERFACE TO DRP

An overview of the IC chip is presented in Fig. 7.1. A frequency synthesizer is combined with a DSP to implement the complete transmitter. The DSP, Texas Instruments TMS320C54x equipped with 28 kilowords of RAM and 128 kilowords of ROM, contains typical peripherals used for cellular applications: timer, API, serial port, and XIO parallel bus interface, including interrupts and wait states. The XIO bus is a dedicated high-speed bidirectional parallel interface of 8-bit address space and 16-bit data registers that couple the digital RF transmitter (DRP$^{TM}$) directly to the DSP. The transmitter registers are mapped into the DSP XIO space and can be accessed using read and write instructions. The DRP is the sole provider of the DSP clock. To avoid injection pulling of the DCO to the

---

*All-Digital Frequency Synthesizer in Deep-Submicron CMOS*, by Robert Bogdan Staszewski and Poras T. Balsara
Copyright © 2006 John Wiley & Sons, Inc.

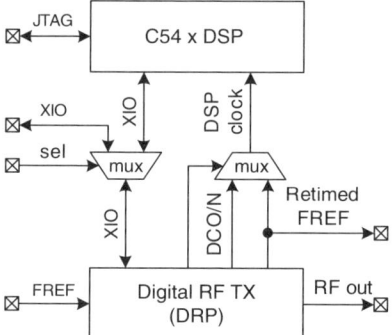

**Figure 7.1** Single-chip digital RF transmitter with a DSP.

$n$th harmonic of the reference frequency, the DSP runs on either the retimed FREF clock by the DCO edge or the down-divided DCO clock. A watchdog timer switches to the FREF clock automatically if the clock failure selected is detected.

## 7.2 TRANSMITTER CORE IMPLEMENTATION

The implementation details of a test chip transmitter are shown in Fig. 7.2. The design is executed based on Texas Instruments' ASIC digital flow with special adjustments due to the analog/RF content. The transmitter core is partitioned into the following blocks:

- *Low-speed digital* (LSD) *superblock* with clocks running on a retimed reference frequency of 13 MHz.
- *High-speed digital* (HSD) *subchip* with clocks much faster than the reference frequency. It contains a variable phase accumulator (running at 2.4 GHz) and $\Sigma\Delta$ dithering of the oscillator tracking bank varactors (running at 600 MHz).
- *Time-to-digital converter* (TDC) *ASIC cell* operating at FREF but with high timing precision requirements. The cell also contains a FREF clock retiming circuit, which operates primarily at 2.4 GHz.
- *DCO + PA ASIC cell* with only digital I/Os, combining a digitally controlled oscillator and class E *power amplifier* (PA) as a single RF module. The PA consists of a digital switch transistor with a matching network built from parasitic components and components external to the IC chip. The RF devices are built using the existing devices available to DSP designers. The planar inductor is constructed using metal layers 3 through 5. The varactors use PPOLY/NWELL MOS structures with fully digital control of the oscillating frequency [50].
- *Control bus interface* (CTL) *superblock*, which allows the transmitter to be controlled through the XIO parallel-port interface.

## 7.2 TRANSMITTER CORE IMPLEMENTATION

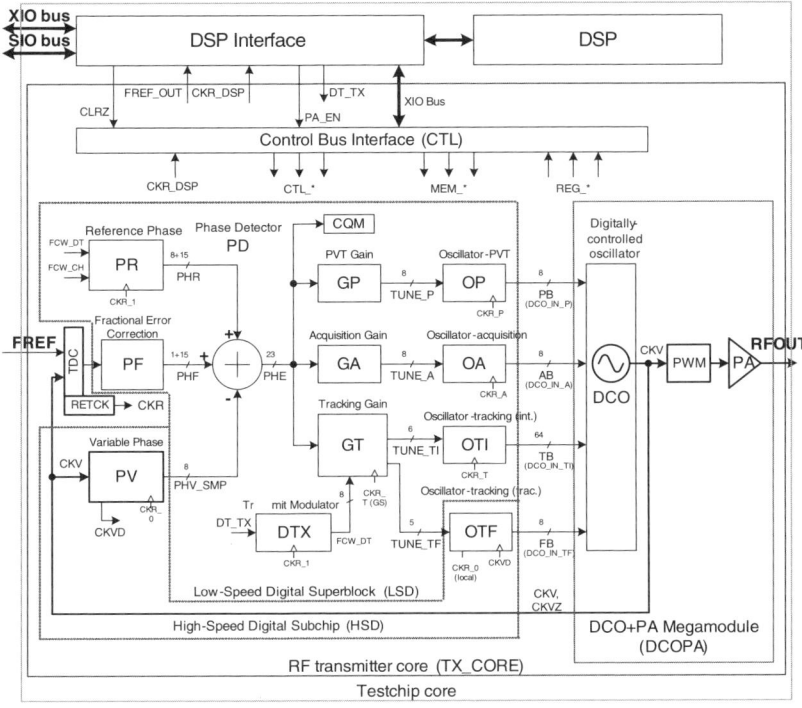

**Figure 7.2** ADPLL-based transmitter core.

The transmitter core is encapsulated by the test chip core, which is the highest level of hierarchy below the final chip level, which contains I/O pads and buffers. The test chip core also contains an interface and wrapper circuitry for the TMS320C54x DSP, commonly used in cellular phones. The control interface block (CTL) allows a transmitter to be controlled from one of several sources: external XIO bus, internal XIO bus from DSP, serial SIO bus, and JTAG interface. The transmitter core has a dedicated 8-bit address space of 16-bit data registers.

All the digital blocks are synthesizable from the *register transfer level* (RTL) subset of VHDL code using a SYNOPSYS design compiler. This includes circuits running with a 2.4-GHz clock, such as variable-phase PV. The blocks are auto-placed and auto-routed using the AVANTI layout software package. The analog and RF blocks are modeled using behavioral VHDL code. Top-level interconnects are described in a structural VHDL, which are synthesized into a netlist without the need to draw schematics.

The implemented transmitter is digitally intensive. The only blocks that follow the established RF/analog design practices internally are DCO and PA, even though at their top I/O level they are characterized using digital flow. These two blocks are vastly different from their conventional counterparts and they seamlessly fit the digital design proposed. Even though they are built of only a handful of

**216** IMPLEMENTATION AND EXPERIMENTAL RESULTS

**Table 7.1 Top-Level Transmitter Core Building Blocks**

| Block Name | Block Abbreviation | Figure |
|---|---|---|
| Reference phase accumulator | PR | 4.65 |
| Variable-phase accumulator | PV | 4.12 |
| Fractional-phase error estimator | PF | 4.13 |
| Time-to-digital converter | TDC | 4.14 |
| FREF clock retiming | RETCK | 4.24 |
| Phase detector | PD | 4.6 |
| Transmitter modulator | DTX | 5.13 |
| PVT gain | GP | 3.9 |
| Acquisition gain | GA | 3.9 |
| Tracking gain | GT | 3.9 |
| PVT oscillator interface | OP | 3.11 |
| Acquisition oscillator interface | OA | 3.12 |
| Tracking oscillator interface | OTI & OTF | 3.16 |
| Digitally controlled oscillator | DCO | 2.14 |
| Pulsewidth modulation | PWM | 5.21 |
| Power amplifier | PA | 5.18 |

transistors, they occupy almost half of the total RF area, as shown in Fig. 7.4. Table 7.1 lists the top-level transmitter core building blocks with their name abbreviations and the figures in which the blocks are described.

The sequencer used to control the operational modes of the synthesizer, such as PVT, acquisition, fast tracking, and tracking, was implemented in the DSP software. It is also possible to step through the operating modes manually by writing external XIO commands. A clock quality monitor (CQM) is used to gather certain statistics about ADPLL operation. It is described in Section 4.14.1.

## 7.3 IC CHIP

The test chip of the BLUETOOTH transmitter described is implemented in a Texas Instruments' 130-nm digital CMOS process. The key process technology parameters are described in Table 1.1, which features a routed digital gate density of 150 equivalent gates per $mm^2$. Figure 7.3 is a die microphotograph of the transmitter test chip. The total silicon dimensions are 3290 μm × 3290 μm. This includes 160 μm dedicated on each side for I/O pads. The companion TMS320C54x DSP, which includes typical peripherals used in 2G cellular phones, occupies 6 $mm^2$ (2430 μm × 2470 μm).

Figure 7.4 is a die microphotograph of the RF transmitter area, which is located in the lower-left corner. Its area is only 0.54 $mm^2$, which is the smallest ever reported. The LC tank inductor occupies a 270 μm × 270 μm square and is clearly discernible as the biggest single component on the entire chip. The synthesized RF output

**Figure 7.3** Micrograph of the BLUETOOTH transmitter test chip.

is buffered to the external pins through a class E power amplifier (PA), which was chosen for its digital-friendly characteristics. This photograph dramatically illustrates the high cost, in terms of digital logic, of conventional RF components in high-density modern CMOS processes and illustrates the benefits of digital implementation of RF synthesizers. This provides additional incentives to minimize

**Figure 7.4** Micrograph of the RF transmitter area in the lower-left corner of the chip.

**Figure 7.5** Evaluation board.

the number of classical RF components and to research novel digitally intensive architectures of RF functions.

The test chip is bonded and encapsulated in an 80-ball 5 × 5 mm MicroStar Junior ball-grid array (BGA) package.

## 7.4 EVALUATION BOARD

Figure 7.5 is a photograph of an evaluation board. The printed circuit board (PCB) is constructed of six layers of standard FR4 material. The test chip is located at the center. The 2.4-GHz RF output, the 13-MHz frequency reference (FREF) input, and its retimed output (BBCLK) are connected using semiprecision *subminiature A* (SMA) connectors, which provide 0- to 10-GHz broadband performance with low reflections and constant 50-$\Omega$ impedance. The connectors on the right-hand side attach the evaluation board to a PC interface board (not shown), whose purpose is to control the test chip by reading and writing its registers by means of a *graphical user interface* (GUI) computer program or through Code Composer$^{TM}$.

## 7.5 MEASUREMENT EQUIPMENT

The RF output and frequency reference input are connected using semiprecision *subminiature A* (SMA) connectors with a characteristic impedance of 50 $\Omega$. The reference input is provided using an HP8662A synthesizer signal generator which has a phase noise of about $-140$ dBc/Hz at a 20-kHz offset. The RF output port is connected to an HP8563E spectrum analyzer. The phase noise of the closed-loop PLL are measured using an HP85671A phase noise measurement utility. The eye-diagram measurement and the modulated spectrum were provided using a Rohde & Schwarz (R&S) FSIQ-7 signal analyzer. The measurement system is set up inside a Faraday cage, which blocks virtually all RF signals present in the ambient environment.

## 7.6 GFSK TRANSMITTER PERFORMANCE

The following figures demonstrate effective and correct operation of the algorithm proposed by displaying an FM-demodulated TX output port with the 1-Mb/s GFSK-modulated 2.4-GHz BLUETOOTH RF signal. An eye diagram of the pseudorandom-modulated data is shown in Fig. 7.6. It was taken with an R&S signal analyzer, which down-converts, FM demodulates, and then plots and uses it to calculate

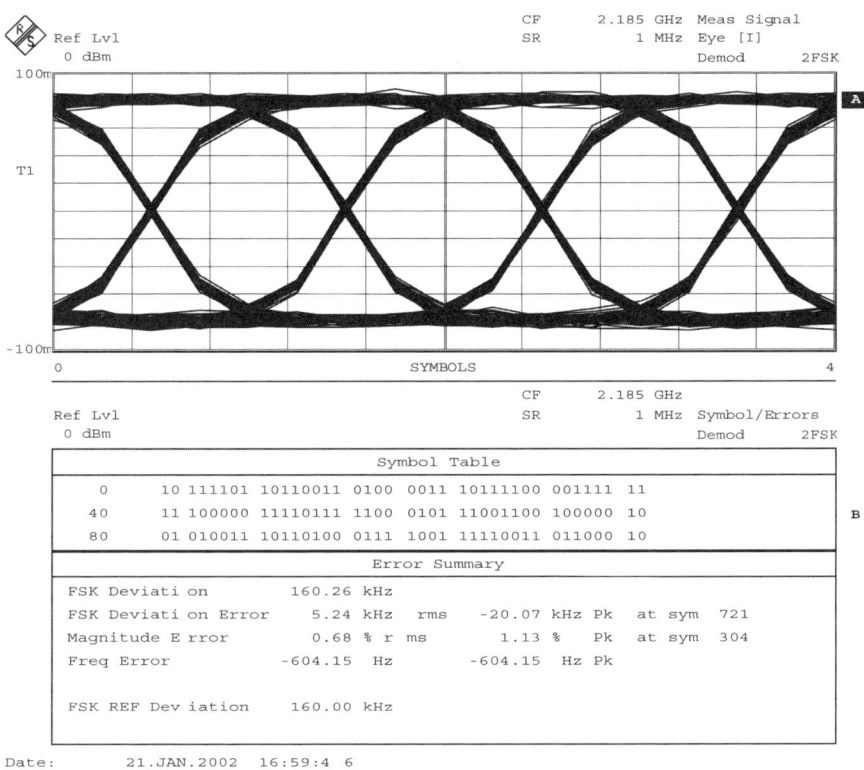

**Figure 7.6** Eye diagram of PN9 pseudorandom data at 2185 MHz, BW = 4 kHz, and room temperature using the statistics measured.

**220** IMPLEMENTATION AND EXPERIMENTAL RESULTS

various statistics. The y-axis is the frequency deviation from the carrier, and the x-axis is the time evolution in 1-μs symbols. The time base of the instrument is synchronized externally to the symbol generator to avoid cycle slipping. The peak-to-peak frequency deviation was measured to be 320.52 kHz, which is very close to the theoretical value of 320 kHz, calculated as a modulation index of 0.32 times the 1-Mb/s data rate frequency. It also demonstrates a very wide eye opening ratio of 86 to 87% (the BLUETOOTH spec is $\geq 80\%$), which is desired for error-free symbol detection and very narrow zero crossings, which are needed for symbol timing and synchronization.

Figure 7.7 shows a similar plot but with a deterministic bit pattern of "111101010000" instead of a pseudorandom sequence. It demonstrates higher-frequency deviation for regions with little ISI ("1111" and "0000") and less frequency deviation for regions with most ISI ("0101"). The ratio of average frequency deviations in those regions is part of the BLUETOOTH qualification tests and should be greater than 80%. The diagram shows that it is easily met with the 84% value.

The eye diagram and TX spectrum in Fig. 7.8 indicate great margin over the BLUETOOTH specification as well as no distortion, due to an imprecise $K_{DCO}$ estimate. This also demonstrates the wide transfer function of the data modulation path. The very wide eye opening shown on the lower plot is desired for error-free

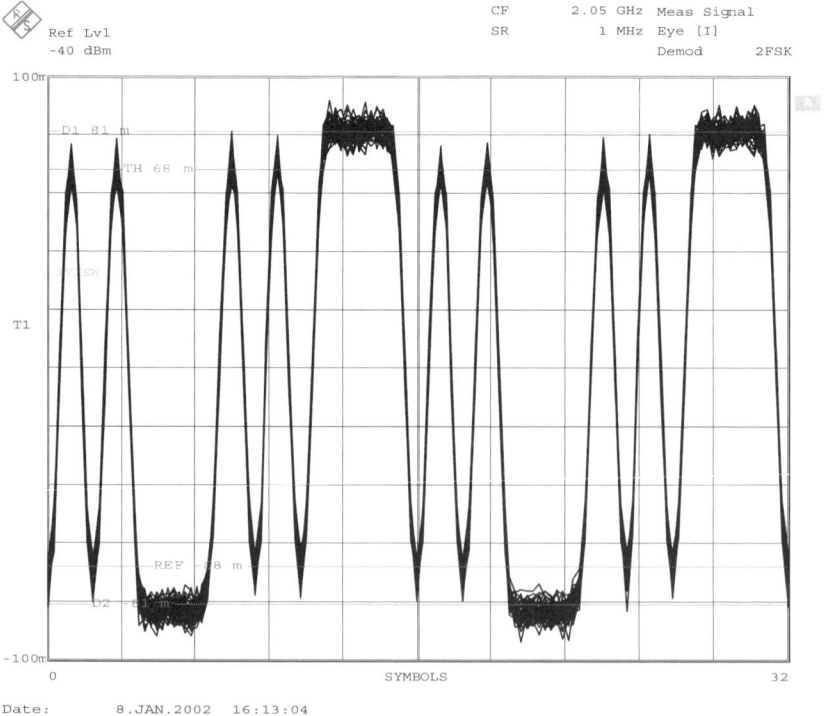

**Figure 7.7** Demodulated diagram of the 111101010000 repetitive pattern. The ratio of $f_{2,\,\text{avg}}$ and $f_{1,\,\text{avg}} = 84\% \geq 80\%$.

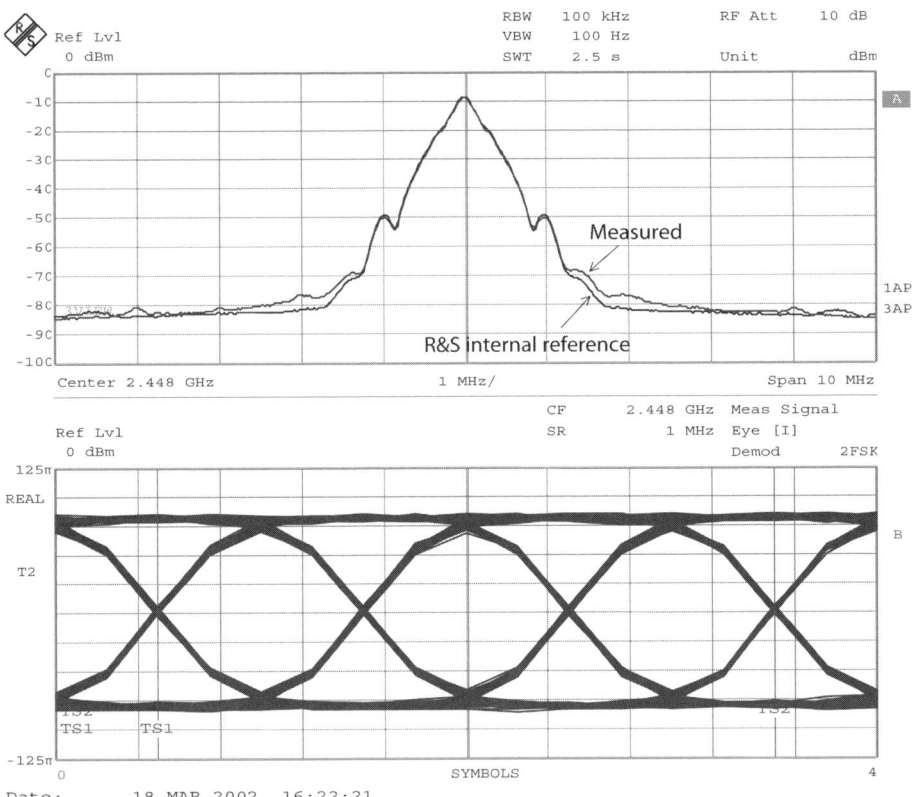

**Figure 7.8** GFSK eye diagram and TX spectrum measured for 1-Mb/s pseudorandom BLUETOOTH data using an R&S FSIQ-7 signal analyzer. Spectrum compared with an instrumentation-quality R&S BLUETOOTH internal reference source.

symbol detection, whereas very narrow zero crossings are desired for symbol timing and synchronization.

## 7.7 SYNTHESIZER PERFORMANCE

Synthesizer performance pertains to a GFSK transmitter operating in its unmodulated state such that only a carrier is present. In this mode, the transmitter noise performance can be evaluated to ensure that it is low enough to maintain an adequate SNR in-band during modulation and to prevent any interference out-of-band. This mode is also active when the channel frequency is changed (either user-initiated or during frequency hopping), since the transmission of data is disabled. The unmodulated state would also be used when the frequency synthesizer acts as a receiver LO.

The closed-loop oscillator output voltage power spectral density $S_x(f)$ defined in Section 1.1.1 is shown in Fig. 7.9 for the wide frequency span of 1 MHz and in

**222** IMPLEMENTATION AND EXPERIMENTAL RESULTS

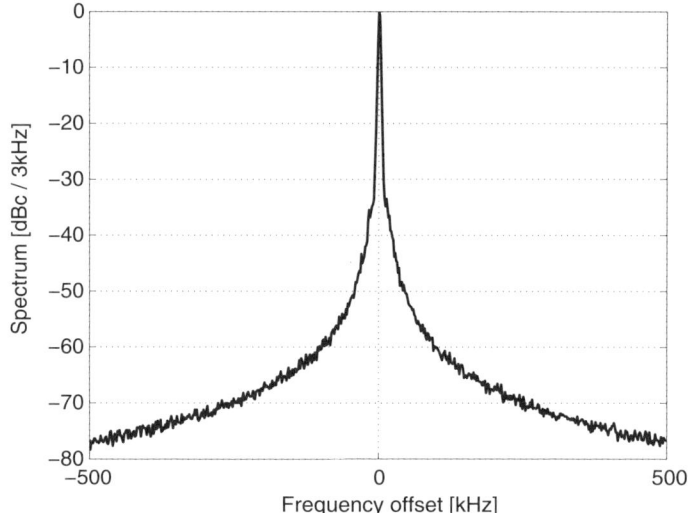

**Figure 7.9** Measured synthesizer output spectrum (wide span); resolution bandwidth = 3 kHz.

Fig. 7.10 for the narrower frequency span of 100 kHz. From Fig. 7.10, the suppression of the close-in phase noise up to about the 10-kHz closed-loop bandwidth is readily visible. The actual phase noise can be estimated, provided that the small-angle criterion is not violated, measuring the relative noise level with respect to the carrier, and compensating for the finite resolution bandwidth

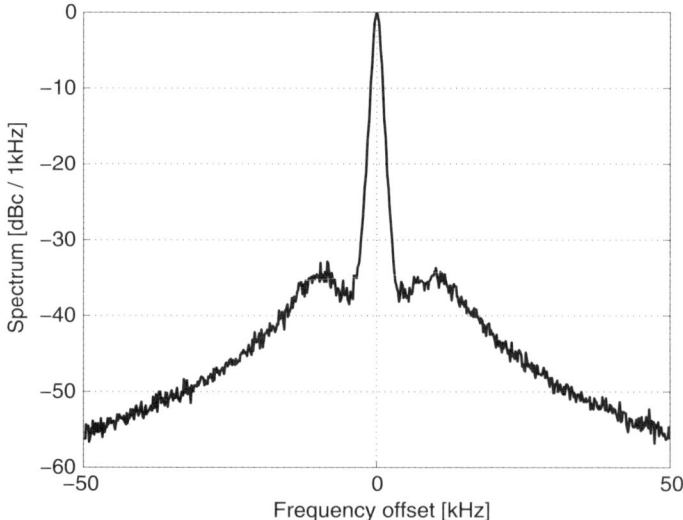

**Figure 7.10** Measured synthesizer output spectrum (narrow span); resolution bandwidth RBW = 1 kHz.

(RBW) of the spectrum analyzer. In this case, the close-in noise level is at $-35$ dBc and the RBW is 1 kHz, so the estimated phase noise is

$$\mathcal{L}\{\Delta f\} = -35\,\text{dBc} - 10\log(1\,\text{kHz}) = -65\,\text{dBc/Hz} \tag{7.1}$$

Figure 7.11 (top) shows the measured synthesizer phase-noise spectrum with a DCO oscillator operating in a type I *all-digital PLL* (ADPLL) with a loop bandwidth

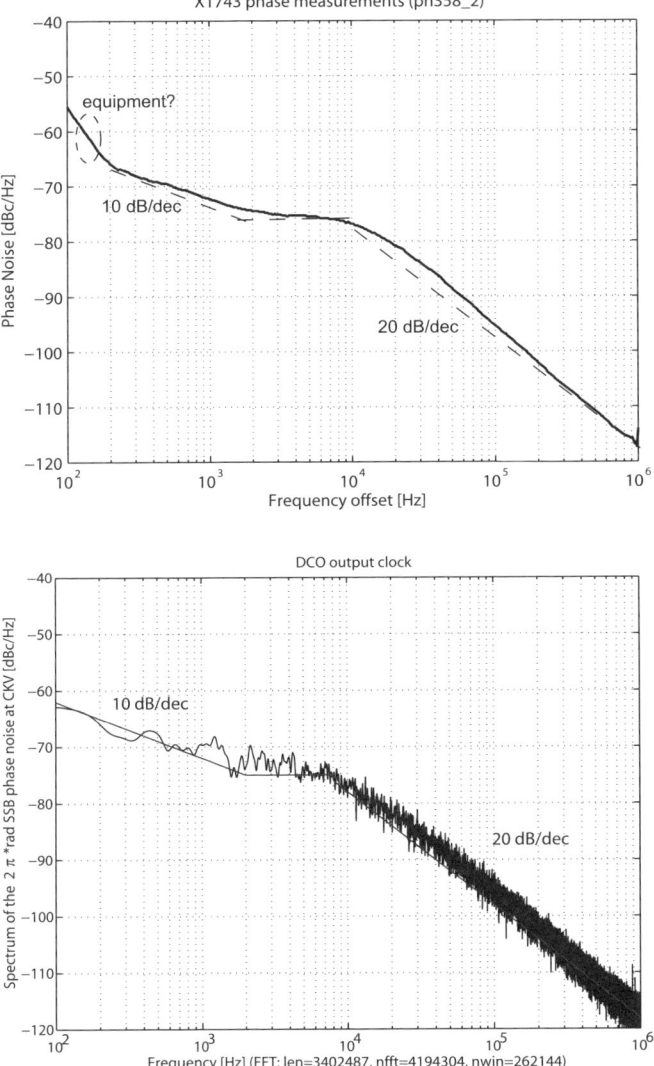

**Figure 7.11** Synthesizer phase noise with the DCO presented: (top) measured using an HP8563E spectrum analyzer with HP85671A phase-noise utility; (bottom) simulated. (From [3], © 2005 IEEE.)

**224** IMPLEMENTATION AND EXPERIMENTAL RESULTS

of 8 kHz. The integrated phase noise in the region 100 Hz to 100 kHz is 2.07° rms. The reference frequency is 13 MHz, and the output frequency is $13 \times 153 = 1989$ MHz. The RF power level is 4 dBm (2.5 mW). The plot concentrates on the up-converted thermal and $1/f$ regions (see Fig. 1.4). The frequency components within the loop filter undergo 20 dB/decade attenuation such that the up-converted thermal noise is flat and the up-converted $1/f$ noise contribution exhibits a 10 dB/decade slope. The phase-noise readout of $-112$ dBc/Hz at 500-kHz offset confirms Eq. 6.7. The simulated phase noise shown in Fig. 7.11 (bottom) closely matches the results measured. This demonstrates the effectiveness of the modeling and simulation methodology described.

## 7.8 SYNTHESIZER SWITCHING TRANSIENTS

Figure 7.12 gives insight into the time-domain operation of an ADPLL. A simple DSP program was written to step the ADPLL through the three operational modes while observing the actual frequency deviation at the RF output. The observation was performed using a Rohde & Schwarz FSIQ-7 signal analyzer, which demodulates the BLUETOOTH RF signal into baseband and plots on the $y$-axis the instantaneous frequency deviation detected. The $y$-axis scale is 300 kHz per grid. The

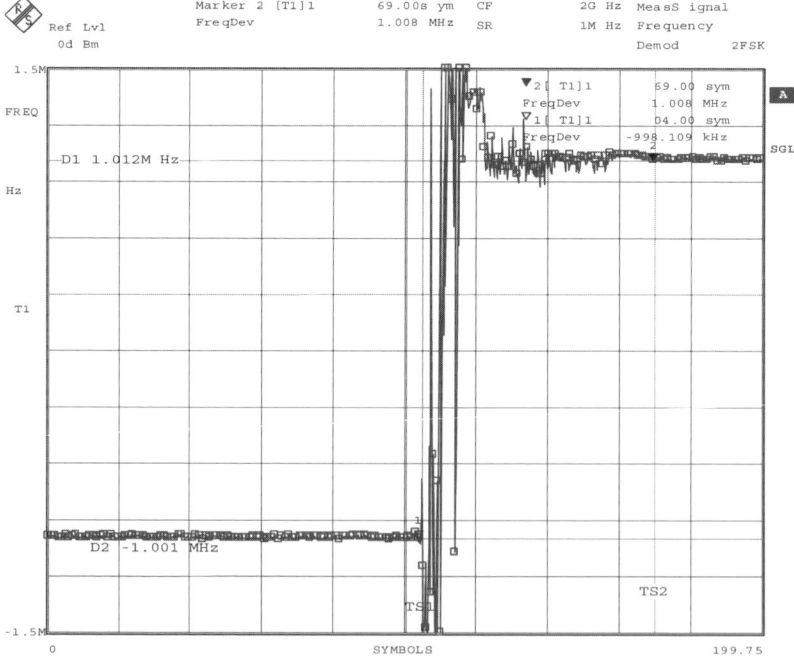

**Figure 7.12** Observed settling time for an ADPLL.

x-axis on the plot is a time progression in units of 1-μs symbols, with a scale of about 20 μs per grid.

The initial frequency deviation was 2 MHz. The following is the predetermined timing sequence: (1) the PVT mode was operational for 5 μs, (2) the acquisition mode was then operational for 25 μs, (3) the fast-tracking mode was then operational for 18 μs, and (4) the regular tracking mode was turned on for the remaining time. The plot shows that in the configuration described above, the settling time is about 50 μs for a 2-MHz jump. The DSP program was not optimized for performance, and it is expected that an acquisition speed similar to that indicated by simulation in Fig. 6.12 could be achieved. The 50-μs figure is still several times better than those reported for conventional RF PLLs.

## 7.9 DSP-DRIVEN MODULATION

As a demonstration of the tight operational integration between a DSP and digital RF, a program was written to perform a software GSM modulation of the transmitter instead of using dedicated hardware for BLUETOOTH modulation.

Impulse response $h[k]$ of the Gaussian filter in the discrete-time domain is expressed as

$$h[k] = \frac{\sqrt{2\pi}}{\sqrt{\ln(2)}} \frac{BT_s}{T_s} \exp\left[-\left(\frac{\sqrt{2\pi}}{\sqrt{\ln(2)}} BT_s \frac{k}{\text{OSR}}\right)^2\right] \tag{7.2}$$

where $B$ is the 3-dB bandwidth, $T_s$ is the symbol period, and $\text{OSR} = f_R/(1/T_S)$ is the symbol oversampling ratio by the reference clock. For GSM, $BT_s = 0.3$ and $T_s = 3.692$ μs and the peak RF frequency deviation is $\pm 67.71$ kHz.

The GSM symbol rate $1/T_S$ is 270.833 kbits/s. At the reference frequency of 13 MHz, the symbol oversampling ratio is exactly 48 ($=13$ MHz$/270.833$ kHz), meaning that 48 FREF samples represent a symbol. A pseudorandom data sequence is being precalculated in software and resides in the RAM. During transmitter modulation, the DSP fetches and adds to the FCW a new sample every six FREF clock cycles, so the actual oversampling ratio is 8, which is quite adequate. The power spectrum of the GMSK-modulated data for GSM is shown in Fig. 7.13. The GSM mask is met at frequency offsets less than 300 kHz from the carrier. Unfortunately, because in this chip the XIO data cannot be fed into the ADPLL in a two-point modulation manner as described in Section 5.1, the higher-frequency spectrum components get filtered by single-pole loop attenuation of 8-kHz bandwidth. It would be quite feasible to predistort the input data accurately by preemphasizing the higher frequencies (the ADPLL transfer function is almost exact, due to its digital nature) to extend the flat part of the transfer function as reported in reference [29]. However, this task was beyond the scope of this book.

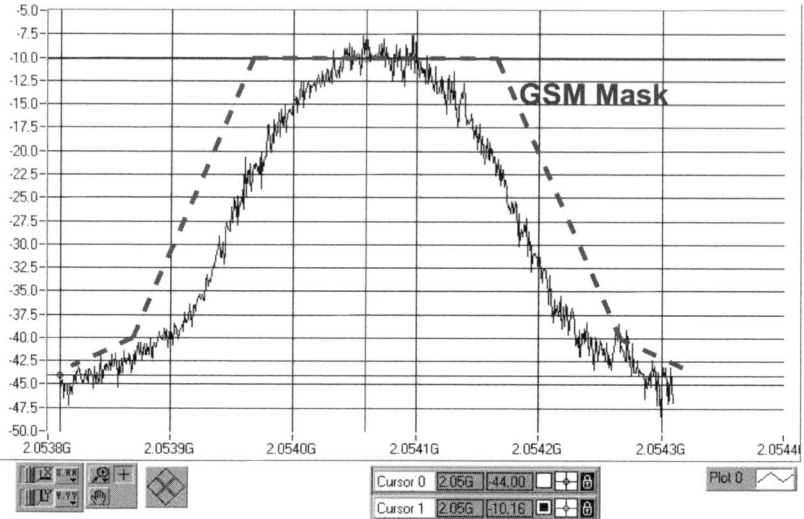

**Figure 7.13** Measured output power spectrum of DSP-driven GSM modulation, with loop attenuation of higher-frequency components.

## 7.10 PERFORMANCE SUMMARY

Table 7.2. summarizes the key transmitter performance parameters measured, some of which are based on the information presented.

The total current consumption is 49.5 mA, as shown in Table 7.3. It includes the DSP under light-to-moderate loading.

**Table 7.2 Key Transmitter Performance Measured**

| | |
|---|---|
| Phase noise | $\leq -114$ dBc/Hz at 500 kHz |
| Spurious tones | $\leq -62.5$ dBc (with antenna filter) |
| DCO frequency pushing | 600 kHz/V |
| PA output power | 4 mW at 50-$\Omega$ load |
| Rms phase error | 2.06° |
| ADPLL settling time | $\leq 50$ μs |

**Table 7.3 Current Consumption for a 1.55-V Supply**

| Circuit | Supply | Current (mA) |
|---|---|---|
| Low-speed digital + DSP | VDD | 12.7 |
| High-speed digital | VDD-HS | 12.8 |
| Oscillator | VDD-OSC | 2.8 |
| RF without an oscillator | VDD-RF | 21.2 |
| Total | | 49.5 |

## 7.11 SUMMARY

In this chapter we presented some implementation details and experimental results and demonstrated the suitability of the all-digital RF frequency synthesizer in the targeted application of a BLUETOOTH transmitter. The eye diagram and modulation spectrum indicate the excellent performance of the transmitter; phase noise and spurious tones of the synthesizer would make it a good choice for the local oscillator of a receiver. The switching speed is extremely fast, making it suitable for modern communication devices utilizing channel hopping. The power consumption is very low, making it ideal for battery-operated mobile units.

# APPENDIX A

# SPURS DUE TO DCO SWITCHING

Let's assume a sinusoidal modulating signal,

$$g(t) = g_{pk} \cos w_m t \tag{A.1}$$

For frequency modulation (FM), the instantaneous frequency $\omega_i$ is the carrier frequency $\omega_c$ modified by the modulating signal times the FM constant $k_f$:

$$\omega_i(t) = \omega_c + k_f g(t) = \omega_c + g_{pk} k_f \cos w_m t \tag{A.2}$$

Defining a new constant called the *peak frequency deviation*,

$$\Delta \omega_{\text{pk}} = g_{pk} k_f \tag{A.3}$$

we can rewrite Eq. A.2 as

$$\omega_i(t) = \omega_c + \Delta \omega_{\text{pk}} \cos w_m t \tag{A.4}$$

The phase of this FM signal is

$$\theta(t) = \omega_c t + \frac{\Delta \omega_{\text{pk}}}{\omega_m} \sin w_m t = \omega_c t + \beta \sin \omega_m t \tag{A.5}$$

*All-Digital Frequency Synthesizer in Deep-Submicron CMOS*, by Robert Bogdan Staszewski and Poras T. Balsara
Copyright © 2006 John Wiley & Sons, Inc.

where

$$\beta = \frac{\Delta\omega_{pk}}{\omega_m} \quad (A.6)$$

is a dimensionless ratio of the peak frequency deviation to the modulating frequency.

The resulting signal is

$$s_{FM}(t) = A\cos(\omega_c t + \beta \sin \omega_m t) \quad (A.7)$$

Equation A.7 could be rewritten after some trigonometric expansion:

$$s_{FM}(t) = A\cos \omega_c t \cos(\beta \sin \omega_m t) - A \sin \omega_c t \sin(\beta \sin \omega_m t) \quad (A.8)$$

For small values of $\beta$ (narrowband FM modulation), we can make the following approximations:

$$\cos(\beta \sin \omega_m t) \approx 1 \quad (A.9)$$
$$\sin(\beta \sin \omega_m t) \approx \beta \sin \omega_m t \quad (A.10)$$

Substituting these into Eq. A.8, we obtain an approximate solution for small $\beta$:

$$s_{NBFM}(t) = A\cos \omega_c t - \beta A \sin \omega_m t \sin \omega_c t \quad (A.11)$$

Expanding Eq. A.11 into phasor form, we have

$$s_{NBFM}(t) = \Re\{Ae^{j(\omega_c t)}(1 + j\beta \sin \omega_m t)\} \quad (A.12)$$

$$= \Re\left\{Ae^{j(\omega_c t)}\left(1 + \frac{1}{2}\beta e^{j\omega_m t} - \frac{1}{2}\beta e^{-j\omega_m t}\right)\right\} \quad (A.13)$$

From Eq. A.12 it is clearly seen that the power spectral density of a narrowband frequency-modulated signal is the continuous-wave carrier frequency plus two sidebands: $\omega_m$ away and $20\log(\frac{1}{2}\beta)$ decibels below the carrier.

## A.1 SPURS DUE TO DCO MODULATION

Let's apply the foregoing analysis of a narrowband FM modulation to our specific case of dithering the oscillating frequency of an $LC$ tank by switching a unit capacitor value through a digital control. When the control signal is high, the capacitor value is $C_{on}$ and the oscillating frequency is

$$f_{osc,h} = \frac{1}{2\pi\sqrt{LC_{on}}} \quad (A.14)$$

Similarly, when the control signal is low, the capacitor value is $C_{\text{off}}$ and the oscillating frequency is

$$f_{\text{osc},l} = \frac{1}{2\pi\sqrt{LC_{\text{off}}}} \tag{A.15}$$

The difference between the high and low oscillation frequencies is

$$\Delta f_{pp} = f_{\text{osc},h} - f_{\text{osc},l} \tag{A.16}$$

The peak angular frequency deviation is

$$\Delta\omega_{\text{pk}} = 2\pi \frac{\Delta f_{pp}}{2} \tag{A.17}$$

The modulating signal frequency is $\omega_m$.

Thus far, the narrowband FM analysis assumed a sinusoidal modulating signal $g(t)$ in Eq. A.1. However, the frequency modulation function of turning a small unit-size capacitor on and off is obviously a rectangular wave [$g_{\text{rect}}(t)$] assumed to have unity peak-to-peak amplitude:

$$\Delta\omega(t) = \Delta\omega_{pp} \cdot g_{\text{rect}}(t) \tag{A.18}$$

with the following zero-mean trigonometric Fourier series decomposition:

$$g_{\text{rect}}(t) = \sum_{n=1}^{\infty} a_n \cos n\omega_m t \tag{A.19}$$

where

$$a_n = 2 \cdot \frac{\tau}{T} \operatorname{sinc} \frac{n\pi\tau}{T} \tag{A.20}$$

is the $n$th coefficient of the trigonometric Fourier series representation, and $\operatorname{sinc} x \equiv (\sin x)/x$. $T = 2\pi/\omega_m$ and $\tau$ are the period and the turn-on time, respectively, of the rectangular wave. For a special case of the symmetric square wave, $\tau = 1/2$ and Eq. A.19 becomes

$$g_{\text{square}}(t) = \sum_{n=1,\text{odd}}^{\infty} \frac{2}{\pi} \frac{1}{n} \sin n\omega_m t \tag{A.21}$$

The modulation index of the $n$th harmonic is now adjusted for the Fourier series decomposition of a rectangular switching wave:

$$\beta_n = \frac{a_n}{n} \frac{\Delta\omega_{pp}}{\omega_m} \tag{A.22}$$

In most cases, especially with a balanced duty cycle, only the fundamental component would really matter.

It should be noted that the frequency-modulating wave $g_{rect}(t)$ in a deep-submicron CMOS process at 2.4 GHz is actually a trapezoidal waveform but with very sharp edge transitions.

**Example 1** Let's consider an example of switching the oscillating frequency by $\Delta f_{pp} = 23$ kHz at a clock rate of 600 MHz. Let's consider the first harmonic in a Fourier series decomposition whose peak amplitude is $2/\pi$ times the square-wave peak-to-peak amplitude. The modulating frequency is one-half of the clock direction switching rate (i.e., $f_m = \omega_m/2\pi = 300$ MHz):

$$\beta_1 = a_1 \frac{\Delta f_{pp}}{f_m} = \frac{2}{\pi} \frac{23 \text{ kHz}}{600 \text{ MHz}/2} = 4.88 \times 10^{-5} \qquad (A.23)$$

This gives rise to spurs 300 MHz away on both sides of the oscillating frequency. Their power level is at

$$20 \log \frac{\beta_1}{2} = -92.3 \text{ dB}$$

relative to the carrier.

The $-92.2$ dBc above ("c" in dBc stands for "relative to the carrier") spur level corresponds to the situation when the DCO is dithered continuously at the highest possible rate. Obviously, this is not a practical case. In fact, $\Sigma\Delta$ dithering will randomize the spurious energy and blur it into the background.

**Example 2** Let's consider now the case of performing the same switching but at the 13 MHz of the FREF frequency:

$$\beta_1 = \frac{2}{\pi} \frac{23 \text{ kHz}}{13 \text{ MHz}/2} = 2.25 \times 10^{-3}$$

and the spur power level is much higher now, at

$$20 \log \frac{\beta}{2} = -59.0 \text{ dB}$$

# APPENDIX B

# GAUSSIAN PULSE-SHAPING FILTER

The Gaussian low-pass filter has a transfer function given by

$$H(f) = \exp(-\alpha^2 f^2) \tag{B.1}$$

The parameter $\alpha$ is related to $B$, the 3-dB bandwidth of the baseband Gaussian shaping filter. It is commonly expressed in terms of a normalized 3-dB bandwidth-symbol time product ($BT_s$):

$$\alpha = \frac{\sqrt{\ln(2)}}{\sqrt{2}} \frac{T_s}{BT_s} \tag{B.2}$$

As $\alpha$ increases, the spectral occupancy of the Gaussian filter decreases and the impulse response spreads over adjacent symbols, leading to increased ISI at the receiver. The impulse response of the Gaussian filter in the continuous-time domain is given by

$$h(t) = \frac{\sqrt{\pi}}{\alpha} \exp\left[-\left(\frac{\pi}{\alpha}t\right)^2\right] \tag{B.3}$$

which could easily be rearranged (Eq. B.4) to reveal its fit with the canonical form of a zero-mean Gaussian random variable with standard deviation $\sigma_h = \alpha/\sqrt{2\pi}$:

$$h(t) = \frac{1}{\sqrt{2\pi}(\alpha/\sqrt{2\pi})} \exp\left[-\frac{t^2}{2(\alpha/\sqrt{2\pi})^2}\right] \tag{B.4}$$

Its integral from $-\infty$ to $\infty$ is, of course, 1.

---

*All-Digital Frequency Synthesizer in Deep-Submicron CMOS*, by Robert Bogdan Staszewski and Poras T. Balsara
Copyright © 2006 John Wiley & Sons, Inc.

Let us now express the Gaussian filter in the discrete-time domain. Let $t_0 = T_s/\text{OSR}$ be an integer oversample of the symbol duration and $t = kt_0$, $k$ being the sample index. The discrete-time impulse response becomes

$$h(kt_0) = \frac{\sqrt{\pi}}{\alpha} \exp\left[-\left(\frac{\pi}{\alpha}kt_0\right)^2\right] \qquad (B.5)$$

Substituting Eq. B.2 and dropping explicit dependence on $t_0$ results in

$$h[k] = \underbrace{\frac{\sqrt{2\pi}}{\sqrt{\ln(2)}} \frac{BT_s}{T_s}}_{h_{\max}} \exp\left[-\left(\frac{\sqrt{2\pi}}{\sqrt{\ln(2)}} BT_s \frac{k}{\text{OSR}}\right)^2\right] \qquad (B.6)$$

The first factor in Eq. B.6 is the peak of the impulse frequency response:

$$h_{\max} = \frac{\sqrt{\pi}}{\alpha} = \frac{\sqrt{2\pi}}{\sqrt{\ln(2)}} \frac{BT_s}{T_s} \qquad (B.7)$$

For BLUETOOTH, with $BT_s = 0.5$ and $T_s = 1$ μs, we obtain $h_{\max} = 1.5054$ MHz. For GSM, with $BT_s = 0.3$ and $T_s = 3.692$ μs, we obtain $h_{\max} = 244.62$ kHz.

For reasons described in Chapter 5, it is more efficient to operate on the *cumulative coefficients*

$$C[k] = \sum_{l=0}^{k-1} h[l] \qquad (B.8)$$

which could be precalculated and stored in a look-up table, with $k = 0 \cdots \text{OSR} - 1$ being the index. The minimum value of $C[k]$ is approximately zero and the maximum value is approximately 1, since the integral of Eq. B.4 is unity.

Figure B.1 shows the impulse $h[k]$, step $C[k]$, and di-bit responses (difference between step and symbol-delayed step responses) of the BLUETOOTH GFSK filter ($BT_s = 0.5$) with a length of three symbols, each symbol oversampled by 8.

Similarly, Fig. B.2 shows the impulse, step, and di-bit responses of the GSM GMSK filter ($BT_s = 0.3$) with a length of four symbols, each symbol oversampled by 8. It reveals much more intersymbol interference (ISI) than in the case of BLUETOOTH.

Figures B.3 and B.4 show the frequency responses of the BLUETOOTH and GSM filters with varying filter lengths of three, four, and five symbols. A filter length of three symbols is completely adequate for precise containment of the

## 234  APPENDIX B: GAUSSIAN PULSE-SHAPING FILTER

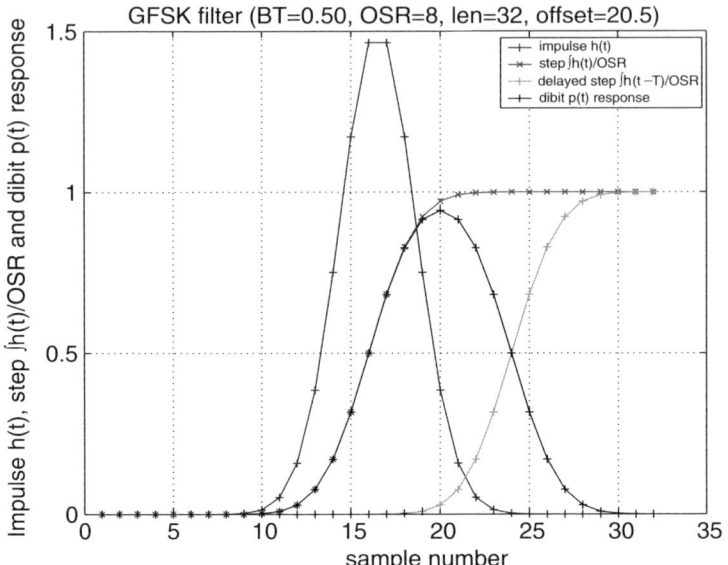

**Figure B.1** Time response of a BLUETOOTH GFSK filter of four-symbol length ($BT_s = 0.3$, OSR $= 8$).

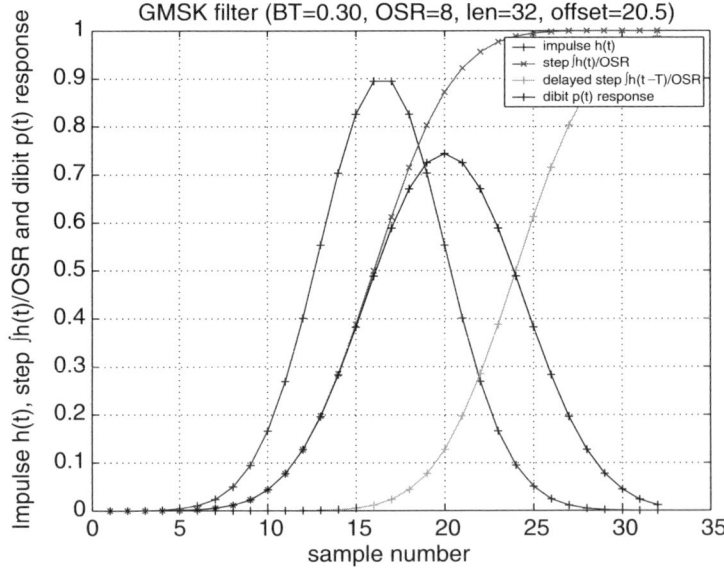

**Figure B.2** Time response of a GSM GMSK filter of four-symbol length ($BT_s = 0.3$, OSR $= 8$).

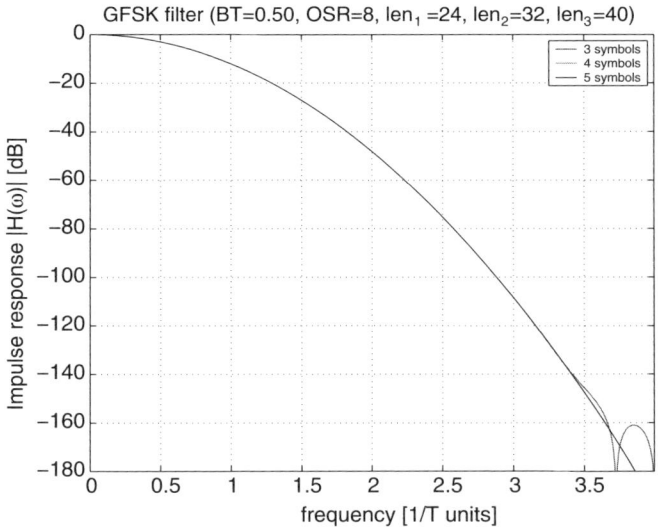

**Figure B.3** Frequency response of a BLUETOOTH GFSK filter for filter lengths of three, four, and five symbols ($BT_s = 0.5$, OSR $= 8$).

modulated output spectrum and sufficient attenuation of frequency components in adjacent channels. However, due to the higher amount of ISI and much tougher requirements for the modulated output spectrum, the GSM-standard filter would require a filter length of at least four symbols.

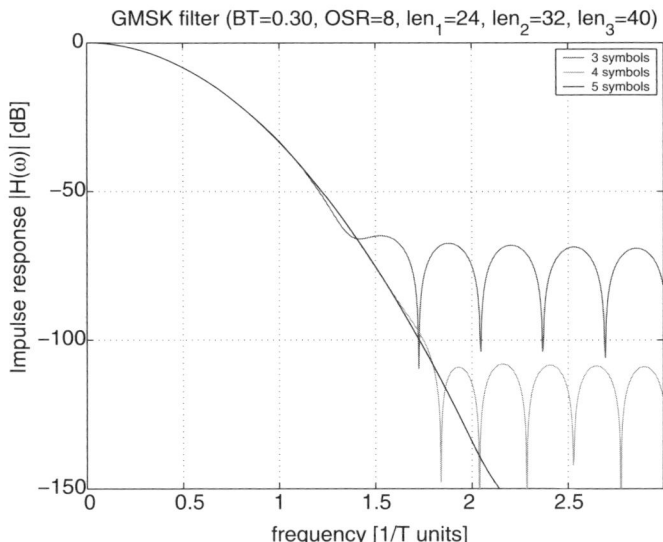

**Figure B.4** Frequency response of a GSM GMSK filter for filter lengths of three, four, and five symbols ($BT_s = 0.3$, OSR $= 8$).

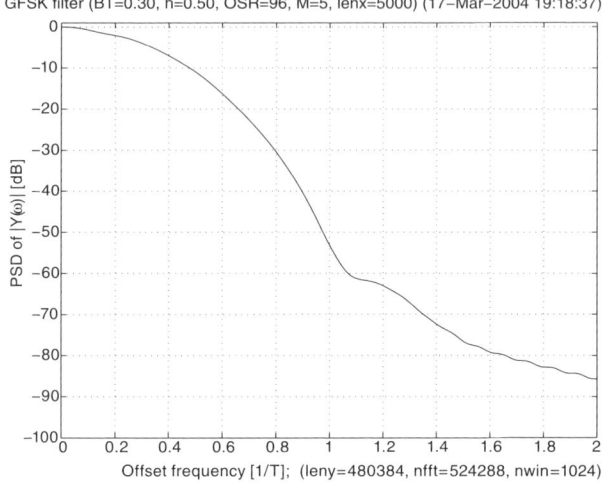

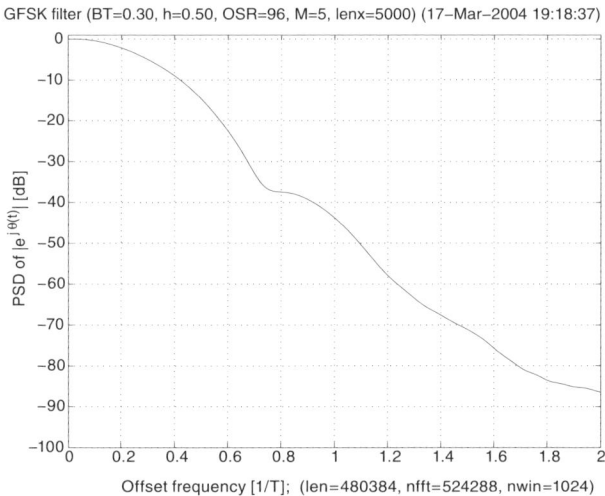

**Figure B.5** Baseband (top) and RF (bottom) spectra of GMSK filter output with pseudorandom input (five-symbol length, $BT_s = 0.3$, OSR $= 96$).

Figure B.5 shows the spectrum of the baseband GMSK filter output FCW and RF port $\Re\{e^{j\theta}\}$ with pseudorandom input data, in which

$$\Delta f[k] = \text{FCW}[k] \frac{f_R}{2^{W_F}} \qquad (\text{B.9})$$

and

$$\theta[k] = \frac{2\pi}{\text{OSR}} \sum_{l=0}^{k-1} \Delta f[k] \qquad (\text{B.10})$$

# APPENDIX C

# VHDL SOURCE CODE

## C.1 DCO LEVEL 2

This VHDL code pertains to the DCO model description in Section 6.5.5. The entity declaration of the level 2 DCO is between lines 18 and 39. The VHDL generics or elaboration-phase parameter constants are declared between lines 19 and 30. The port declaration of the block's I/O signals is between lines 31 and 39. The behavioral architecture describing the block starts at line 42. Lines 43 through 49 declare internal signals. Lines 56 through 61 describe the DCO varactor merging operation on the left of Fig. 6.10. The period is calculated and checked for upper and lower bounds between lines 67 and 81. Finally, the period-controlled oscillator (PCO) engine is instantiated between lines 87 and 98.

```
1   --------------------------------------------------
2   --
3   -- Digitally controlled Oscillator, dco_l2.vhd
4   --
5   --------------------------------------------------
6   -- (C) Robert B. Staszewski, Texas Instruments Inc
7   --------------------------------------------------
8
9   library ieee;
10    use ieee.std_logic_1164.all;
11
```

*All-Digital Frequency Synthesizer in Deep-Submicron CMOS*, by Robert Bogdan Staszewski and Poras T. Balsara
Copyright © 2006 John Wiley & Sons, Inc.

**238** APPENDIX C: VHDL SOURCE CODE

```vhdl
12  library rf;
13    use rf.components_common.src_pco_c;
14
15  entity dco_12 is
16    generic (
17      DCO_PER_0      : time := 417 ps;   -- period of center freq.
18      DCO_PEROFF_LIM : time := 83 ps;    -- maximum period deviation
19      DCO_TIME_RES   : time := 1 fs;     -- finest time resolution
20      DCO_QUANT_P    : positive := 402;
21      DCO_QUANT_A    : positive := 80;
22      DCO_QUANT_T    : positive := 4;    -- (DCO_TIME_RES units)
23      DCO_INIT_DLY   : time := 0 ns;     -- initial oscillation delay
24      DCO_WRMS       : time := 0 ps;     -- accumulative jitter (wander)
25      DCO_JRMS       : time := 0 ps;     -- nonaccumulative jitter
26      SEED           : integer := -1     -- time-based, if < 0
27    );
28    port (
29      dco_in_p  : in integer;
30      dco_in_a  : in integer;
31      dco_in_ti : in integer;
32      dco_in_tf : in integer;
33      mat_pdev  : out time := 0 ns;      -- DCO period deviation
34      ckv       : out std_logic          -- digitized DCO clock output
35    );
36  end;
37  ----------------------------------------------------------------
38
39  architecture behav of dco_12 is
40    signal mat_quant_p: integer;
41    signal mat_quant_a: integer;
42    signal mat_quant_ti: integer;
43    signal mat_quant_tf: integer;
44    signal mat_quant_ls: integer;                 -- low-speed terms
45    --
46    signal mat_per: time := DCO_PER_0;   -- period of oscillation
47  begin
48
49  ----------------------------------------------------------------
50  -- Calculate the tuning input (w/o tracking fractional part)
51  ----------------------------------------------------------------
52
53    mat_quant_p  <= dco_in_p * DCO_QUANT_P;
54    mat_quant_a  <= dco_in_a * DCO_QUANT_A;
55    mat_quant_ti <= dco_in_ti * DCO_QUANT_T;
56    mat_quant_tf <= dco_in_tf * DCO_QUANT_T;
57    mat_quant_ls <= mat_quant_p + mat_quant_a + mat_quant_ti;
58
59  ----------------------------------------------------------------
60  -- Calculate the period
61  ----------------------------------------------------------------
62
63    process (mat_quant_tf, mat_quant_ls)
64      variable mat_pdev_var: time;
```

```
65    begin
66      mat_pdev_var := DCO_TIME_RES * (mat_quant_tf + mat_quant_ls);
67      --
68      -- limit the oscillation period
69      if mat_pdev_var > DCO_PEROFF_LIM then
70        mat_pdev_var := DCO_PEROFF_LIM;
71      elsif mat_pdev_var < -DCO_PEROFF_LIM then
72        mat_pdev_var := -DCO_PEROFF_LIM;
73      end if;
74      --
75      mat_pdev <= mat_pdev_var;
76      mat_per  <= DCO_PER_0 - mat_pdev_var;
77    end process;
78
79    ---------------------------------------------------------------
80    -- Computationally efficient period-controled oscillator
81    ---------------------------------------------------------------
82
83    xpco: src_pco_c
84      generic map (
85        INIT_DELAY => DCO_INIT_DLY,
86        WANDER_RMS => DCO_WRMS,
87        JITTER_RMS => DCO_JRMS,
88        SEED       => SEED
89      )
90      port map (
91        period0 => mat_per,
92        clk     => ckv,
93        clk2    => open
94      );
95
96  end;
97  ---------------------------------------------------------------
98  -- end of dco_12.vhd
```

## C.2   PERIOD-CONTROLLED OSCILLATOR

This VHDL code pertains to the PCO referred to in Section C.1. Input port "period0" of type (line 28) controls the oscillator period during the next cycle. The "smp" internal signal is used to control event activity and to establish the next timestamp (line 82). It is also used to schedule rise and fall times of the clock (line 77). With each timestamp, jitter and wander values are added at lines 58 to 67 and 68 to 76, respectively.

```
1   ---------------------------------------------------------------
2   --
3   -- Period-controlled Oscillator, src_pco_c.vhd
4   --
5   -- Includes jitter.
6   -- Includes wander (random walk).
7   -- Uses foreign-C subroutine for Gaussian random number generator.
```

```vhdl
 8  --
 9  ----------------------------------------------------------------
10  -- (C) Robert B. Staszewski, Texas Instruments Inc
11  ----------------------------------------------------------------
12
13  library ieee;
14    use ieee.std_logic_1164.all;
15    use ieee.math_real.all;
16  library rf;
17    use rf.sub_rand.all;
18
19  entity src_pco_c is
20    generic (
21      INIT_DELAY  : time    := 0 ns;
22      WANDER_RMS  : time    := 0 ps;          -- set to >0 at the top level
23      JITTER_RMS  : time    := 0 ps;          -- set to >0 at the top level
24      SEED        : integer := -1             -- time-based seed, if < 0
25    );
26  port (
27    period0 : in time;
28    en      : in std_logic := '1';
29    clk     : out std_logic := '0';
30    clk2    : out std_logic := '0'
31    );
32  end entity;
33  ----------------------------------------------------------------
34
35  architecture behav of src_pco_c is
36    signal smp: bit := '0';                                     -- tick
37    -- internal measurement signals
38    signal period_s: time := 0 ns;
39    signal tdiff_s: time := 0 ns;      -- diff between actual and ref sample
40  begin
41    process (smp, en) is
42      variable initial: boolean := true;
43      variable tref: time := 0 ns;             -- reference sample time
44      variable jitter: time := 0 ns;        -- instantaneous jitter value
45      variable jitter_prev: time := 0 ns;
46      variable wander: time := 0 ns;        -- instantaneous wander value
47      variable period: time := 0 ns;          -- the current clock period
48      variable s1: integer := SEED;
49      variable randvar: real;
50    begin
51      if not initial and en='1' then
52        -- find time difference between actual and referenced samples
53        tdiff_s <= now - tref;
54        tref := tref + period0;
55        -- adjust the next VC0 period
56        period := period0;
57        if JITTER_RMS /= 0 ns then
58          -- add Gaussian-distributed jitter
59          sub_randn(randvar);
60          jitter := randvar * JITTER_RMS;
```

```
61          if abs(jitter) >= period/2 then        -- check if an outlier
62            jitter := 0 ns;
63          end if;
64          period := period + jitter - jitter_prev;
65          jitter_prev := jitter;
66        end if;
67        if WANDER_RMS /= 0 ns then
68          -- add Gaussian-distributed wander
69          sub_randn(randvar);
70          wander := randvar * WANDER_RMS;
71          if abs(wander) >= period/2 then        -- check if an outlier
72            wander := 0 ns;
73          end if;
74          period := period + wander;
75        end if;
76        clk <= '1', '0' after period/2;         -- clock with 50% duty cycle
77        -- half-rate clock
78        if smp = '0' then  clk2 <= '0';
79        else               clk2 <= '1';
80        end if;
81        smp <= not smp after period;
82        period_s <= period;
83      else
84        sub_randomize_i(s1);
85        sub_randomize_o(s1);
86        period := period0;                      -- the initial time period is period0
87        tref := INIT_DELAY;                     -- mark the first transition as reference
88        clk <= '0';                             -- the initial clock output is low
89        clk2 <= '0';                            -- the initial clock/2 output is low
90        smp <= transport '1' after INIT_DELAY;  -- first transition
91        initial := false;
92        period_s <= period;
93      end if;
94    end process;
95  end architecture;
96  -------------------------------------------------------------------
97  -- end of src_pco_c.vhd --
```

## C.3 TACTICAL FLIP-FLOP

This VHDL code pertains to the flip-flop model description in Section 6.6. The flip-flop is powered-up at a random logic low or high state with equal probability. This is performed between lines 53 and 62. In the course of normal operation, upon the rising edge of the clock (line 65), the input level is sampled (line 69) and activated on the complementary output ports (lines 81 and 87) after an appropriate delay of TD_Q. Metastability detection between the D data input and CLK is performed on line 67. If the data are changing within a certain window T_SU before the clock, an X will be produced at the output.

The "translate_on" and "translate_off" pragma directives hide the nonsynthesizable behavioral code from synthesis tools but keep it visible for the VHDL simulator. The "dc_script_begin" and "dc_script_end" pragma directives convey special non-VHDL information to the SYNOPSYS Design Compiler. In this case, they specify the exact flip-flop and library location to be used.

```
1  ----------------------------------------------------------------
2  --
3  -- Single-bit register with complementary outputs.
4  -- (Detects metastability.)
5  -- (Randomizes the power-up value of the registers.)
6  --
7  ----------------------------------------------------------------
8  -- (C) Robert B. Staszewski, Texas Instruments Inc
9  ----------------------------------------------------------------
10
11 library ieee;
12   use ieee.std_logic_1164.all;
13
14 -- pragma translate_off
15 library rf;
16   use rf.sub_rand.all;
17 -- pragma translate_on
18
19 entity DTT01 is
20    -- pragma translate_off
21    generic (
22      SEED : integer := -1;                    --time-based, if < 0
23      T_SU : time := 0 ps;
24      TD_Q : time := 0 ps
25    );
26    -- pragma translate_on
27    port (
28      D   : in std_logic;
29      CLK : in std_logic;
30      Q   : out std_logic;
31      QZ  : out std_logic.
32    );
33    -- pragma dc_script_begin
34    -- set_register_type -exact -flip_flop DTT01
35    -- remove_attribute GS40_DTT01_W_115_1.35_CORE.db/DTT01 dont_use
36    -- pragma dc_script_end
37 end;
38 ----------------------------------------------------------------
39
40 architecture rtl of DTT01 is
41    signal qq: std_logic;
42 begin
43
44    process (CLK)
45      -- pragma translate_off
46      variable s1: integer := SEED;
47      variable initial: boolean := true;
48      variable randvar: natural;
49      -- pragma translate_on
```

APPENDIX C: VHDL SOURCE CODE   **243**

```
50    begin
51      -- pragma translate_off
52      if initial then
53        sub_randomize_i(s1);
54        sub_randomize_o(s1);
55        sub_randb(randvar);
56        if randvar=0 then
57          qq <= '0';
58        else
59          qq <= '1';
60        end if;
61        initial := false;
62      elsif now > 0 ns then
63        -- pragma translate_on
64        if CLK'event and CLK='1' then
65          -- pragma translate_off
66          if D'last_event >= T_SU then
67            -- pragma translate_on
68            qq <= D;
69            -- pragma translate_off
70          else
71            qq <= 'X';
72          end if;
73          -- pragma translate_on
74        end if;
75      -- pragma translate_off
76      end if;
77      -- pragma translate_on
78    end process;
79
80    Q <= qq
81    -- pragma translate_off
82    after TD_Q
83    -- pragma translate_on
84    ;
85
86    QZ <= not qq
87    -- pragma translate_off
88    after TD_Q
89    -- pragma translate_on
90    ;
91
92    end;
93    ----------------------------------------------------------------
94    -- end of DTT01.vhd
```

## C.4  TDC PSEUDO-THERMOMETER OUTPUT DECODER

This VHDL code pertains to the TDC output vector decoder model described in Section 6.6. The 48-bit input vector from the TDC is inspected bit by bit for "forced unknown" (X) on line 76. If detected, a logic level 0 (line 78) or 1 (line 79) is used instead with

**244** APPENDIX C: VHDL SOURCE CODE

equal probability. Thus a resolved input vector is used by the main process (lines 99 through 152). Lines 107 through 123 perform a combinatorial logic (terminating on a latch, line 105) of priority decoding by detecting the earliest occurrence of rising and falling digital edge transitions. The half-period calculation for the period inversion estimation is carried out between lines of 140 and 144.

```
 1  ----------------------------------------------------------------
 2  --
 3  -- Thermometer-code decoder of time-to-digital converter, pf_dec.vhd
 4  --
 5  -- (Combinatorial logic)
 6  -- (Includes edge-skipping compensation)
 7  --
 8  ----------------------------------------------------------------
 9  -- (C) Robert B. Staszewski, Texas Instruments Inc
10  ----------------------------------------------------------------
11
12  library ieee;
13    use ieee.std_logic_1164.all;
14    use ieee.numeric_std.all;
15
16  -- pragma translate_off
17  library rf;
18    use rf.sub_rand.all;
19  -- pragma translate_on
20
21  entity pf_dec is
22    generic (
23      -- pragma translate_off
24      TD_Q  : time    := 0 ps;
25      SEED  : integer := -1;              -- time-based, if < 0
26      -- pragma translate_on
27      SELQ  : integer := 4;   -- TDC_Q index used for edge selection
28      DTDC  : integer := 48;              -- latched TDC array bus width
29      WTDC  : integer := 6                -- decoded TDC output bus width
30    );
31    port (
32      tdc_q    : in std_logic_vector (DTDC downto 1);
33      ckr      : in std_logic;                    -- level-sensitive latch
34      tdc_rise : out unsigned (WTDC-1 downto 0);-- quant. rise delta
35      tdc_skip : out std_logic;            -- skip one full CKV cycle
36      tdc_hper : out unsigned (WTDC-1 downto 0)-- quant. half-period
37    );
38    -- pragma dc_script_begin
39    -- set_driving_cell -cell IV120 all_inputs()
40    -- set_max_fanout 1 {tdc_q}
41    -- create_clock -name CKR -period 20.0 find(port,ckr)
42    -- set_input_delay 0.3 --clock CKR {tdc_q}
43    -- set_fix_multiple_port_nets -all
44    -- pragma dc_script_end
45  end;
46  ----------------------------------------------------------------
```

```vhdl
architecture rtl of pf_dec is
 signal q1: std_logic_vector (DTDC downto 1);
 constant SLV_0: std_logic_vector (SELQ downto 1):= (others=>'0');
begin

------------------------------------------------------------------
-- Resolve the TDC X'es
------------------------------------------------------------------

process (tdc_q)
    -- pragma translate_off
    variable s1: integer := SEED;
    variable initial: boolean := true;
    variable randvar: natural;
    -- pragma translate_on
    variable q: std_logic_vector (DTDC downto 1);
begin
    -- pragma translate_off
    if initial then
      sub_randomize_i(s1);
      sub_randomize_o(s1);
      initial := false;
    else
    -- pragma translate_on
      -- resolve X'es
      for k in 1 to DTDC loop
        -- pragma translate_off
        if tdc_q(k)='X' then
          sub_randb(randvar);
          if randvar = 0  then q(k) := '0';
          elsif randvar=1 then q(k) := '1';
          end if;
        else
        -- pragma translate_on
          q(k) := tdc_q(k);
        -- pragma translate_off
        end if;
        -- pragma translate_on
      end loop;
      --
      q1 <= q;
    -- pragma translate_off
    end if;
    -- pragma translate_on
end process;

------------------------------------------------------------------
-- Decode and latch
------------------------------------------------------------------

process (ckr, q1)
    variable rise: integer range DTDC-1 downto 0;
```

```
100    variable fall: integer range DTDC-1 downto 0;
101    variable half_period: integer range DTDC-1 downto 0;
102      variable skip: std_logic;
103  begin
104    if ckr='1' then
105      -- digital rising transition detector
106      rise := 0; - in case not found
107      for k in 2 to DTDC loop
108        if q1(k-1)='1' and q1(k)='0' then
109          rise := k-1;
110          exit;
111        end if;
112      end loop;
113      -- digital falling transition detector
114      fall := 0;    -- in case not found
115      for k in 2 to DTDC loop
116        if q1(k-1)='0' and q1(k)='1' then
117          fall := k-1;
118          exit;
119        end if;
120      end loop;
121      --
122      tdc_rise <= to_unsigned(rise, WTDC)
123      -- pragma translate_off
124      after TD_Q
125      -- pragma translate_on
126      ;
127      if q1(SELQ downto 1) = SLV_0 then
128        skip := '1';
129      else
130        skip := '0';
131      end if;
132      --
133      tdc_skip <= skip
134      -- pragma translate_off
135      after TD_Q
136      -- pragma translate_on
137      ;
138      -- calculate the oscillator clock instantaneous half-period
139      if rise > fall then
140        half_period := rise - fall;
141      else
142        half_period := fall - rise;
143      end if;
144      --
145      tdc_hper <= to_unsigned(half_period, WTDC)
146      -- pragma translate_off
147      after TD_Q
148      -- pragma translate_on
149      ;
150    end if;
151  end process;
152  end;
153  ----------------------------------------------------------------
154  -- end of pf_dec.vhd
```

# REFERENCES

1. R. B. Staszewski, Digital deep-submicron CMOS frequency synthesis for RF wireless applications, Ph.D. dissertation, University of Texas at Dallas, Aug. 2002.
2. R. B. Staszewski and S. Kiriaki, Top-down simulation methodology of a 500 MHz mixed-signal magnetic recording read channel using standard VHDL, *Proceedings of the Behavioral Modeling and Simulation Conference*, sec. 3.2, Oct. 1999.
3. R. B. Staszewski, C. Fernando, and P. T. Balsara, Event-driven simulation and modeling of phase noise of an RF oscillator, *IEEE Transactions on Circuits and Systems I*, vol. 52, no. 4, pp. 723–733, Apr. 2005.
4. B. Razavi, *RF Microelectronics*, Prentice Hall, Upper Saddle River, NJ, 1998.
5. J. Craninckx and M. Steyaert, *Wireless CMOS Frequency Synthesizer Design*, Kluwer Academic, Norwell, MA, 1998.
6. T. H. Lee, *The Design of CMOS Radio-Frequency Integrated Circuits*, Cambridge University Press, Cambridge, 1998.
7. T. C. Weigandt, B. Kim, and P. R. Gray, Analysis of timing jitter in CMOS ring oscillators, *Proceedings of the IEEE Symposium on Circuits and Systems*, pp. 27–30, 1994.
8. V. Reinhardt, K. Gould, K. McNab, and M. Bustamante, A short survey of frequency synthesizer techniques, *Proceedings of the 40th Annual Frequency Control Symposium*, pp. 355–365, May 1986.
9. J. Tierney, C. M. Radar, and B. Gold, A digital frequency synthesizer, *IEEE Transactions on Audio Electroaccoustics*, vol. 19, pp. 48–57, Mar. 1971.
10. L. K. Tan and H. Samueli, A 200 MHz quadrature digital synthesizer/mixer in 0.8 μm CMOS, *IEEE Journal of Solid-State Circuits*, vol. 30, no. 3, pp. 193–200, Mar. 1995.

*All-Digital Frequency Synthesizer in Deep-Submicron CMOS*, by Robert Bogdan Staszewski and Poras T. Balsara
Copyright © 2006 John Wiley & Sons, Inc.

11. W. F. Egan, *Frequency Synthesis by Phase Lock*, Wiley, New York, 2000.
12. W. F. Egan, *Phase Lock Basics*, Wiley, New York, 1998.
13. S. T. Lee, S. J. Fang, D. J. Allstot, A. Bellaouar, A. R. Fridi, and P. A. Fontaine, A quad-band GSM-GPRS transmitter with digital auto-calibration, *IEEE Journal of Solid-State Circuits*, vol. 39, no. 12, pp. 2200–2214, Dec. 2004.
14. A. N. Hafez and M. I. Elmasry, A low power monolithic subsampled phase-locked loop architecture for wireless transceivers, *Proceedings of the IEEE Symposium on Circuits and Systems*, vol. 2, pp. 549–552, May–June 1999.
15. B. Razavi, Challenges in the design of frequency synthesizers for wireless applications, *Proceedings of the Custom Integrated Circuits Conference*, pp. 395–402, 1997.
16. K. Muhammad, R. B. Staszewski, and P. T. Balsara, Challenges in integrated CMOS transceivers for short distance wireless, *Proceedings of the Great Lakes Symposium on VLSI*, Mar. 2001.
17. I. Elahi, Robust receiver design using digitally intensive techniques to overcome analog impairments, Ph.D. dissertation, Department of Electrical Engineering, University of Texas at Dallas, Nov. 2005.
18. T. S. Rappaport, *Wireless Communications: Principles and Practice*, Prentice Hall, Upper Saddle River, NJ, 1996.
19. M. Bopp et al., A DECT transceiver chip set using SiGe technology, *Proceedings of the IEEE Solid-State Circuits Conference*, sec. MP4.2, pp. 68–69, 447, Feb. 1999.
20. B. Zhang and P. Allen, Feed-forward compensated high switching speed digital phase-locked loop frequency synthesizer, *Proceedings of the IEEE Symposium on Circuits and Systems*, vol. 4, pp. 371–374, 1999.
21. F. M. Gardner, Charge-pump phase-locked loops, *IEEE Transactions on Communications*, vol. 28, pp. 1849–1858, Nov. 1980.
22. D. H. Wolaver, *Phase-Locked Loop Circuit Design*, Prentice Hall, Englewood Cliffs, NJ, 1993.
23. W. B. Wilson, U. K. Moon, K. R. Lakshmikumar, and L. Dai, A CMOS self-calibrating frequency synthesizer, *IEEE Journal of Solid-State Circuits*, vol. 35, no. 10, pp. 1437–1444, Oct. 2000.
24. I. C. Hwang, S. H. Song, and S. W. Kim, A digitally controlled phase-locked loop with a digital phase-frequency detection for fast acquisition, *IEEE Journal of Solid-State Circuits*, vol. 36, no. 10, pp. 1574–1581, Oct. 2001.
25. T. P. Kenny, T. A. Riley, N. M. Filiol, and M. A. Copeland, Design and realization of a digital delta-sigma modulator for fractional-$N$ frequency synthesis, *IEEE Transactions on Vehicular Technology*, vol. 48, no. 2, pp. 510–521, Mar. 1999.
26. B. Miller and R. J. Conley, A multiple modulator fractional divider, *IEEE Transactions on Instrumentation and Measurement*, vol. 40, no. 3, pp. 578–583, June 1991.
27. T. Riley, M. Copeland, and T. Kwasniewski, Delta–sigma modulation in fractional-$N$ frequency synthesis, *IEEE Journal of Solid-State Circuits*, vol. 28, no. 5, pp. 553–559, May 1993.
28. Y. Matsua, K. Uchimura, A. Iwata, T. Kobayashi, M. Ishikawa, and T. Yoshitome, A 16-bit oversampling A/D conversion technology using triple integration noise shaping, *IEEE Journal of Solid-State Circuits*, vol. 22, pp. 921–929, Dec. 1987.
29. M. H. Perrott, T. Tewksbury, and C. Sodini, A 27-mW CMOS fractional-$N$ synthesizer using digital compensation for 2.5-Mb/s GFSK modulation, *IEEE Journal of Solid-State Circuits*, vol. 32, no. 12, pp. 2048–2060, Dec. 1997.

## REFERENCES

30. W. T. Bax and M. A. Copeland, A GMSK modulator using a $\Delta\Sigma$ frequency discriminator-based synthesizer, *IEEE Journal of Solid-State Circuits*, vol. 36, no. 8, pp. 1218–1227, Aug. 2001.
31. H. Brugel and P. F. Driessen, Variable bandwidth DPLL bit synchronizer with rapid acquisition implemented as a finite state machine, *IEEE Transactions on Communications*, vol. 42, pp. 2751–2759, Sept. 1994.
32. J. Dunning, G. Garcia, J. Lundberg, and E. Nuckolls, An all-digital phase-locked loop with 50-cycle lock time suitable for high performance microprocessors, *IEEE Journal of Solid-State Circuits*, vol. 30, pp. 412–422, Apr. 1995.
33. R. E. Best, *Phase Locked Loops: Design, Simulation and Applications*, 3rd ed., McGraw-Hill, New York, 1997.
34. T. Y. Hsu, B. J. Shieh, and C. Y. Lee, An all-digital phase-locked loop (ADPLL)-based clock recovery circuit, *IEEE Journal of Solid-State Circuits*, vol. 34, pp. 1063–1073, Aug. 1999.
35. M. Olivieri and A. Trifiletti, An all-digital clock generator firm-core based on differential fine-tuned delay for reusable microprocessor cores, *Proceedings of the IEEE Symposium on Circuits and Systems*, vol. 4, pp. 638–641, 2001.
36. A. Kajiwara and M. Nakagawa, A new PLL frequency synthesizer with high switching speed, *IEEE Transactions on Vehicular Technology*, vol. 41, no. 4, pp. 407–413, Nov. 1992.
37. T. H. Lee, CMOS RF: (still) no longer an oxymoron, *Proceedings of the Symposium on Radio Frequency Integrated Circuits*, pp. 3–6, 1999.
38. A. A. Abidi, Wireless transceivers in CMOS IC technology: the new wave, *Proceedings of the Symposium on VLSI Technology*, pp. 151–158, 2000.
39. J. N. Burghartz, M. Hargrove, C. S. Webster, et al., RF potential of a 0.18-$\mu$m CMOS logic device technology, *IEEE Transactions on Electron Devices*, vol. 47, no. 4, pp. 864–870, Apr. 2000.
40. H. Iwai, CMOS technology for RF applications, *Proceedings of the 22nd International Conference on Microelectronics*, vol. 1, pp. 27–34, May 2000.
41. J. T. Wu, M. J. Chen, and C. C. Hsu, A 2 V 900 MHz CMOS phase-locked loop, *Proceedings of the IEEE Symposium on VLSI Circuits*, pp. 52–53, June 1998.
42. Q. Huang, P. Orsatti, and F. Piazza, GSM transceiver front-end circuits in 0.25-$\mu$m CMOS, *IEEE Journal of Solid-State Circuits*, vol. 34, no. 3, pp. 292–303, Mar. 1999.
43. J. L. Tham, M. A. Margarit, B. Pregardier, C. Hull, R. Magoon, and F. Carr, A 2.7-V 900-MHz/1.9-GHz dual-band transceiver IC for digital wireless communication, *IEEE Journal of Solid-State Circuits*, vol. 34, no. 3, pp. 286–291, Mar. 1999.
44. B. Murmann and B. E. Boser, *Digitally Assisted Pipeline ADCs: Theory and Implementation*, Kluwer Academic, Norwell, MA, 2004.
45. GS40 0.11-$\mu$m CMOS standard cell/gate array, in *Texas Instruments Application Specific Integrated Circuits Macro Library Summary, Version 1.0*, Jan. 2001.
46. *Specification of the BLUETOOTH System, Version 1.1*, www.bluetooth.com, Feb. 22, 2001.
47. G. K. Dehng, C. Y. Yang, J. M. Hsu, and S-I. Liu, A 900-MHz 1-V CMOS frequency synthesizer, *IEEE Journal of Solid-State Circuits*, vol. 35, no. 8, pp. 1211–1214, Aug. 2000.

48. *The National Technology Roadmap for Semiconductors*, Semiconductor Industries Association, San Jose, CA, 1997.
49. N. K. Verghese, T. J. Schmerbeck, and D. J. Allstot, *Simulation Techniques and Solutions for Mixed-Signal Coupling in Integrated Circuits*, Kluwer Academic, Norwell, MA, 1995.
50. R. B. Staszewski, C.-M. Hung, D. Leipold, and P. T. Balsara, A first multigigahertz digitally controlled oscillator for wireless applications, *IEEE Transactions on Microwave Theory and Techniques*, vol. 51, no. 11, pp. 2154–2164, Nov. 2003.
51. C. L. Huang and N. D. Arora, Measurements and modeling of MOSFET $I$–$V$ characteristics with polysilicon depletion effect, *IEEE Transactions on Electron Devices*, vol. 40, no. 12, pp. 2330–2337, Dec. 1993.
52. C.-M. Hung, B. A. Floyd, N. Park, and K. O. Kenneth, Fully integrated 5.35-GHz CMOS VCOs and prescalers, *IEEE Transactions on Microwave Theory and Techniques*, vol. 49, no. 1, pp. 17–22, Jan. 2001.
53. E. Hegazi, J. Rael, and A. Abidi, *The Designer's Guide to High-Purity Oscillators*, Kluwer Academic, Norwell, MA, 2005.
54. J. B. Shyu, G. C. Temes, and F. Krummenacher, Random error effects in matched MOS capacitors and current sources, *IEEE Journal of Solid-State Circuits*, vol. 19, pp. 948–955, Dec. 1984.
55. S. H. Lee and B. S. Song, Digital-domain calibration of multistep analog-to-digital converters, *IEEE Journal of Solid-State Circuits*, vol. 27, pp. 1679–1688, Dec. 1992.
56. R. B. Staszewski, D. Leipold, K. Muhammad, and P. T. Balsara, Digitally controlled oscillator (DCO)-based architecture for RF frequency synthesis in a deep-submicrometer CMOS process, *IEEE Transactions on Circuits and Systems II*, vol. 50, no. 11, pp. 815–828, Nov. 2003.
57. M. H. Perrott, M. D. Trott, and C. G. Sodini, A modeling approach for $\Sigma$-$\Delta$ fractional-$N$ frequency synthesizers allowing straightforward noise analysis, *IEEE Journal of Solid-State Circuits*, vol. 37, no. 8, pp. 1028–1038, Aug. 2002.
58. R. B. Staszewski, D. Leipold, and P. T. Balsara, Just-in-time gain estimation of an RF digitally-controlled oscillator for digital direct frequency modulation, *IEEE Transactions on Circuits and Systems II*, vol. 50, no. 11, pp. 887–892, Nov. 2003.
59. T. H. Lee and A. Hajimiri, Oscillator phase noise: a tutorial, *IEEE Journal of Solid-State Circuits*, vol. 35, no. 3, pp. 326–336, Mar. 2000.
60. A. Hajimiri and T. H. Lee, A general theory of phase noise in electrical oscillators, *IEEE Journal of Solid-State Circuits*, vol. 35, no. 3, pp. 326–336, Feb. 1998.
61. A. Hajimiri and T. H. Lee, *The Design of Low Noise Oscillators*, Kluwer Academic, Norwell, MA, 1999.
62. J. C. Candy and G. C. Temes, Oversampling methods for A/D and D/A conversion, in *Oversampling Delta-Sigma Data Converters*, IEEE Press, New York, 1991.
63. R. E. Radke, A. Eshraghi, and T. S. Fiez, A 14-bit current-mode $\Sigma$–$\Delta$ DAC based upon rotated data weighted averaging, *IEEE Journal of Solid-State Circuits*, vol. 35, no. 8, pp. 1074–1084, Aug. 2000.
64. F. M. Gardner, Interpolation in digital modems, part I: fundamentals, *IEEE Transactions on Communications*, vol. 41, no. 3, pp. 501–507, Mar. 1993.

65. R. B. Staszewski and P. T. Balsara, Phase-domain all-digital phase-locked loop, *IEEE Transactions on Circuits and Systems II*, vol. 52, no. 3, pp. 159–163, Mar. 2005.
66. P. Dudek, S. Szczepanski, and J. Hatfield, A high-resolution CMOS time-to-digital converter utilizing a Vernier delay line, *IEEE Journal of Solid-State Circuits*, vol. 35, no. 2, pp. 240–247, Feb. 2000.
67. R. B. Staszewski, K. Muhammad, D. Leipold, C.-M. Hung, Y.-C. Ho, J. L. Wallberg, C. Fernando, K. Maggio, R. Staszewski, T. Jung, J. Koh, S. John, I. Y. Deng, V. Sarda, O. Moreira-Tamayo, V. Mayega, R. Katz, O. Friedman, O. E. Eliezer, E. de-Obaldia, and P. T. Balsara, All-digital TX frequency synthesizer and discrete-time receiver for BLUETOOTH radio in 130-nm CMOS, *IEEE Journal of Solid-State Circuits*, vol. 39, no. 12, pp. 2278–2291, Dec. 2004.
68. B. N. Nikolic, V. G. Oklobdzija, V. Stajonovic, W. Jia, J. Chiu, and M. Leung, Improved sense-amplifier-based flip-flop: design and measurements, *IEEE Journal of Solid-State Circuits*, vol. 35, no. 6, pp. 876–884, June 2000.
69. T. J. Gabara, G. J. Cyr, and C. E. Stroud, Metastability of CMOS master/slave flip-flops, *IEEE Transactions on Circuits and Systems II*, vol. 39, no. 10, pp. 734–740, Oct. 1992.
70. C. Brown and K. Feher, Measuring metastability and its effect on communication signal processing systems, *IEEE Transactions on Instrumentation and Measurement*, vol. 46, no. 1, pp. 61–64, Feb. 1997.
71. R. B. Staszewski, K. Muhammad, and P. Balsara, A 550-Msample/s 8-tap FIR digital filter for magnetic recording read channels, *IEEE Journal of Solid-State Circuits*, vol. 35, pp. 1205–1210, Aug. 2000.
72. K. Muhammad, R. B. Staszewski, and P. T. Balsara, Speed, power, area, and latency tradeoffs in adaptive FIR filtering for PRML read channels, *IEEE Transactions on VLSI Systems*, vol. 9, no. 1, pp. 42–51, Feb. 2001.
73. R. B. Staszewski, K. Muhammad, and P. T. Balsara, A constrained asymmetry LMS algorithm for PRML disk drive read channels, *IEEE Transactions on Circuits and Systems II*, vol. 48, pp. 793–798, Aug. 2001.
74. B. Razavi, Design of monolithic phase-locked loops and clock recovery circuits: a tutorial, in *Monolithic Phase-Locked Loops and Clock Recovery Circuits: Theory and Design*, IEEE Press, New York, 1996.
75. T. Riley, N. Filiol, Q. Du, and J. Kostamovaara, Techniques for in-band phase noise reduction in $\Sigma\Delta$ synthesizers, *IEEE Transactions on Circuits and Systems II*, vol. 50, no. 11, pp. 794–803, Nov. 2003.
76. E. Duvivier, G. Puccio, S. Cipriani, L. Carpineto, P. Cusinato, B. Bisanti, F. Galant, F. Chalet, F. Coppola, S. Cercelaru, N. Vallespin, J. Jiguet, and G. Sirna, A fully integrated zero-IF transceiver for GSM-GPRS quad-band application, *IEEE Journal of Solid-State Circuits*, vol. 38, no. 12, pp. 2249–2257, Dec. 2003.
77. F. Spagna, Phase locked loop using delay compensation techniques, *Proceedings of the IEEE Symposium on Computers and Communications*, pp. 417–423, 2000.
78. F. M. Gardner, *Phaselock Techniques*, Wiley, New York, 1979.
79. T. M. Almeida and M. S. Piedade, High performance analog and digital PLL design, *Proceedings of the IEEE Symposium on Circuits and Systems*, vol. 4, pp. 394–397, 1999.
80. R. B. Staszewski, G. Shriki, and P. T. Balsara, All-digital PLL with ultrafast acquisition, *Proceedings of the IEEE Asian Solid-State Circuits Conference*, Taipei, Taiwan, sec. 11-7, pp. 289–292, Nov. 2005.

81. J. Lee and B. Kim, A 200 MHz low jitter adaptive bandwidth PLL, *Proceedings of the IEEE Solid-State Circuits Conference*, sec. WA20.1, pp. 346–347, 477, Feb. 1999.
82. H. Sato, K. Kato, and T. Sase, A fast pull-in PLL IC using two-mode pull-in technique, *Electronics and Communications in Japan*, pt. 2, vol. 75, no. 3, pp. 41–50, 1992.
83. R. B. Staszewski, D. Leipold, and P. T. Balsara, Direct frequency modulation of an ADPLL for BLUETOOTH/GSM with injection pulling elimination, *IEEE Transactions on Circuits and Systems II*, vol. 52, no. 6, pp. 339–343, June 2005.
84. B. Razavi, A study of injection pulling and locking in oscillators, *Proceedings of the 2003 IEEE Custom Integrated Circuits Conference*, pp. 305–312, Sept. 2003.
85. T. Sowlati, C. A. Salama, J. Sitch, G. Rabjohn, and D. Smith, Low voltage, high efficiency GaAs class E power amplifiers for wireless transmitters, *IEEE Journal of Solid-State Circuits*, vol. 30, no. 10, pp. 1074–1080, Oct. 1995.
86. N. J. Kasdin, Discrete simulation of colored noise and stochastic processes and $1/(f^{\alpha})$ power law noise generation, *Proceedings of the IEEE*, vol. 8, no. 5, pp. 802–827, May 1995.
87. S. R. Norsworthy, D. A. Rich, and T. R. Viswanathan, A minimal multibit digital noise shaping architecture, *Proceedings of 1996 IEEE International Symposium on Circuits and Systems*, pp. 5–8, 1996.
88. W. H. Press, S. A. Teukolsky, W. T. Vetterling, and B. P. Flannery, *Numerical Recipes in C*, 2nd ed., Cambridge University Press, Cambridge, 1994.
89. U.-K. Moon, K. Mayaram, and J. T. Stonick, Spectral analysis of time-domain phase jitter measurements, *IEEE Transactions on Circuits and Systems II*, vol. 49, pp. 321–327, May 2002.
90. A. Zanchi, A. Bonfanti, S. Levantino, and C. Samori, General SSCR vs. cycle-to-cycle jitter relationship with application to the phase noise in PLL, *Proceedings of the Southwest Symposium on Mixed-Signal Design*, pp. 32–37, Feb. 2001.
91. A. J. Acosta, A. Barriga, M. Valencia, M. Bellido, and J. L. Huertas, Modeling of real bistables in VHDL, *Proceedings of the European Design Automation Conference*, pp. 460–465, Sept. 1993.

# INDEX

Acquisition mode, 40–41, 44–45, 147
Acquisition time, 16
Additive white Gaussian noise (AWGN), 193
ADPLL (all-digital PLL), 23, 76
  clock names, 81
  clock retiming, 103
  difference mode, 85
  direct modulation, 159
  fractional error correction, 82
  fractional phase, 91
  frequency control word (FCW), 78, 116, 149, 156
  frequency response, 115
    type I, 116
    type II, 130
  higher-order, 133, 135
  integer-domain operation, 86
  linear model, 119
  loop
    bandwidth, 113, 117, 128
    delay, 116
    filter, 115
    proportional-integral control, 127
  loop gain, 109
    attenuation, 110
    damping, 110
    gear shifting, 142, 153
    programmable, 113
    proportionality factor, 111
  metastability, 79, 100, 103
  modeling, 189
    arithmetic, 84
  noise, 119
    DCO noise, 119
    TDC noise, 120
  nonlinear differential term, 139
  performance, 221
    switching transient, 224
  phase
    accumulators, 86, 89
    comparison, 79
    detection, 83
    estimate, 81
    reference, 78, 82, 86, 113
    resolution, 113
    variable, 82, 89
  phase error, 82, 111
    excess, 111, 149–151
    instantaneous variation, 139
    excess, 111
  phase noise spectrum, 121, 126, 222–223

*All-Digital Frequency Synthesizer in Deep-Submicron CMOS*, by Robert Bogdan Staszewski and Poras T. Balsara
Copyright © 2006 John Wiley & Sons, Inc.

ADPLL (all-digital PLL) (*Continued*)
  phase-domain
    architecture, 112
    structure, 24
    operation, 77, 79
  power spectral density, 222, 229
  proportional-integral control, 127
  quantization error, 80
  reference clock (FREF), 1, 76–77
    retiming, 79, 100
  reference phase, 77–78
  retimed reference (CKR), 79, 103, 113
  $s$-domain model, 115, 118
  scaling factor, 110
  stability analysis, 137
  TDC (time-to-digital converter), 91
  transfer function
    closed-loop, 116, 131
    open-loop, 116, 130–131
  transmitter core, 215
  type I, 79, 112
  type II, 127, 130
  variable phase, 77–78
  $z$-domain model, 119, 132
  zero-phase restart, 149
    hitless, 143, 150
ADPLL synthesizer(s), 24
ADPLL transmitter, 156
  core, 215
  direct modulation, 157
    transfer function, 162
Amplitude control, 183
Amplitude control word (ACW), 187
Amplitude modulation, 177–178, 185
  digital pulse slimmer, 178
  phase adjustment, 179, 182
Analog
  intensive, xiii
  PLL, 17
  synthesis, vi
  voltage resolution, xiii
Arithmetic, 84
  binary-weighted, 35, 64
  encoding, 61
  fixed-point, 113
  incrementer, 89
  modulo, 84, 87
  redundant system, 43
  signed representation, 50, 61
  wordlength limit, 86, 111, 113
ASIC design flow, xiv, 214
Asynchronous retiming, 104
Auto-place and route, 215

Banerjee figure of merit (BFM), 121
Bit error rate (BER), 12
BLUETOOTH, xiv, 9, 12–13, 42, 152, 216, 219, 225
  center channel, 40, 46
  extended data rate (EDR), 177
  frequency deviation, 40, 173
  frequency resolution, 40, 50
  modeling, 193–194
  modulation index, 165
  operating modes, 40
  test chip, 216
    evaluation board, 218
    performance, 219

Cartesian, 185
Center frequency, 41, 44, 61–62
Channel
  hopping, 9, 16
  select, 41, 45
Charge pump, 17–18
CKR (retimed reference clock), 79, 103, 113
CKV (DCO output), 77, 95
Class E power amplifier, 175–176
Clock divider
  fractional-$N$, 8
  integer-$N$, 17
  noise, 200
  pulse swallower, 18
  $\Sigma\Delta$ modulated, 20
Clock quality monitoring (CQM), 140
Clock retiming, 103
  majority-voting, 107
  metastability, 107
  mid-edge selector, 107
Closed-loop transfer function, 116, 131
CMOS, xiii, 25, 27
Constant-envelope modulation, 178
Continuous-time filtering, 186
CORDIC algorithm, 185
Cycle skipping, 95

Damping, 110, 131
Data modulation, 40, 156
Data rate, 165
DCO (digitally-controlled oscillator), 27, 30
  acquisition, 40–41, 44–45
  ASIC cell, 47
  center frequency, 41, 44, 61–62
  digital
    control, 33–34, 37
    control bits, 34, 37
    word, 39
  digital-to-frequency, 30, 39

## INDEX

dithering, 34, 43, 56–57
dynamic range, 40, 43
frequency
   deviation, 40, 49, 208
   natural, 61–62
   nominal, 47
   modulation, 46
   offset, 61
   resolution, 40, 43–44, 49
   tuning, 45, 42
   tuning word retiming, 55
gain ($K_{DCO}$), 53, 116
   estimation, 53, 141, 158
   normalization, 54
jitter, 55
LC tank, 35
   differential varactor, 37
   varactor, 31
   varactor array, 33, 38
   varactor banks, 39, 41–42
narrowband, 39
normalized (nDCO), 52, 54
operating modes, 40
   PVT mode, 40
   acquisition mode, 41
   tracking mode, 42
oscillator core, 37
oscillator tuning word (OTW), 31,
   50, 53, 64, 164
perturbations, 55
phase accumulation, 74
phase noise, 119, 123, 125
PVT calibration mode, 40, 42
   look-up table, 42
PVT variations, 40
schematic, 38
spurs, 65, 229
time-domain model, 47
timing deviation (TDEV), 49–50
tracking, 40–42, 45
   bank encoding, 43
   VHDL model, 201, 237, 239
DCO gain, 53, 116
   estimation, 141, 158
   just-in-time calculation, 164
   normalization, 54
DCO output clock (CKV), 77, 95
Deep-submicron CMOS, xiii, 25
   process parameters, 27
Delay control, 180
Delay line, 180
   control, 179
   resolution, 180
Digital amplitude modulation, 177–178

Digital baseband, xiii, 28
Digital pulse slimmer, 178, 182
Digital radio, 28
Digital RF processor (DRP$^{TM}$), xvi, 28, 213
Digital transmit filter, 167
Digital, xiii
   ASIC, xiv
   CMOS, xiii, 25
   digitally intensive, xiii, 27, 215
   edge transition times, xiii, 2
   gate density, 26
   switching noise, 29
   testing cost, 26
   time-domain resolution of, xiii, 26
Digital-to-RF-amplitude conversion
   (DRAC), 185
Digitally-controlled oscillator (DCO), 27, 30
Direct modulation, 156
   compensating input, 159
   direct input, 159
   feed-forward, 159
   modulation data, 157
   perturbation, 159
   residual error, 160
   spectral replicas, 158
Discrete-time modulating signal, 157
Discrete-time oscillator, 30
Dithering, 34, 43, 56
   noise shaping, 125
   phase noise, 122, 125
   $\Sigma\Delta$, 21, 181, 183
Divider circuit, 18
DSP-coupled transmitter, 213
DSP-driven modulation, 225
DSP-DRP interface, 213
Dynamic element matching (DEM), 70

EDGE, 177
Edge pulling force, 115
Error
   attenuation, 110
   fractional-period, 91
   tracking, 82
   phase, 82, 110, 113
Evaluation board, 218
Excess phase, 150

FCW (frequency command word), 1, 78,
   116, 156
FIR filter, 133, 196
Fixed-point, 113
Flicker noise, 4, 25
   modeling, 195
      antialiasing filter, 197–198

Flicker noise (*Continued*)
　first-order IIR, 196
　time-domain, 196
Flip-flop, 101
　delays, 100
　metastability, 100, 203
　modeling, 204
　pulse generator, 102
　resolution time, 100
　sense amplifier, 102
　SR latch, 102
　uncertainty, 101
　VHDL, 242
Following edge, 92
Fractional error
　correction, 82, 113
　cycle skipping, 95
　estimate, 94
Fractional phase, 91, 94
Fractional spurs, 19–20, 98
Fractional-$N$ divider, 8, 18
　$\Sigma\Delta$ modulated, 20
　fractional spurs, 20
　phase interpolation, 20
　programmable, 8
　time-averaged ratio, 19, 34
Fractional-period error, 91, 94
FREF (reference clock), 1, 76–77
Frequency
　beating, 114
　center, 41, 44, 61–62
　deviation, 48
　　instantaneous, 207, 209
　　peak, 111, 228–229
　division ratio, 78
　drift, 110
　hopping, 9
　modulation (FM), 15, 157
　　narrowband, 8, 22
　natural, 43–44, 61–62
　nominal, 47
　offset, 111
　perturbation, 110, 144
　quantization, 110
　resolution, 40, 43–44, 49, 97
　resonant, 35
　tuning, 42, 45
Frequency control word (FCW), 1, 78, 116, 156
Frequency modulation (FM), 15, 157, 162
Frequency reference (FREF), 1, 76–77
Frequency response, 115
　type I, 116
　type II, 130

Frequency synthesizer, xiv, 1, 5
　acquisition time, 16
　channel hopping, 9, 16
　direct analog, 6
　direct digital (DDFS), 6
　　digital reconstruction, 7
　dithering, 21
　fast acquisition, 16
　fractional-$N$ divider, 8, 18
　frequency control word (FCW), 1, 78, 116, 156
　frequency reference (FREF), 1, 76–77
　fully digital, xv
　fully monolithic, xv
　hybrid, 8
　indirect (PLL), 7
　integrateable, xv, 10
　local oscillator, 10
　loop bandwidth, 16
　mobile communications, 16
　RF, xiv, 9
　switching speed, 9
　techniques, 5
　tuning bandwidth, 10
Frequency transfer function, 110

Gain
　DCO ($K_{DCO}$), 53
　estimation, 141, 158, 160, 165
　just-in-time calculation, 164
　loop, 109
　programmable, 113
　proportional, 113
　normalization, 54
Gaussian filter, 168
Gaussian frequency shift keying (GFSK), 15, 167, 235
　modulation index, 165
　pulse-shaping, 167
Gaussian low-pass filter, 232
Gear shifting, 142, 153
　acquisition, 147
　autonomous, 144
　IIR filter pole, 147
　multiple, 145
　perturbation, 144
　phase-error trajectory, 145
　settline time, 153
　tracking, 144
GSM, 9, 13, 135, 173, 219, 225, 235–236

High-rate dithering, 34, 56
High-speed digital, 214

Higher-order ADPLL, 133
  attenuation factor, 135
  FIR filter, 133
  IIR filter, 133
  pole location, 134
  single-pole IIR filter, 134
Hitless, 143, 149–150

IIR filter, 133, 196
  four-pole, 135
  pole location, 134
  single-pole, 134
Incrementer, 89
  partitioned, 90
Injection locking, 11
Injection pulling, 11, 213, 114
Instantaneous phase resolution, 113
Integer-domain operation, 86
Integer-$N$ divider, 17
Intermediate frequency (IF), 11
Interpolative transmit filter, 172
Intersymbol interference (ISI), 13, 169, 220, 233–234

Jitter, 5
  absolute, 194
  accumulative, 194
  DCO, 55
  modeling, 192
Just-in-time gain calculation, 164

LC tank, 35
  capacitance(s)
    binary-weighted, 35–36
    effective switchable, 36
    perturbation, 55
    single cell, 36
  frequency, 36
    resonant, 35
    natural, 43–44
  inductor, 45
    area, 216
  matching, 36, 39–40, 70
  varactors, 35, 37, 42
Local oscillator, xiv, 10
Loop
  bandwidth, 113, 117, 128, 211
  delay, 116
  filter, 8, 17, 115
  gain, 109
    gear shifting, 142, 153
    programmable, 113
    proportional, 113
Low-pass filter, 10–11, 232
Low-speed digital, 214

MASH $\Sigma\Delta$ modulator, 21, 66, 69
Mean time between failures (MTBF), 103, 105
Metastability, 24, 79, 100, 103, 203
  modeling, 203
Mid-edge detector, 104
Mobile RF systems, 9
  evaluation criteria, 9
  RF synthesizers, 16
Modeling and simulation, 189, 223
  data samples, 191–192
  data storage, 192
  DCO, 201, 237, 239
  flip-flop, 205
  frequency deviation, 207–208
  MATLAB models, 190
  metastability, 203
  noise modeling, 192
    random numbers, 192
  phase noise, 192, 209
    AWGN, 193
    clock divider effect, 200
    flicker noise, 195
    oscillator wander, 194
    period deviation, 193
    timing jitter, 192
  Simulink, 190
  SPICE-based, 189
  SPW, 190
  stream processing, 192
  time-domain model
    DCO, 50
    nDCO, 73
    VHDL, 202
  Verilog, 190
  VHDL, 190, 237
    abstraction levels, 191
    computation time, 190
    event-driven, 190
Modulating data, 157
Modulation index, 165
Monolithic, xv
MOS varactor, 31
  behavior of, 31
  C-V curve, 32
  PMOS, 32
Multirate, 110

Narrowband frequency modulation, 8, 22, 39
Natural frequency, 43–44, 61–62
nDCO (normalized DCO), 52, 54, 61
  active mode, 64
  arithmetic encoding, 61
  at reset, 62

nDCO (Normalized DCO) (Continued)
  control logic, 61
  dithering
    discrete-time, 57
    update clock, 59
  dynamic element matching, 70
  gain, 53
    estimation, 53
    normalization, 54, 60
    paths, 60
    PVT independent, 54
  mode switchover, 64
  normalized tuning word (NTW), 54, 73
  operational frequency resolution, 66
  oscillate failure, 62
  oscillator interface, 61
  spurs, 65
  time deviation (TDEV), 74
  time-domain model, 73
  tracking, 64
    control, 67
    fractional bits, 64, 66, 69
    high-speed dithering, 64
    integer bits, 64, 69
    MASH $\Sigma\Delta$, 21, 66, 69
    $\Sigma\Delta$ modulator, 65
    switching matrix, 68
    thermometer code, 64
  varactor rearrangement, 71
  zero-phase restart (ZPR), 64, 149
Noise, 2, 119
  digital switching, 29
  flicker, 4, 25, 195
  floor, 193, 209
  ideal oscillator, 2
  immunity, 119
  jitter, 5
  modeling, 192, 209
  phase noise, 2–3, 8, 119, 125, 209, 222
  power spectrum, 3, 126
  $\Sigma\Delta$ quantization, 20
  spectral noise density, 3–4
  spurious tones, 4–5
  thermal, 4, 38, 210
Nominal frequency, 47
NTW (normalized tuning word), 54, 73

Open-loop transfer function, 116, 130
Oscillator
  ideal, 2
  local, 10

OTW (oscillator tuning word), 31, 50, 53, 64, 164
  control bit, 34, 37, 42
  digital word, 39

Packet, 164
Period deviation, 193
Phase
  accumulation, 74, 78, 86
  comparison, 79
  detection, 81, 83
    signal names, 83
    structure, 83
  detector, 84
  differentiation, 185
  domain, 77
  drift, 110
  estimate, 81
  interpolation, 20
  resolution, 113
Phase error, 82, 111
  dynamic range, 111
  excess, 149–150
  instantaneous variations, 139
  steady-state, 111
  variability, 145
Phase modulation, 157, 185
Phase noise, 2–3, 8, 119, 125, 209, 222
  DCO dithering, 125
  floor, 209
  jitter, 5, 192
  modeling, 192, 209
  power, 121, 123
  sources, 119
  spectral noise density, 3–4
  spectrum, 3, 121, 126, 211
    measured, 222–223
  TDC resolution, 120–121
  thermal, 38, 210
Phase-domain ADPLL, 112
Phase-domain operation, 77, 79
Phase-domain PLL, 24
Phase-locked-loop (PLL), 5, 8, 17, 76
  charge-pump, 17–18
  fractional-$N$ divider, 8, 18–19
  integer-$N$ divider, 17
  loop filter, 8, 17
  pulse swallower, 18
Phase/frequency detector (PFD), 7, 17
Polar modulator (transmitter), 14, 177, 184, 187
  amplitude control word, 183
  amplitude modulation, 178
  phase modulation, 185
  power amplifier, 185

quantization noise, 185
   signals, 185
Power amplifier, 10, 175, 188
   bond wire inductor, 176
   class E, 175–176
   digital control, 183, 185
   load resistance, 176
   matching network, 175
   power dissipation, 176
   RF choke, 175
Power spectral density (PSD), 3–4, 222, 229
Preceding edge, 92
Pre-power amplifier, 175, 177
Priority detection, 92
Programmable loop gain, 113
Proportional loop gain, 113
Proportional-integral control, 127
Pseudo-thermometer code, 92
Pulse shaping, 13
   filter, 167, 173
Pulse swallower, 18
Pulsewidth modulation, 178
   dead zone, 178
   delay control, 179
   delay line, 180
   delay resolution, 180
PVT (process-voltage-temperature)
   calibration mode, 40, 45
   independent, 54
   variations, 16

Quadrature amplitude modulation (QAM), 12, 182, 184
Quantization
   effects, 97
   error, 80, 82, 123
   noise, 110, 122
   resolution, 91

Receiver, 11
   direct down-conversion, 12
   local oscillator, 11
   low-noise amplifier, 11
   zero-IF, 11
Reference clock (FREF), 1, 76–77
Reference
   edge estimation, 93
   feedthrough, 112
   frequency, 1, 76–77
   phase, 77–78, 82
      accumulator, 86, 110, 150
      modulo arithmetic, 87
      signal, 113
   retiming, 100, 103

Reset
   synchronous, 151
   system-wide, 106
Retimed reference clock (CKR), 79, 103, 113
RF choke, 175
RF processes, 22
RF synthesizer, 9

$s$-domain model, 115, 118–119
Scaling factor, 110
Sense amplifier flip-flop, 102
$\Sigma\Delta$ modulator, 20, 35, 125
   accumulator cell, 67
   dithering, 21, 181
   MASH, 21, 66, 69
   quantization noise, 21
Signal-to-noise ratio (SNR), 12
Single-chip radio, 26, 28
Spectral replicas, 158
Spectral shaping, 13
Spurs, 4–5, 65, 98, 113–114, 229
   fractional, 19
Stability, 110, 137
Staggered clock, 92
Stream processing, 192
Switching transients, 224
Symbol
   deviation, 208
   period, 169
Synchronous phase domain, xiv
Synchronous reset, 151
System integration, 27
System-on-chip (SoC), 189
System-wide reset, 106

TDC (time-to-digital converter), 91
   averaging, 95
   edge skip, 95
   fractional error, 94
   inverter, 91
   number of taps, 92
   phase noise, 120–121
   pseudo-thermometer code, 92
   quantization
      effects, 97
      noise, 122
   resolution, 91
      effects, 97, 120
   signal names, 96
   staggered clock, 92
   transition position, 92
   Vernier delay line, 92
   VHDL model, 244

Test chip, 214
  die microphotograph, 217
  evaluation board, 218
Thermal noise, 4, 38, 210
Thermometer code, 64, 92
TI TMS320C54x, 213
Time-causal, 92
Timing deviation (TDEV), 49–50, 74
Tracking mode, 40–42, 44–45, 144, 147
  bits, 65
  control, 67
  type II, 128
Transceiver, xiii
Transition timestamp, 30, 77
Transmitter, 10, 156
  bandlimited channel, 13
  bit coder, 167
  building blocks, 216
  core, 215
  D/A converters, 10
  data rate, 165
  DCO gain, 158
  estimation circuit, 160
    estimation flowchart, 166
    forced frequency deviation, 164
    just-in-time calculation, 164
    measurement variance, 166
  digital amplitude control, 183
  digital amplitude modulation, 177
  digitally intensive, 215
  digital-to-RF-amplitude conversion
      (DRAC), 185
  direct modulation, 157
    compensating feed, 159, 162
    direct feed, 159, 162
    improvement, 162
    modulation index, 165, 169
    transfer function, 161
  direct up-conversion, 10
  dithering, 183
  Gaussian frequency shift keying
      (GFSK), 15, 167
  intersymbol interference (ISI), 13, 169,
      171–172
  local oscillator, 10
  low-pass filter, 10
  mixer-based, 10
  packet, 164
  performance, 226, 219
  polar modulation, 14, 177, 184
  power amplifier, 10, 175, 177
  power regulation, 181
  pulse shaping, 13
    filter, 168

quadrature amplitude modulation
    (QAM), 12, 182, 184
spectral shaping, 13
symbol period, 169
transmit filter, 167
  baseband clock, 168
  coefficients, 169
  curve storage, 171
  Gaussian filter, 168
  interpolative filter, 172
  look-up table, 169
  output curves, 171
  output data FCW, 170
  oversampling, 168
  structure, 170
  trigger sequence, 170
Tuning word, 111
Two-pole system, 131
Type I ADPLL, 79, 112, 115
  closed-loop transfer function, 116
  frequency response, 116
  loop bandwidth, 117
  magnitude response, 116
  open-loop transfer function, 116
  phase noise spectrum, 223
Type II ADPLL, 127
  closed-loop transfer function, 131
  fast tracking, 128
  filter, 127
  frequency response, 130
  loop bandwidth, 128
  loop noise performance, 132
  magnitude response, 130
  open-loop transfer function, 130
  phase error, 128
  proportional-integral
      control, 127
  steady-state error, 127

UI (unit interval), 74, 77

Varactor, 31
  array, 33, 38
    dynamic element matching
        (DEM), 70
    matching, 39
    rearrangement, 71
  differential, 37
  digital control, 33, 35, 37
  matching, 36
Variable phase, 77–78, 82
  accumulator, 82, 86, 89, 109
  modulo arithmetic, 87
  signal, 113

Vernier delay line, 92
VHDL, 190, 237
   synthesizable RTL, 215
Voltage-controlled oscillator
     (VCO), 8, 62

Wander, 194
Watchdog timer, 214
Wideband CDMA
     (WCDMA), 177
Wireless
  channel, 9

transmitter, 164
transceiver, xiv
Wordlength limit, 86, 111, 113

$z$-domain model, 119, 132
Zero IF, 11
Zero-phase restart (ZPR), 64, 143, 149
   discontinuities, 149
   excess phase error, 149–150
   hitless, 149–150
   simulations, 151
   synchronous reset, 150